Energy Efficiency and Indoor Environment Quality

Energy Efficiency and Indoor Environment Quality

Editor

Roberto Alonso González Lezcano

MDPI • Basel • Beijing • Wuhan • Barcelona • Belgrade • Manchester • Tokyo • Cluj • Tianjin

Editor
Roberto Alonso González
Lezcano
Universidad CEU San Pablo
Spain

Editorial Office
MDPI
St. Alban-Anlage 66
4052 Basel, Switzerland

This is a reprint of articles from the Special Issue published online in the open access journal *Energies* (ISSN 1996-1073) (available at: https://www.mdpi.com/journal/energies/special_issues/indoor_environment).

For citation purposes, cite each article independently as indicated on the article page online and as indicated below:

LastName, A.A.; LastName, B.B.; LastName, C.C. Article Title. *Journal Name* **Year**, *Volume Number*, Page Range.

ISBN 978-3-0365-2506-8 (Hbk)
ISBN 978-3-0365-2507-5 (PDF)

Contents

About the Editor

Roberto Alonso González Lezcano is a Tenured Professor at the Department of Architecture and Design in the field of Building Systems, within the Institute of Technology at Universidad CEU San Pablo in Madrid. He is accredited by ANECA as a Senior Professor, and between 2003 and 2015, he undertook a six-year research period with CNEAI. He is the coordinator of the subject Mechanical Systems and of the Laboratory of Building Systems within Universidad CEU San Pablo, as well as the coordinator of the postgraduate course in Energy Efficiency and Mechanical Systems in Buildings. He is a member of the PhD Programs "Health Science and Technology" and "Composition, History and Techniques pertaining to Architecture and Urbanism", where he is the coordinator of the Construction, Innovation and Technology line and coordinator of the training course "Methodology of technical and statistical experimentation". He is the lecturer of the master's course in Quality in Construction and the postgraduate course in Overall Management of Buildings and Services at the Universidad Politécnica de Madrid. He is also coordinator of the Wind Energy section within the master's degree program in Renewable Energy at the Institute of Technology (Universidad CEU San Pablo), as well as a lecturer of Advanced Manufacturing Processes in the master's degree program in Industrial Engineering at the Universidad Europea de Madrid. In addition, he lectures in Mechanics of Continuous Media and Theory of Structures, Electromechanics and Materials at seven Spanish universities. He has published 29 books, 20 of which in the last five years, and is the coordinator and co-author of the collections *ABC of Building Systems* (Munilla-Lería Eds.), consisting of five volumes, and *Building Systems in Building Design* (Asimétricas Eds.), consisting of seven volumes. He has published 58 articles in the last five years concerning areas of energy, systems, and mechanics of continuous media (41 JCR and 17 SJR) and he is the editor of several books on issues of sustainability, energy efficiency and indoor environment quality, indexed in the Book Citation Index, WoS, Scopus, and SPI, among others. He was director of the PhD theses "Efficiency of ventilation in residential environments for the promotion of health of its occupants: impact on architectural indoor design" and "Energy simulation as a prognostic tool in architecture: design of passive strategies in different climatic zones of Spain", which he obtained in June 2018 and September 2020, respectively, Cum Laude, with mention of an International PhD. He is also the Director/Co-director of three PhD theses currently in progress related to energy efficiency, comfort, and indoor air quality, entitled "Influence of occupants on energy consumption in multi-family dwellings in Madrid: Characterization of their habits", "Building Systems in efficient office buildings" and "Energy simulation as a forecast tool in architecture: The importance of the solar reflectance index in the design stage of residential buildings in Spain". He is a member of the Editorial Board and Review Board of several journals indexed in JCR, as well as a Guest Editor of Special Issues and Topic Editor for JCR-indexed MDPI journals.

energies

MDPI

Article

Real-Time Reconstruction of Contaminant Dispersion from Sparse Sensor Observations with Gappy POD Method

Zheming Tong [1,2,*,†] and Yue Li [1,2,†]

1 State Key Laboratory of Fluid Power and Mechatronic Systems, Zhejiang University, Hangzhou 310027, China
2 School of Mechanical Engineering, Zhejiang University, Hangzhou 310027, China
* Correspondence: tzm@zju.edu.cn
† These authors contributed equally to this work.

Received: 25 March 2020; Accepted: 11 April 2020; Published: 15 April 2020

Abstract: Real-time estimation of three-dimensional field data for enclosed spaces is critical to HVAC control. This task is challenging, especially for large enclosed spaces with complex geometry, due to the nonuniform distribution and nonlinear variations of many environmental variables. Moreover, constructing and maintaining a network of sensors to fully cover the entire space is very costly, and insufficient sensor data might deteriorate system performance. Facing such a dilemma, gappy proper orthogonal decomposition (POD) offers a solution to provide three-dimensional field data with a limited number of sensor measurements. In this study, a gappy POD method for real-time reconstruction of contaminant distribution in an enclosed space is proposed by combining the POD method with a limited number of sensor measurements. To evaluate the gappy POD method, a computational fluid dynamics (CFD) model is utilized to perform a numerical simulation to validate the effectiveness of the gappy POD method in reconstructing contaminant distributions. In addition, the optimal sensor placement is given based on a quantitative metric to maximize the reconstruction accuracy, and the sensor placement constraints are also considered during the sensor design process. The gappy POD method is found to yield accurate reconstruction results. Further works will include the implementation of real-time control based on the POD method.

Keywords: gappy proper orthogonal decomposition; sparse sensor observations; contaminant distribution; reconstruction; CFD

---

1. Introduction

Automatic control of the HVAC system plays a significant role in improving indoor air [1–4] and reducing building energy consumption [5–8]. Real-time estimation of contaminant distribution inside any enclosed space could provide immediate feedback to the control of ventilation systems and, thus, is of great significance. However, the temporal evolution of contaminant distribution is characterized by complex nonlinear dynamics. It is challenging to reconstruct the spatiotemporal distribution of contaminants for real-time control of the ventilation system [9].

There are three main approaches to constructing indoor field data: spatial data interpolation, physics-based simulation, and the data-driven approach. Generally, ordinary Kriging is the most widely used spatial interpolation method [10] and could produce an effective estimate of an indoor thermal map [11] and pollutant distributions [11–13] based on sensor measurements. However, the accuracy of the interpolation is strongly dependent on sensor placements and the number of sensors. Typically, a large number of sensors are required to achieve adequate spatial resolution [11]. On another hand, physics-based models such as computational fluid dynamics (CFD) [14–16], fast fluid dynamics (FFD) [17–22], and zonal models [23–25] can provide three-dimensional field data for an indoor environment with only a small number of boundary conditions that are, however often difficult to obtain in real applications. Even though CFD have been widely employed for predicting the spatial variation of indoor environmental variables [26], the large amount of computational time makes real-time prediction extremely difficult and, thereby, unlikely to achieve real-time control. To overcome the large computational cost, the FFD model is proposed by solving the continuity equation and unsteady Navier-Stokes equations with a time-splitting approach [21]. In addition, the computational speed of FFD could be further accelerated by using parallel and GPU computing [19]. Considering its reasonable accuracy and high computational efficiency, the FFD method has been extensively studied for fast simulation of transient flow [18,19] and for inverse design of the ventilation parameters [20]. Although the reduction in computing time is significant, it is still computationally costly to implement, especially for real-time control. Different from CFD and FFD models, the zonal models are easily incorporated into the control system [25] and require very little computation time, because they only solve energy and mass balance equations at each thermal zone. However, based on the assumption that all quantities are uniform at each zone, zonal models are unable to provide detailed spatial information for a room space [23]. The third category is the data-driven approach. The artificial neural network (ANN) is one of the most widely used data-driven approaches for fast estimation of an indoor environment [27–30]. ANN models the nonlinear relationships between boundary conditions and environmental variables by training the data obtained from CFD simulations. However, an ANN model is very prone to over-fitting [12]. Moreover, as a black box model, it is difficult to interpret the results predicted by an ANN model.

Recently, the proper orthogonal decomposition (POD) has been extensively studied as a major class of data-driven approaches for dynamics reconstruction, since it provides a powerful tool to identify low-order modes of a system. Compared with ANN, POD provides a more interpretative reconstruction result by capturing the dominant physics-based coherent structures. At present, POD has wide application for dynamic reconstruction in the context of fluid mechanics [31,32], structural dynamics [33], ocean engineering [34,35], etc. As for the application in fluid dynamics, POD utilizes a few dominant POD bases for representation of the extracted flow features [36]. Moreover, to achieve active flow control in practice, sensor measurements must be introduced to the POD framework. Thus, gappy POD has been recently developed as an extension of the POD method by combining the POD model and sensor measurements [37]. Given a set of POD modes, gappy POD is able to reconstruct fluid dynamics with a limited number of sensor measurements through linear regression.

POD has recently been introduced to the field of built environment for fast reconstruction of the indoor temperature field [38–43] and airflow pattern [38,44]. For example, Allery et al. estimated the air flow pattern in a two-dimensional ventilated cavity by a POD-Galerkin construction [44]. Sempey et al. obtained reduced-order models of temperature distribution in an air-conditioned room under steady flow conditions based on the POD-Galerkin method [42]. Phan et al. applied the POD model for investigation of the temperature distribution in a data center by using the POD-Galerkin method and further investigated the effects of design parameters on the reconstruction accuracy [43]. Li et al. established a reduced-order model for fast estimation of the indoor thermal map by using Galerkin projection and further developed a real-time controller based on the reduced order model [41]. Tallet et al. used a POD model for real-time reconstruction of indoor temperature distribution based on sensor measurements [38]. Meyer et al. combined the POD method with the linear stochastic estimation method for reconstruction of the fluctuating thermal field in an experimental room based on real-time sensor measurements [40]. Jiang et al. fuse the input parameters of a CFD simulation and sparse sensor observations for reconstruction of a steady indoor thermal field [39]. However, gappy POD, which has been confirmed to perform well in turbulent flow sensing [37,45], has seldom been used for reconstructing the transient variations of the indoor environment.

Sensor measurements are critical to the estimation and control of the indoor environment. Considering the nonuniform distribution of indoor contaminants, a large number of sensors are required to guarantee satisfactory HVAC performance, which, however, is usually cost-expensive. Thus, it is of great significance to extrapolate the three-dimensional distribution of indoor air contaminants based on limited sensor measurements. In this paper, a gappy POD method for the real-time reconstruction of contaminant distribution in an enclosed space is proposed by combining the POD method with a limited number of sensor measurements. Even though gappy POD has been confirmed to perform well in the field of turbulent flow sensing, the effects of gappy POD on the real-time reconstruction of contaminant distributions remain to be investigated. To evaluate the effects of gappy POD for the reconstruction of indoor environments, CFD models are utilized to perform the simulation study to validate the reconstruction results. The article is organized as follows. Section 2 briefly introduces the gappy POD method and the algorithm for optimal sensor placement. The reconstruction results of the contaminant distribution are presented in Section 3. Finally, the main conclusions are presented in Section 4.

2. Materials and Methods

2.1. Standard POD

The spatiotemporal distribution of contaminants is characterized by complex nonlinear dynamics. The POD method provides a powerful tool to decompose the high-dimensional dynamics to basic POD modes for representation of the coherent structures for contaminant distribution. The snapshot method was introduced by Sirovich et al. for determination of the basic POD modes [46]. The snapshots X are formed by collecting the time-series data.

$$\boldsymbol{X} = \begin{bmatrix} \boldsymbol{x}(t_1) & \boldsymbol{x}(t_2) & \cdots & \boldsymbol{x}(t_k) & \cdots & \boldsymbol{x}(t_m) \end{bmatrix} \tag{1}$$

where $\boldsymbol{x}(t_k)$ is a vector of n elements representing the spatial distribution of the contaminants at a time t_k. The snapshot $\boldsymbol{x}(t_k)$ could be obtained from CFD simulation results. The correlation matrix R of size m × m is defined by

$$\boldsymbol{R} = \frac{1}{m}\boldsymbol{X}^T\boldsymbol{X} \tag{2}$$

Then, the eigenvectors $\boldsymbol{\varphi} = \begin{bmatrix} \varphi^1, \varphi^2 & \cdots & , \varphi^m \end{bmatrix}$ of the correlation matrix R, and its corresponding eigenvalues $\boldsymbol{\lambda} = \begin{bmatrix} \lambda^1, \lambda^2 & \cdots & , \lambda^m \end{bmatrix}$ are computed. The basic POD modes $\boldsymbol{\phi}$ is given by a linear combination of snapshots

$$\boldsymbol{\phi}^j = \frac{\boldsymbol{X}\boldsymbol{\varphi}^j}{\|\boldsymbol{X}\boldsymbol{\varphi}^j\|_2} \tag{3}$$

where $\boldsymbol{\varphi}^j$ denotes the jth eigenvector. The eigenvalue is ordered in terms of the importance of their corresponding POD modes ($\lambda^1 > \lambda^2 > \lambda^3 \quad \cdots \quad > \lambda^m$). The lager eigenvalue means that the corresponding basic POD modes plays a more important role in describing the contaminant distribution. By using the first p eigenvectors, the contaminant concentration could be reconstructed as

$$\boldsymbol{x} \approx \sum_{i=1}^{p} \boldsymbol{b}_i \boldsymbol{\Phi}^i \tag{4}$$

where b_i are temporal coefficients of the ith POD basis mode, and x is the vector representing the estimation of contaminant distribution in the enclosed space. For POD applications, the most important issue is the determination of the coefficients for basic POD modes.

2.2. Gappy POD

Gappy POD provides an effective method for the calculation of POD coefficients based on sparse sensor observations. The mask vector H of the n elements records the sensor locations. If the sensor is located on the ith grid point, then the ith element in H is equal to one; otherwise, the ith element in H is equal to zero. The incomplete vector x of the n elements describes the contaminant concentration at n locations corresponding to sensor locations but also has some elements missing. The full vector $\widetilde{x}$ of the n elements represents the reconstructed contaminant concentration. The gappy POD could reconstruct the full vector $\widetilde{x}$ from the incomplete vector x. The POD coefficient could be obtained by solving a linear regression in Equation (5).

$$M\widetilde{b} = f \tag{5}$$

where

$$M_{ij} = \left((H, \Phi^i), (H, \Phi^j)\right) \tag{6}$$

and

$$f_i = \left(x, (H, \Phi^i)\right) \tag{7}$$

The coefficient $\widetilde{b}$ could be obtained by solving Equation (5), and the spatial distribution of contaminants could be immediately reconstructed through Equation (4).

2.3. Sensor Placement

According to Equation (5), the POD coefficient is obtained through data regression. The condition number of matrix M is a good proxy for evaluating the solution accuracy of Equation (5). The smaller condition number could contribute to the higher solution stability for Equation (5). The optimal sensor placement for the N sensors could be determined by the following equations:

$$\begin{gathered} \min \kappa(M) \\ s.t.\ K(i,1)\ \epsilon\ \{0,1\},\ i = 1,2,\ldots,n \\ \textstyle\sum_{j=1}^{n} K(i,1) = N \end{gathered} \tag{8}$$

where $\kappa(M)$ is the condition number of M.

The sensor placement could be obtained based on a greedy algorithm to minimize $\kappa(M)$ [37]. Firstly, loop over all possible placement points, evaluate the condition number of M for each point, and choose the point that minimizes $\kappa(M)$ to locate the first sensor. Then, determine the location of the next sensor that could minimize the condition number $\kappa(M)$. Repeat the previous steps until all the N sensors are placed in appropriate locations. For sensors with placement constraints, the number of possible sensor placement points would be fewer, and the only change we have to make is to loop over locations within location constraints for the optimal sensor placement.

2.4. Numerical Model

In the present study, the turbulent model based on Reynolds-Averaged Navier Stokes (RANS) is adopted. According to the previous studies focusing on gaseous contaminant dispersion in an enclosed space, the Eulerian method is able to provide an effective solution for simulating an indoor concentration distribution [26,47]. Therefore, the Eulerian method is used to predict the contaminant distributions in this study. Moreover, it is assumed that there is no chemical reaction during the gaseous contaminant dispersion process.

The experimental results given by Yuan et al. were utilized to validate the reliability of our numerical model [48]. The experiments were conducted in a test chamber with a size of 5.16 m × 3.65 m × 2.43 m (length × width × height) under the displacement ventilation. The total flow rate was 0.05 m^3/s. The supply inlet was placed at the middle of the side wall near the floor, and the exhaust device of size

0.43 m × 0.43 m was fixed at the center of the ceiling. Two heated manikins, two computers, and six lamps were used to simulate heat sources. SF_6 was used to simulate contaminants from occupants, and the contaminant source was placed above the simulated occupants.

Figure 1 gives the comparisons of the velocity and dimensionless SF_6 concentration profiles between the numerical results and the experimental data. It could be observed that the agreement between the CFD results and the experimental data is reasonably good. Discrepancies between the simulated results and experimental values could be observed in the upper part of the space. Additionally, the discrepancies could also be found in the simulation study by Yuan et al. [49], who conducted this experiment and explained that this was because of recirculating flows in the upper part of the space. Moreover, it should be noted that the uncertainties for the experimental data are 0.01 m/s and 10% for the SF_6 concentration [49].

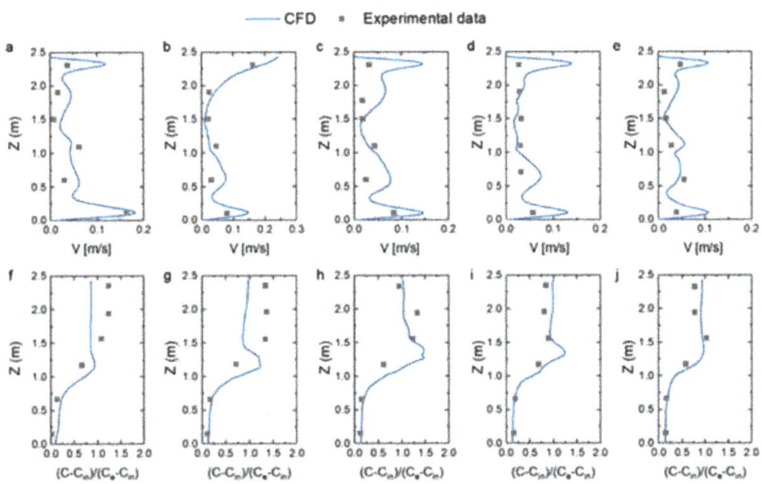

Figure 1. Validation of the numerical model. (**a**)–(**e**) Validation of the airflow model. (**f**)–(**j**) Validation of the contaminant transport model.

2.5. Model Setup

In this study, the case is set in an enclosed space with dimensions of 4 m (L) × 3 m (W) × 2.5 m (H). Both airflow inlet and outlet are located on the left wall with dimensions of 0.6 m × 0.3 m. The inlet is located in the upper part of the left-side wall of 0.3 m beneath the ceiling, while the outlet is located in the lower part of the left-side wall of 0.3 m above the floor. A gaseous contaminant source is located in the center of the room. Detailed configurations of the ventilation system are shown in Figure 2. Moreover, the sensor locations are assumed to be constrained on the left and the back wall.

The reconstruction of the contaminant distribution is decomposed into two steps: the offline stage and online stage (Figure 3). In the offline stage, the snapshots are decomposed into basic POD modes, and the optimal sensor placement is determined. In the online stage, the temporal coefficients for POD modes are obtained based on the sparse sensor measurements through linear regression and then are applied for reconstruction of the contaminant distribution through a linear combination of the dominant POD modes. The POD-based online-offline algorithm has already been applied to many cases for optimal control of the dynamic systems, including optimal control of the water flooding reservoir [50] and optimal control of the indoor temperature [38,41]. The detailed procedure is described as follows:

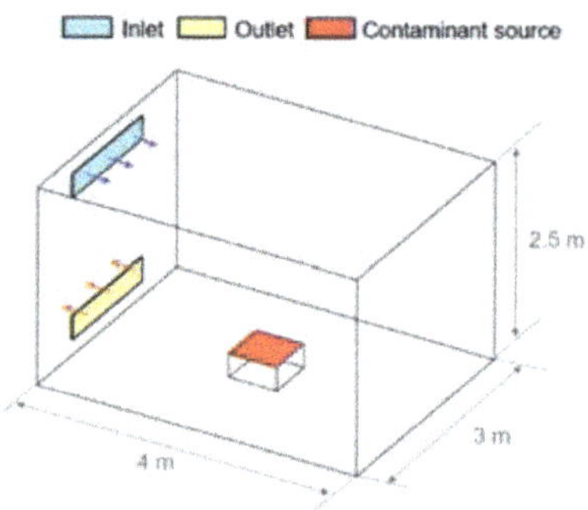

Figure 2. Geometry of the three-dimensional enclosed space.

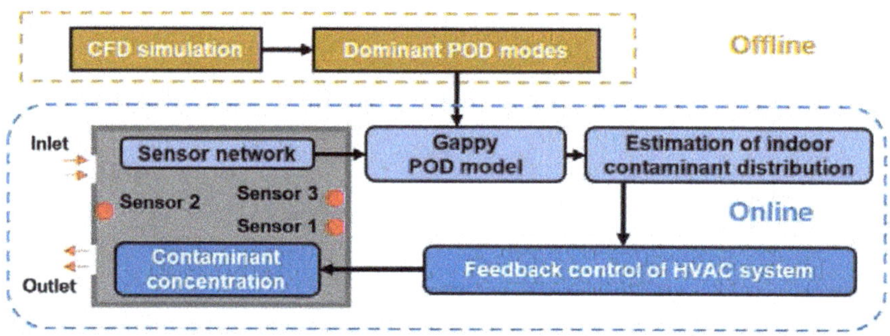

Figure 3. Flow diagram illustrating the control process for indoor air quality management. POD: proper orthogonal decomposition.

(1) Collect a set of snapshots with a combination of different inlet velocities and contaminant source strengths. In case I, we first obtain the steady contaminant distribution with the CFD model for an inlet velocity of 1 m/s, and the gaseous contaminant is steadily released with a source strength of 1 mL per cubic meter per second. Using the steady contaminant distribution as the initial condition, the inlet velocity experiences a step change from 1 m/s to 2 m/s, and the source strength remains unchanged. The simulation results are recorded as snapshots until the next steady state is reached. In summary, the snapshots are recorded every 1 s, and 180 snapshots are obtained during the response process in case I. Case II follows a similar procedure as case I; the only difference between case I and case II is that the gaseous contaminant is steadily released with a source strength of 2 times higher than the source strength in case I. A sum of 320 snapshots are recorded for case II. Finally, a snapshot matrix combining the snapshots in both case I and case II is formed.

(2) Decompose the snapshots to obtain the POD modes, and then the optimal senor placement can be determined by minimizing the condition number in Equation (8). Moreover, it should be noted that the number of sensors should be more than the number of the dominant POD modes to make sure that the optimal algorithm for sensor placement is not an underdetermined problem.

(3) In the online stage, the temporal coefficients of the POD modes are determined based on the real-time measurements from the sparse sensors. The CFD simulations in the test case are used to provide sensor observations for implementing gappy POD reconstruction and to validate the reconstruction results. In particular, it is noteworthy that the gappy POD in the online step does not rely on any CFD simulation, and the CFD simulation here is used to perform a simulation study to evaluate the gappy POD model. The test case is set when the contaminant is steadily released at a source strength of 1.5 times higher than the source strength in case I. We first obtain the steady contaminant distribution with the CFD model for an inlet velocity of 1.2 m/s, and then, the inlet velocity

is subject to a 0.4 m/s positive step to 1.6 m/s. The gappy POD method can be applied for reconstruction of the contaminant distribution in this test case, because the airflow pattern in our test case shares similar coherent structures with the snapshots. Moreover, Sempey et al. confirmed that the POD method performs well in reconstructing system dynamics with boundary conditions different from the conditions used for constructing snapshots [42].

(4) Reconstruct the spatiotemporal distribution of the contaminants based on the basic POD modes and the corresponding temporal coefficients.

3. Results

3.1. POD Decomposition

The snapshots of the contaminant distributions are of great importance for characterizing the system dynamics, and the dominant POD modes are built based on the snapshot ensemble. In detail, the contaminant is released from the center of the enclosed space, and the concentration is highest around the contaminant source. The contaminant distribution evolves with time as a response to the step increase of the inlet velocity from 1 m/s to 2 m/s.

A total of 500 POD modes and 500 corresponding eigenvalues are obtained based on the POD decomposition process. The exponential decay of the normalized eigenvalues is demonstrated in Figure 4. The normalized eigenvalues are calculated by dividing the sum of the total 500 eigenvalues and are regarded as a significant parameter for evaluating the percentage of energy contained in their corresponding POD modes and for measuring the importance of the POD modes. The normalized eigenvalues in Figure 4 are ordered in terms of their ability to describe the spatiotemporal distribution of the contaminants. It could be observed that the first few modes account for most of the system's energy. For example, the first, second, and third POD mode accounts for 66.63%, 17.83%, and 4.26% of the total system energy, respectively. In addition, the first six POD modes contain about 95% of the total system energy, while the first 16 POD modes contain about 99% of the total system energy. As the normalized eigenvalue decreases, the corresponding POD mode will exhibit less meaningful spatial structures and contribute less to reconstructing the contaminant dispersion process.

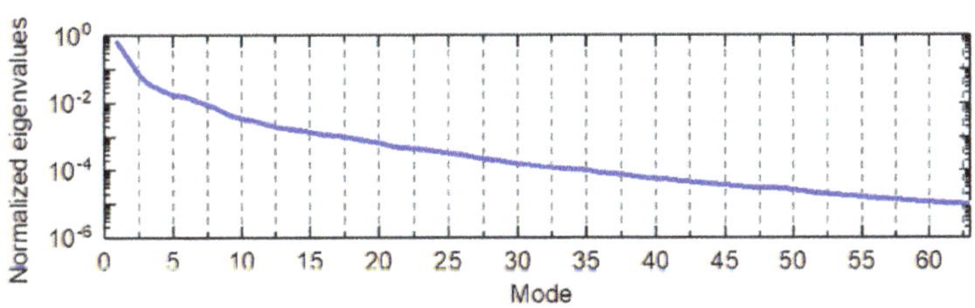

Figure 4. The exponential decay of the normalized eigenvalue. The eigenvalue is normalized by dividing the sum of the 500 eigenvalues. The first 64 POD modes contain more than 99.9% of the total system energy.

3.2. Sensor Placement

It is challenging to optimize the sensor placement to maximize the reconstruction accuracy, since there are thousands of potential locations for sensor placement. A quantitative framework is used to determine the sensor locations by minimizing the condition number of the M matrix in Equation (8). The sensor locations are assumed to be constrained on the left and the back wall. For example, the optimal sensor placements of 20 sensors with location constraints is demonstrated in Figure 5. The potential sensor location could be in any location of the approximately 3000 grids on the left and the back wall. The optimization strategy is to pick up 20 points from the 3000 potential locations for the placement of sensors that could minimize the condition number of M in Equation (8) based on the greedy algorithm. It should be noted that increasing the POD mode number or increasing the sensor number would both result in a higher computational cost for finding the optimal sensor location. However, the computation of the optimal sensor placement is in the off-line stage and would not affect the speed for the real-time contaminant reconstruction in the online stage. Moreover, it can be observed that the sensors with location constraints are always kept away from the contaminant source. This can benefit the sensors by protecting them from long-term exposure to high-concentration contaminants and can improve the sensors' service lives.

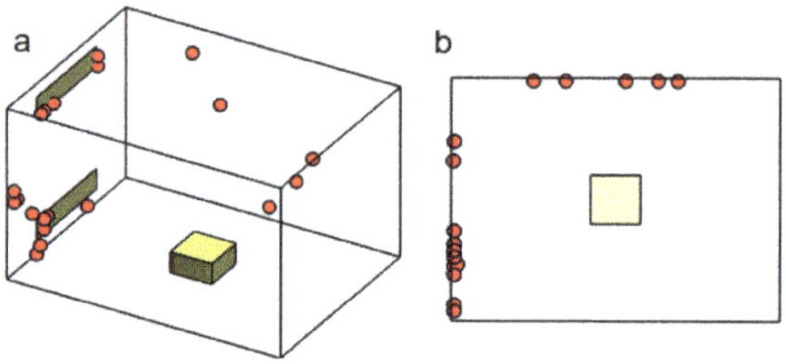

Figure 5. Optimal sensor placements: (**a**): isometric view and (**b**): top view.

3.3. *Gappy Flow Reconstruction*

Quantifying the indoor air quality plays an important role in analyzing indoor occupant exposure [51]. Figure 6 demonstrates the estimation of the contaminant distribution based on 16 dominant POD modes with sensor placement constraints. It should be noted that the reconstruction accuracy is significantly limited by the number of dominant POD modes chosen for estimation of the contaminant distribution, and the first 16 modes contain more than 99% of the total system energy. As shown in Figure 6, the gappy POD method exhibits a high reconstruction accuracy and performs extremely well in estimating a contaminant concentration higher than 0.02 ppm. This is because 16 POD modes could capture a sufficiently detailed structure for reconstruction of the contaminant dispersion process. Moreover, the difference between gappy POD reconstruction and CFD simulation can be observed at t = 20 s, because it is difficult to reconstruct the contaminant distribution during this period with dramatical variations in the airflow field.

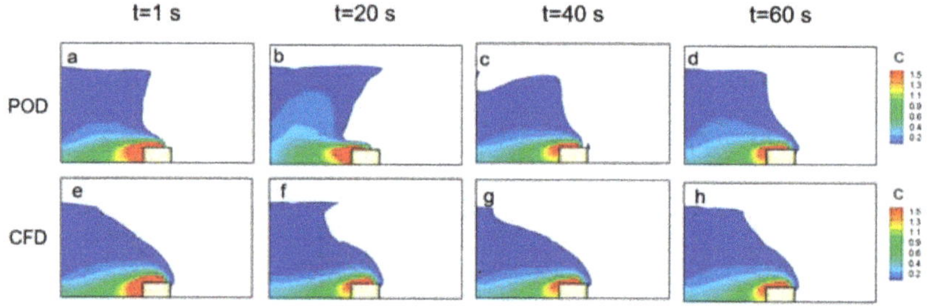

Figure 6. Comparison of the estimated contaminant distributions (ppm) at $x = 1.5$ m by the POD method and by CFD simulation at time 1s, 20 s, 40 s, and 60 s. (**a**)–(**d**): POD reconstruction; (**e**)–(**h**): CFD simulations.

For further evaluation of the gappy POD method, comparison of the estimated contaminant concentration along the line $x = 1.5$ m, $y = 1$ m is conducted (Figure 7). The difference between the POD reconstruction and CFD simulation could be observed at $t = 20$ s due to the slightly different airflow pattern from the snapshots, which could also be observed in Figure 6. However, the gappy POD performs well in most conditions, because the airflow pattern in our test case shares similar coherent structures with snapshots in most conditions. Moreover, Sempey et al. confirmed that the

POD method performs well in reconstructing system dynamics with boundary conditions different from the conditions used for constructing snapshots [42].

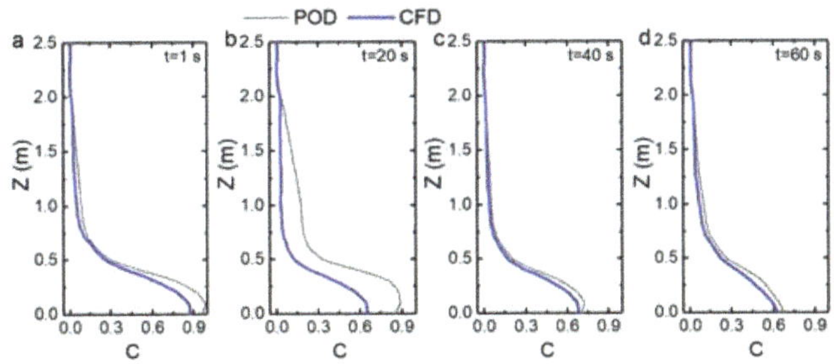

Figure 7. Comparison of the estimated contaminant concentrations (ppm) along the line $x = 1.5$ m, $y =$ 1 m at time (**a**): $t = 1$ s; (**b**): $t = 20$ s; (**c**): $t = 40$ s; (**d**): $t = 60$ s.

The reconstruction accuracy could be improved by increasing the dominant POD modes. This is because more POD modes can contribute to more detailed information for the description of the pollutant distribution. In our study, we found that 16 dominant POD modes are sufficient for estimation of the contaminant distribution in the enclosed space with sensor placement constraints.

4. Discussions and Conclusions

Real-time estimation of the contaminant distribution is essential for ventilation control and indoor air quality management. Moreover, it is known that automatic control is of great importance for improving a systems' energy efficiency in engineering applications [52–55]. However, it is challenging to reconstruct spatiotemporal distributions of contaminants in an enclosed space, due to the ununiform distribution and nonlinear variations of indoor contaminants. As an essential part of active control, the sensor measurements only provide limited information about contaminant concentrations near sensor locations. Usually, a large number of sensors are required to guarantee the satisfactory performance of HVAC control, which might result in high costs for maintaining and conducting a sensor network in practice. Facing such a dilemma, the gappy POD offers a solution to provide real-time contaminant distributions with a limited number of sensor measurements.

In this study, a gappy POD method for the real-time reconstruction of pollutant distribution in an enclosed space is proposed by combining the POD method with a limited number of sensor measurements. In fact, the spatial distribution of indoor contaminants is often represented by a combination of a few dominant patterns, and this inherent property enables reconstruction of the contaminant distribution with sparse sensor observations. Moreover, our study gives the optimal sensor placement based on a quantitative metric in order to maximize the reconstruction accuracy. The sensor placement constraints are also considered during the sensor design process. It should be noted that the reconstruction accuracy is significantly limited by the number of dominant POD modes chosen for estimation of the pollutant distributions. For example, the first six POD modes only contain about 95% of the total system energy, while the first 16 POD modes contain about 99% of the total system energy. According to our study, reconstruction based on 16 POD modes are sufficient for accurate reconstruction of the pollutant distribution in the enclosed space.

For feedback control of the HVAC, gappy POD is able to provide a reliable estimation of three-dimensional data for the indoor environment with a high fidelity and low computational cost. Considering the low-order nature of the POD model, a closed-loop control might be achieved by controlling the coefficient of the POD modes, and the estimated POD coefficients could provide the necessary information for driving the HVAC actuators. Further work will include real-time control of the dynamic system based on the gappy POD method.

Author Contributions: Y.L.: conceptualization, methodology, and writing—original draft preparation. Z.T.: methodology, project administration and writing—review and editing. All authors have read and agreed to the published version of the manuscript.

Funding: This work was supported by the National Natural Science Foundation of China (51708493); Zhejiang Provincial Natural Science Foundation (LR19E050002); the National Key R&D Program of China (grant no. 2019YFB2004604); the Zhejiang Province Key Science and Technology Project (2018C01020, 2018C01060, and 2019C01057); and the Youth Funds of State Key Laboratory of Fluid Power & Mechatronic Systems (SKLoFP_QN_1804).

Conflicts of Interest: The authors declare no conflict of interest.

References

1. Lim, E.; Ito, K.; Sandberg, M. Performance evaluation of contaminant removal and air quality control for local ventilation systems using the ventilation index Net Escape Velocity. *Build. Environ.* **2014**, *79*, 78–89. [CrossRef]
2. Kolokotsa, D.; Pouliezos, A.; Stavrakakis, G.; Lazos, C. Predictive control techniques for energy and indoor environmental quality management in buildings. *Build. Environ.* **2009**, *44*, 1850–1863. [CrossRef]
3. Lee, S.; Hwangbo, S.; Kim, J.T.; Yoo, C.K. Gain scheduling based ventilation control with varying periodic indoor air quality (IAQ) dynamics for healthy IAQ and energy savings. *Energy Build.* **2017**, *153*, 275–286. [CrossRef]
4. Tong, Z.; Chen, Y.; Malkawi, A.; Liu, Z.; Freeman, R.B. Energy saving potential of natural ventilation in China: The impact of ambient air pollution. *Appl. Energy* **2016**, *179*, 660–668. [CrossRef]
5. Cao, Y.; Du, J.; Soleymanzadeh, E. Model predictive control of commercial buildings in demand response programs in the presence of thermal storage. *J. Clean. Prod.* **2019**, *218*, 315–327. [CrossRef]
6. King, J.; Perry, C. *Smart Buildings: Using Smart Technology to Save Energy in Existing Buildings*; Amercian Council for an Energy-Efficient Economy: Washington, DC, USA, 2017.
7. Hu, S.; Hoare, C.; Raftery, P.; O'Donnell, J. Environmental and energy performance assessment of buildings using scenario modelling and fuzzy analytic network process. *Appl. Energy* **2019**, *255*, 113788. [CrossRef]
8. Li, K.; Su, H. Forecasting building energy consumption with hybrid genetic algorithm-hierarchical adaptive network-based fuzzy inference system. *Energy Build.* **2010**, *42*, 2070–2076. [CrossRef]
9. Manohar, K.; Brunton, B.W.; Kutz, J.N.; Brunton, S.L. Data-Driven Sparse Sensor Placement for Reconstruction: Demonstrating the Benefits of Exploiting Known Patterns. *IEEE Contr. Syst. Mag.* **2018**, *38*, 63–86.
10. Burrough, P.A.; McDonnell, R.; McDonnell, R.A.; Lloyd, C.D. *Principles of Geographical Information Systems*; Oxford University Press: Oxford, UK, 2015.
11. Huang, Y.; Shen, X.; Li, J.; Li, B.; Duan, R.; Lin, C.H.; Chen, Q. A method to optimize sampling locations for measuring indoor air distributions. *Atmos. Environ.* **2015**, *102*, 355–365. [CrossRef]
12. Li, J.; Li, H.; Ma, Y.; Wang, Y.; Abokifa, A.A.; Lu, C.; Biswas, P. Spatiotemporal distribution of indoor particulate matter concentration with a low-cost sensor network. *Build. Environ.* **2018**, *127*, 138–147. [CrossRef]
13. Cheng, K.-C.; Tseng, C.-H.; Hildemann, L.M. Using Indoor Positioning and Mobile Sensing for Spatial Exposure and Environmental Characterizations: Pilot Demonstration of PM2. *5 Mapping. Environ. Sci. Technol. Lett.* **2019**, *6*, 153–158. [CrossRef]
14. Kosutova, K.; van Hooff, T.; Vanderwel, C.; Blocken, B.; Hensen, J. Cross-ventilation in a generic isolated building equipped with louvers: Wind-tunnel experiments and CFD simulations. *Build. Environ.* **2019**, *154*, 263–280. [CrossRef]
15. Aganovic, A.; Steffensen, M.; Cao, G. CFD study of the air distribution and occupant draught sensation in a patient ward equipped with protected zone ventilation. *Build. Environ.* **2019**, *162*, 106279. [CrossRef]
16. Liu, W.; Liu, D.; Gao, N. CFD study on gaseous pollutant transmission characteristics under different ventilation strategies in a typical chemical laboratory. *Build. Environ.* **2017**, *126*, 238–251. [CrossRef]
17. Zuo, W.; Chen, Q. Real-time or faster-than-real-time simulation of airflow in buildings. *Indoor Air* **2009**, *19*, 33. [CrossRef]
18. Zuo, W.; Chen, Q. Fast and informative flow simulations in a building by using fast fluid dynamics model on graphics processing unit. *Build. Environ.* **2010**, *45*, 747–757. [CrossRef]

19. Zuo, W.; Chen, Q. Simulations of Air Distributions in Buildings by FFD on GPU. *HVAC&R Res.* **2010**, *16*, 785–798.
20. Xue, Y.; Liu, W.; Wang, Q.; Bu, F. Development of an integrated approach for the inverse design of built environment by a fast fluid dynamics-based generic algorithm. *Build. Environ.* **2019**, *160*, 106205. [CrossRef]
21. Liu, W.; Jin, M.; Chen, C.; You, R.; Chen, Q. Implementation of a fast fluid dynamics model in OpenFOAM for simulating indoor airflow. *Numer. Heat Transf. Part A Appl.* **2016**, *69*, 748–762. [CrossRef]
22. Jin, M.; Zuo, W.; Chen, Q. Improvements of Fast Fluid Dynamics for Simulating Air Flow in Buildings. *Numer. Heat Transf. Part B Fundam.* **2012**, *62*, 419–438. [CrossRef]
23. Megri, C.A.; Haghighat, F. Zonal modeling for simulating indoor environment of buildings: Review, recent developments, and applications. *HVAC&R Res.* **2007**, *13*, 887–905.
24. Jiru, E.T.; Haghighat, F. Modeling ventilated double skin façade—A zonal approach. *Energy Build.* **2008**, *40*, 1567–1576. [CrossRef]
25. Song, Z.; Murray, B.T.; Sammakia, B. A dynamic compact thermal model for data center analysis and control using the zonal method and artificial neural networks. *Appl. Therm. Eng.* **2014**, *62*, 48–57. [CrossRef]
26. Chen, Q. Ventilation performance prediction for buildings: A method overview and recent applications. *Build. Environ.* **2009**, *44*, 848–858. [CrossRef]
27. Song, Z.; Murray, B.T.; Sammakia, B. Airflow and temperature distribution optimization in data centers using artificial neural networks. *Int. J. Heat Mass Transf.* **2013**, *64*, 80–90. [CrossRef]
28. Cao, S.-J.; Ren, C. Ventilation control strategy using low-dimensional linear ventilation models and artificial neural network. *Build. Environ.* **2018**, *144*, 316–333. [CrossRef]
29. Zhou, L.; Haghighat, F. Optimization of ventilation system design and operation in office environment, Part I: Methodology. *Build. Environ.* **2009**, *44*, 651–656. [CrossRef]
30. Zhou, L.; Haghighat, F. Optimization of ventilation systems in office environment, Part II: Results and discussions. *Build. Environ.* **2009**, *44*, 657–665. [CrossRef]
31. Oberleithner, K.; Sieber, M.; Nayeri, C.N.; Paschereit, C.O.; Petz, C.; Hege, H.C.; Wygnanski, I. Three-dimensional coherent structures in a swirling jet undergoing vortex breakdown: Stability analysis and empirical mode construction. *J. Fluid Mech.* **2011**, *679*, 383–414. [CrossRef]
32. Lengani, D.; Simoni, D.; Pichler, R.; Sandberg, R.D.; Michelassi, V.; Bertini, F. Identification and quantification of losses in a LPT cascade by POD applied to LES data. *Int. J. Heat Fluid Flow* **2018**, *70*, 28–40. [CrossRef]
33. Kumar, N.; Burton, T.D. Use of random excitation to develop POD based reduced order models for nonlinear structural dynamics. In Proceedings of the ASME 2007 International Design Engineering Technical Conferences and Computers and Information in Engineering Conference, Las Vegas, NV, USA, 4–7 September 2007; American Society of Mechanical Engineers Digital Collection: New York, NY, USA, 2007.
34. Cao, Y.; Zhu, J.; Luo, Z.; Navon, I.M. Reduced-Order Modeling of the Upper Tropical Pacific Ocean Model using Proper Orthogonal Decomposition. *Comput. Math. Appl.* **2006**, *52*, 1373–1386. [CrossRef]
35. Luo, Z.; Zhu, J.; Wang, R.; Navon, I.M. Proper orthogonal decomposition approach and error estimation of mixed finite element methods for the tropical Pacific Ocean reduced gravity model. *Comput. Methods Appl. Mech. Eng.* **2007**, *196*, 4184–4195. [CrossRef]
36. Brunton, S.L.; Kutz, J.N. *Data-Driven Science and Engineering: Machine Learning, Dynamical Systems, and Control*; Cambridge University Press: Chambridge, UK, 2019.
37. Willcox, K. Unsteady flow sensing and estimation via the gappy proper orthogonal decomposition. *Comput. Fluids* **2006**, *35*, 208–226. [CrossRef]
38. Tallet, A.; Allery, C.; Allard, F. POD approach to determine in real-time the temperature distribution in a cavity. *Build. Environ.* **2015**, *93*, 34–49. [CrossRef]
39. Jiang, C.; Soh, Y.C.; Li, H. Two-stage indoor physical field reconstruction from sparse sensor observations. *Energy Build.* **2017**, *151*, 548–563. [CrossRef]
40. Meyer, R.D.; Tan, G. Provide detailed and real-time indoor environmental information using POD–LSE and limited measurements. *Energy Build.* **2014**, *73*, 59–68. [CrossRef]
41. Li, K.; Su, H.; Chu, J.; Xu, C. A fast-POD model for simulation and control of indoor thermal environment of buildings. *Build. Environ.* **2013**, *60*, 150–157. [CrossRef]
42. Sempey, A.; Inard, C.; Ghiaus, C.; Allery, C. Fast simulation of temperature distribution in air conditioned rooms by using proper orthogonal decomposition. *Build. Environ.* **2009**, *44*, 280–289. [CrossRef]

43. Phan, L.; Lin, C.-X. Reduced order modeling of a data center model with multi-Parameters. *Energy Build.* **2017**, *136*, 86–99. [CrossRef]
44. Allery, C.; Béghein, C.; Hamdouni, A. Applying proper orthogonal decomposition to the computation of particle dispersion in a two-dimensional ventilated cavity. *Commun. Nonlinear Sci. Numer. Simul.* **2005**, *10*, 907–920. [CrossRef]
45. Bui-Thanh, T.; Damodaran, M.; Willcox, K. Proper Orthogonal Decomposition Extensions for Parametric Applications in Compressible Aerodynamics. In Proceedings of the 21st AIAA Applied Aerodynamics Conference, Orlando, FL, USA, 23–26 June 2003; American Institute of Aeronautics and Astronautics: Reston, VA, USA, 2003.
46. Sirovich, L. Turbulence and the dynamics of coherent structures. *I. Coherent structures. Q. Appl. Math.* **1987**, *45*, 561–571.
47. Ye, X.; Kang, Y.; Yang, F.; Zhong, K. Comparison study of contaminant distribution and indoor air quality in large-height spaces between impinging jet and mixing ventilation systems in heating mode. *Build. Environ.* **2019**, *160*, 106159. [CrossRef]
48. Yuan, X.; Chen, Q.; Glicksman, L.R.; Hu, Y.; Yang, X. Measurements and computations of room airflow with displacement ventilation. *Ashrae Trans.* **1999**, *105*, 340.
49. Stull, R.B. Review of non-local mixing in turbulent atmospheres: Transilient turbulence theory. *Bound.-Lay. Meteorol.* **1993**, *62*, 21–96. [CrossRef]
50. Sun, X.-H.; Xu, M.-H. Optimal control of water flooding reservoir using proper orthogonal decomposition. *J. Comput. Appl. Math.* **2017**, *320*, 120–137. [CrossRef]
51. Tong, Z.; Chen, Y.; Malkawi, A.; Adamkiewicz, G.; Spengler, J.D. Quantifying the impact of traffic-related air pollution on the indoor air quality of a naturally ventilated building. *Environ. Int.* **2016**, *89*, 138–146. [CrossRef]
52. Tong, S.; Huang, Y.; Tong, Z.; Cong, F. A novel short-frequency slip fault energy distribution-based demodulation technique for gear diagnosis and prognosis. *Int. J. Adv. Robot. Syst.* **2020**, *17*, 1729881420915032. [CrossRef]
53. Tong, S.; Cheng, Z.; Cong, F.; Tong, Z.; Zhang, Y. Developing a grid-connected power optimization strategy for the integration of wind power with low-temperature adiabatic compressed air energy storage. *Renew. Energy* **2018**, *125*, 73–86. [CrossRef]
54. Cheng, Z.; Tong, S.; Tong, Z. Bi-directional nozzle control of multistage radial-inflow turbine for optimal part-load operation of compressed air energy storage. *Energy Convers. Manag.* **2019**, *181*, 485–500. [CrossRef]
55. Tong, Z.M.; Xin, J.G.; Tong, S.G.; Yang, Z.Q.; Zhao, J.Y.; Mao, J.H. Internal flow structure, fault detection, and performance optimization of centrifugal pumps. *J. Zhejiang Univ. Sci. A* **2020**, *21*, 85–117. [CrossRef]

energies

MDPI

Review

Energy Efficiency Evaluation Based on Data Envelopment Analysis: A Literature Review

Tao Xu, Jianxin You, Hui Li and Luning Shao *

School of Economics and Management, Tongji University, Shanghai 200092, China; xutao1007@tongji.edu.cn (T.X.); yjx2256@tongji.edu.cn (J.Y.); lihui@mail.tongji.edu.cn (H.L.)

* Correspondence: shaoluning@tongji.edu.cn

Received: 7 June 2020; Accepted: 7 July 2020; Published: 10 July 2020

Abstract: The importance and urgency of energy efficiency in sustainable development are increasing. Accurate assessment of energy efficiency is of considerable significance and necessity. The data envelopment analysis (DEA) method has been widely used to study energy efficiency as a total factor efficiency assessment method. In order to summarize the latest research on DEA in the field of energy efficiency, this article first analyzes the overall situation of related literature published in 2011–2019. Subsequently, the definition, measurement and evaluation variables of energy efficiency are introduced. After that, this article reviews the current DEA model and its extension models and applications based on different scenarios. Finally, considering the shortcomings of the existing DEA model, possible future research topics are proposed.

Keywords: energy efficiency; data envelopment analysis; literature review; future research

1. Introduction

Energy efficiency is a major global issue that plays an essential role in achieving sustainable development. Although the use of clean energy is gradually increasing, about 80% of global energy consumption is still fossil fuels, such as oil and natural gas, and about 50% of power generation depends on coal resources [1]. As a result, the public, researchers and governments are paying more attention to this issue. It is of considerable significance to evaluate the energy efficiency of different regions and sectors, not only can help identify differences in energy efficiency, but also to provide a quantitative basis for improving efficiency [2]. Patterson [3] first proposed the concept of energy efficiency, considering that it means using fewer resources at the same output, and gave four indicators of energy efficiency measurement. According to this definition, the indicators that measure energy efficiency can be divided into economic energy efficiency and physical energy efficiency.

To the best of our knowledge, the energy efficiency measured by different definitions and indicators varies widely. In order to measure energy efficiency more accurately, many scholars have studied the measurement of energy efficiency. Among them, Hu and Wang [4] proposed the concept of total factor energy efficiency (TFEE), which was widely recognized. TFEE believes that a single energy input cannot produce any output, which means that energy must be combined with other factors (such as labor and capital) to produce output. Based on the TFEE framework, energy efficiency is defined as the ratio of the target energy input to the actual input required at a particular output level. The proposal of TFEE effectively makes up for the shortcomings of traditional single-factor energy efficiency evaluation and has significant enlightening effects on subsequent research.

Early TFEE assessment methods consider that the input of production factors such as labor, capital, and energy will ultimately yield a single output, which is generally expressed as the gross domestic product (GDP). As fossil fuels dominate the world's energy consumption structure, environmental pollution is becoming increasingly severe. Therefore, a large number of studies have gradually incorporated ecological issues into energy efficiency evaluation [5]. It also means that energy efficiency is a significant issue related to the coordinated development of economy, energy and the environment.

With regard to energy efficiency measurement methods, there are parametric and non-parametric methods. Parametric Stochastic Frontier Analysis (SFA) requires the assumption of production function [6], which may have a multicollinearity problem [7]. The Non-parametric DEA method can better deal with the efficiency evaluation of decision-making units under the complicated situation of multiple inputs–outputs and has been widely used to evaluate the TFEE. DEA was first proposed in 1978 as a mathematical programming method for determining the relative effectiveness of homogeneous decision-making units (DMUs) [8]. Zhu, et al. [9] pointed out that DEA is a data-oriented method for evaluating the efficiency of a set of homogeneous DMUs. Compared with previous efficiency evaluation methods, DEA does not need to build a production function, which means that it can better deal with the efficiency of DMUs.

In the existing research, a large number of studies are conducted from the perspective of theory and application based on the data of countries, regions, industries and enterprises. In order to sort out the latest research results of DEA in the field of energy efficiency, this article reviews different DEA models used in energy efficiency evaluation, including basic DEA models and their extensions in different scenarios. Future research directions of DEA-based energy efficiency evaluation methods have also been proposed.

In the rest of this article, the basics of the literature published by DEA Energy Efficiency are presented in Section 2. Section 3 introduces the definition of energy efficiency and input and output variables. Sections 4 and 5 review the DEA-based energy efficiency assessment models and applications, respectively. Conclusions and future research discussions are illustrated in Section 6.

To produce clear descriptions, all the acronyms mentioned in this paper are listed in the nomenclatures.

2. DEA-Based Energy Efficiency Publications

This section first analyzes the publication years, journals and authors. Subsequently, through a visual analysis of the keywords and their evolutionary context, the basic situation of research in the field of energy efficiency evaluation based on DEA can be discovered.

2.1. Number of Publications, Journal Distribution and Authors

In order to analyze the latest research in the field of DEA energy efficiency evaluation, this article searches the relevant articles in the core database of the web of science. The search method used in this article is for the literature with the subject "energy efficiency" and the titles containing "DEA", as well as a publication period of 2011–2019. In the search results, the "article" type was selected, and 281 articles were selected in the end.

Based on the published papers retrieved, this article first analyzes the journal distribution of the literature. Journals with more than six published articles are shown in Figure 1. Among them, the journal with the most published articles is *Energy Economics*, with 26 published articles. Besides, the number of journals published in *Sustainability* and *Energy* exceeded 20. *Energies* and *Journal of Cleaner Production* have also published more than 15 articles.

Figure 1. Journal distribution and number of publications.

Figure 2 shows the number of papers published in the field of DEA energy efficiency from 2011 to 2019. As can be seen from Figure 2, the number of papers published since 2011 has shown a clear upward trend. The quantity reached 61 articles in 2019. This also indicates that the issue of energy efficiency is getting more and more attention from researchers.

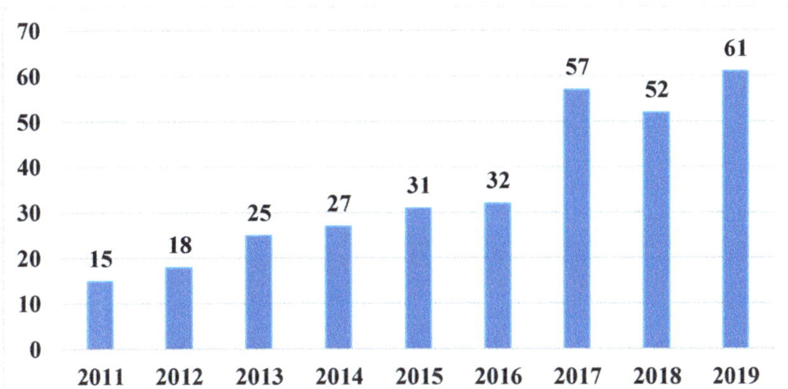

Figure 2. Number of publications from 2011 to 2019.

2.2. Keyword Evolution Analysis

In order to further analyze the research progress of DEA in the field of energy efficiency, this article uses Citespace to visualize the keywords and their research evolutionary context between 2010 and 2019. The context of keyword evolution is shown in Figure 3.

It can be found from Figure 3 that related research is mainly carried out from theory and applications. Theoretical analysis is primarily based on the DEA method to improve the evaluation model to evaluate the energy efficiency better.

Combined with the development and evolution of the DEA model, with the deepening of the research, the construction of the DEA-based energy efficiency evaluation model is more in line with the actual situation. These models have changed from traditional radial models (CCR, BCC) to radial SBM models, from a single output to considering undesirable output, and from simple static structures to dynamic models of complex network structures. Besides, it can be seen in Figure 3 that the primary research objects of the existing literature are regional energy efficiency (country, province, and city), industrial energy efficiency (manufacturing and agriculture), and company energy efficiency (power plant). The detailed theory and application of DEA in energy efficiency will be further discussed in the following article.

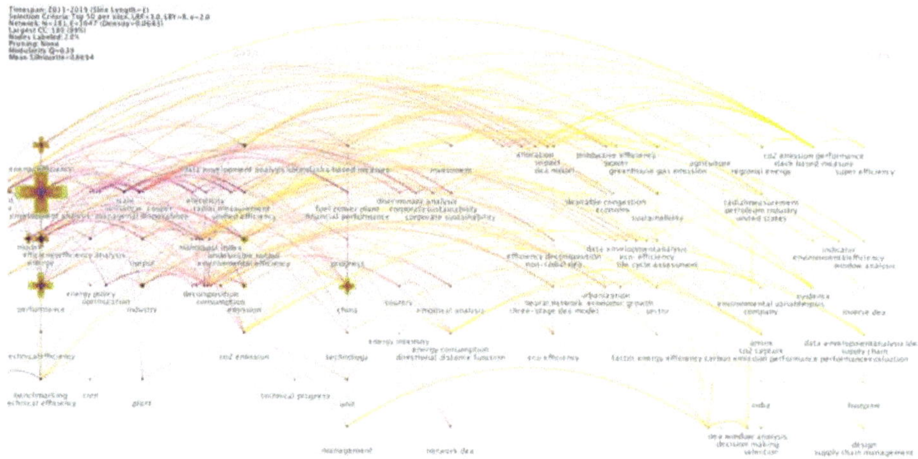

Figure 3. The context of keyword evolution.

3. Energy Efficiency Definition and Input–Output Variables

3.1. Energy Efficiency Definition

As traditional energy efficiency measurement methods ignore other inputs, the concept of TFEE proposed by Hu and Wang [4] has been widely accepted. In the TFEE framework, it is believed that energy itself cannot produce any output and must be put together with other factors to produce output. The TFEE index incorporates energy, labor, and capital into the input system to generate economic output. Energy efficiency is defined as the ratio of target energy input to actual energy input, as shown below.

$$0 \leq \frac{Target\ Energy\ Input}{Actual\ Energy\ Input} \leq 1$$

3.2. Input–Output Variables

Selecting the appropriate input and output variables is an important step in the evaluation of TFEE using the DEA model. Although there have been many studies on energy efficiency analysis, there is still no unique standard for selecting input and output variables. In conventional energy efficiency measures, energy is used as a single input to generate GDP [3]; Hu and Wang [4] introduced labor and capital as input into the energy efficiency evaluation system for the first time, to evaluate the energy efficiency of 29 provinces and cities in China. By adopting the same variables, Honma and Hu [10] calculated the energy efficiency of a Japanese region from 1993 to 2003. Zhang, et al. [11] evaluated the energy efficiency of 27 developing countries.

In production activities, as the energy consumption structure is still dominated by fossil energy, a large number of carbon emissions, wastewater, and waste gas generated by the input of traditional energy have a serious impact on the environment. In this context, carbon emissions, waste gas, etc., are included as undesirable outputs when evaluating energy efficiency [12]. For example, Zhang, Sun and Huang [5] used carbon emissions as undesirable output and GDP as a desirable output when evaluating the energy efficiency of CDM member states. Li and Lin [13], Zhang and Choi [14], and Wang, et al. [15] also used GDP and carbon emissions as output variables when evaluating energy efficiency in 30 provinces and cities in China. Makridou, et al. [16] and Feng and Wang [17] also use the above variables to evaluate the efficiency of high energy consuming industries in EU countries and Chinese provincial-level industrial sectors.

Through the above analysis, it can be seen that there is a common system when using DEA for TFEE evaluation. Without loss of generality, energy, capital and labor are used as inputs. The main reason for introducing labor and capital is that energy itself cannot produce any output, and it must be put together with other factors to produce the output [4]. Regarding output variables, GDP is generally used as the expected output. Considering the impact of energy consumption on the environment, carbon emissions are typically used as undesirable output.

The variables will also vary depending on the different subjects. For example, when evaluating the energy efficiency of industries or enterprises, the input–output variables should be consistent with their actual production processes. Wu, et al. [18] selected passenger seats, transportation energy consumption, fixed assets, and transportation mileage as inputs. Passenger turnover and carbon emissions are selected as outputs to evaluate the energy efficiency of China's transportation sector. Zhang and Choi [14] used the amount of power generation as the expected output when assessing the energy efficiency of 252 power plants in China.

4. Construction of DEA-Based Models in Energy Efficiency Evaluation

Since DEA was proposed in 1978, it has been widely used in the efficiency evaluation of multiple input–output problems. Since then, there has been continuous research to expand the DEA model based on different theoretical and realistic backgrounds. Many models have also been proposed, including radial and non-radial, static and dynamic, single structure and network structure models, etc. [19]. In the field of energy efficiency assessment, due to the existence of undesirable carbon emissions, modeling of energy efficiency, including undesirable outputs, has also attracted the attention of researchers. This section reviews the energy efficiency methods from the perspective of basic and extended DEA methods.

4.1. The Theoretical Basis of DEA

Data Envelopment Analysis (DEA) is a data-oriented method for estimating the full factor efficiency of homogeneous decision units (DMUs) [19]. DEA can obtain the weight of a set of optimal input and output variables through optimization methods based on objective data of the evaluation object, and determine the efficiency level of DMU in the form of input and output ratio [8]. The basic logic of the DEA method is to construct a set of homogeneous DMU convex combinations based on input–output data to obtain an efficient production frontier. By comparing the actual input–output data of the DMU with the projection data of the frontier, the DMU is evaluated by relative efficiency [20,21]. Since the DEA method was proposed, it has been widely used in different fields [9,19]. As to the DEA method, many scholars have also proposed new concepts and models, such as the cross-efficiency model, the super-efficiency model, SBM model, network structure model, etc. [22–24].

Charnes, Cooper and Rhodes [8] first proposed the DEA theory method. In the later DEA literature, the first DEA model they created, namely the CCR model, was named after the first letters of the three surnames of Charnes, Cooper and Rhodes. The CCR model assumes constant returns to scale (CRS). In the DEA method, the evaluated objects are represented by DMUs. Each DMU produces R outputs by inputting M production factors. In model (1), x_{mo}, y_{ro} represents the m-th input and r-th output of the DMU, respectively, and u_m, v_r are the weights of the corresponding output and input variables. The input–output efficiency of the o-th DMU can be obtained from model (1). According to the weights of input and output variables, the CCR model can be understood as turning the multi-input–output problems into a virtual single input–output one. For a specific DMU, this single efficiency is measured by the ratio of virtual output to input. The linear programming of the CCR model can be expressed as maximizing the efficiency of a specific DMU with the condition that the efficiency of all DMUs does not exceed 1.

$$
\begin{aligned}
&\max \frac{\sum_{m=1}^{M} u_m y_{mo}}{\sum_{r=1}^{R} v_r x_{ro}} \\
&s.t. \frac{\sum_{m=1}^{M} u_m y_{mn}}{\sum_{r=1}^{R} v_r x_{rn}} \leq 1; \\
&u \geq 0; v \geq 0;\ m = 1,2,\ldots,M; \\
&r = 1,2,\ldots,R;\ n = 1,2,\ldots,N
\end{aligned}
\tag{1}
$$

Since model (1) is a nonlinear programming one, let $t = \frac{1}{\sum_{r=1}^{R} v_r x_{ro}}, \mu = tu, v = tv$, model (1) can be transformed into an equivalent linear programming model, as shown in model (2). Model (2) is called the DEA multiplier form, and its dual model can be represented by model (3), known as the envelope form. The CCR model assumes that the returns to scale are unchanged, and the technical efficiency obtained includes the scale efficiency component, so it is called comprehensive technical efficiency [8]. According to the way efficiency is measured, DEA models can also be divided into input-oriented, output-oriented, and non-oriented.

$$
\begin{aligned}
&\max \sum_{m=1}^{M} \mu_m y_{mo} \\
&s.t. \sum_{m=1}^{M} \mu_m y_{mn} - \sum_{r=1}^{R} v_m x_{mn} \leq 1 \quad m = 1,2,\ldots,M \\
&\sum_{r=1}^{R} v_m x_{mo} = 1 \quad r = 1,2,\ldots,R \\
&v \geq 0; \mu \geq 0; n = 1,2,\ldots N
\end{aligned}
\tag{2}
$$

$$
\begin{aligned}
&\min \theta \\
&s.t. \sum_{n=1}^{N} \lambda_n x_{mn} \leq \theta x_{mo}\ m = 1,2,\ldots,M \\
&\sum_{n=1}^{N} \lambda_n y_{rn} \geq y_{ro} \quad r = 1,2,\ldots,R \\
&\lambda_n \geq 0, n = 1,2,\ldots,N \\
&j = 1,2,\ldots N
\end{aligned}
\tag{3}
$$

As mentioned above, many scholars have also proposed various new concepts and models in the field of the DEA method, and many of those novel methods are utilized to evaluate energy efficiency. In this paper, some basic or extended DEA models for assessing the efficiency of energy performance will be summarized. To better understand these models more visually and clearly, the main features of these models as well as their application scenario are presented in Table 1.

Table 1. The main features and application scenario of DEA.

Type	Model	Main Features and Application Scenario
Basic DEA	CCR DEA	CCR model assumes constant returns to scale, so that the results of the model include scale efficiencies
	BCC DEA	The BCC model assumes variable returns to scale, and the results obtained are referred to as pure technical efficiency
	SBM DEA	As a non-oriented efficiency evaluation model, SBM can be better used to deal with the slack improvement of the input and output variables

Table 1. *Cont.*

Type	Model	Main Features and Application Scenario
Extended DEA	Undesirable DEA	A good solution which can solve the problem of bad output in the production process
	Network DEA	By constructing a DEA model with intermediate output, the network DEA can open the black box to show more detailed efficiency in the production process
	Dynamic DEA	Dynamic DEA can effectively analyze the efficiency of different stages and the changes in efficiency, which can be achieved through the window DEA method or the Malmquist index
	Game Cross-Efficiency DEA	Game Cross-Efficiency DEA can analyze the situation of competition between DMUs, which is helpful to improve the accuracy of efficiency assessment
	Meta-Frontier DEA	Meta-Frontier DEA can solve the difference between different decision-making units due to technical or resource problems, and improve the accuracy of evaluation

4.2. Energy Efficiency Evaluation Model Based on Basic DEA

4.2.1. CCR-Based Evaluation Model

Based on the input-oriented CCR model, Hu and Wang [4] proposed a TFEE evaluation model, as shown in model (4). It is also assumed that there are N DMUs, and the production process of each DMU is the input of energy and other M production factors, resulting in R outputs. The energy input of the o-th DMU is denoted by e_o. x_{mo}, y_{ro} represents the non-energy input and expected output of the DMU respectively.

$$
\begin{aligned}
& \min\theta \\
& s.t. \sum_{n=1}^{N} \lambda_n e_n \leq \theta e_o \\
& \sum_{n=1}^{N} \lambda_n x_{mn} \leq \theta^* x_{mo} \quad m = 1,2,\ldots,M \\
& \sum_{n=1}^{N} \lambda_n y_{rn} \geq y_{ro} \quad r = 1,2,\ldots,R \\
& \lambda_n \geq 0, n = 1,2,\ldots,N
\end{aligned}
\tag{4}
$$

The optimal solution can be obtained by solving model (4), which represents the efficiency of the o-th DMU. However, the above model cannot directly obtain the target energy input. Thus, Ali and Seiford [25] proposed a two-stage method so that the slack variables of input and output can be obtained. The slack variable of energy input can be expressed by s_e^-. The TFEE based on the CCR model can be calculated by formula (5).

$$
CCR - based \quad TFEE = \frac{\theta \times e_o - s_e^-}{e_o} = 1 - \frac{(1-\theta) \times e_o + s_e^-}{e_o} \tag{5}
$$

4.2.2. BCC-Based Evaluation Model

The CCR model assumes that the scale effect of production technology is maintained, but in fact, not all DMUs are in the optimal production scale state. Therefore, the efficiency calculated by the CCR model includes scale efficiency. Subsequently, Banker, Charnes and Cooper [21] proposed a DEA model that considers returns to scale, known as the BCC model. The BCC model is based on variable returns to scale, and the technical efficiency derived excludes the effects of returns on the scale, so it is called pure technical efficiency. The BCC model adds a constraint $\sum_{n=1}^{N} \lambda_n = 1$ to the CCR model; λ is the linear combination coefficient of input–output variables. The BCC-based energy efficiency evaluation model is shown in model (6), and the meaning of the variables in model (6) is consistent with their counterparts in model (4).

$$
\begin{aligned}
& \min \theta \\
& s.t. \sum_{n=1}^{N} \lambda_n e_n \leq \theta e_o \\
& \sum_{n=1}^{N} \lambda_n x_{mn} \leq \theta x_{mo} \quad m = 1,2,\ldots,M \\
& \sum_{n=1}^{N} \lambda_n y_{rn} \geq y_{ro} \quad r = 1,2,\ldots,R \\
& \sum_{n=1}^{N} \lambda_n = 1 \\
& \lambda_n \geq 0, n = 1,2,\ldots,N \\
& BCC-based \quad TFEE = \frac{\theta \times e_o - s_e^-}{e_o} = 1 - \frac{(1-\theta)\times e_o + s_e^-}{e_o}
\end{aligned}
\tag{6}
$$

4.2.3. SBM-Based Evaluation Model

The CCR and BCC models improve the inefficient DMU by reducing (increasing) all inputs (or outputs) proportionally, which are called radial models. In fact, for the inefficient DMU, the gap between the current state and the effective target value also includes part of the slack improvement. Therefore, Tone [26] proposed the Slack-Based Model (SBM), a non-oriented efficiency evaluation model, which is used to deal with the slack improvement of the input and output. The SBM has a higher discriminating ability than the CCR and BCC model. The energy efficiency evaluation model based on the non-oriented SBM is shown in model (7).

In model (7), the optimal value of the objective function is the efficiency value of the unoriented SBM. s_e^- represents the slack variable of the energy input, S_m^- and S_m^+ represent the slack variable of the m-th input and r-th output. The remaining variables are consistent with those in the model (6). According to the Charnes–Cooper transformation, the abovementioned nonlinear SBM can be transformed into a linear model for solving.

$$
\begin{aligned}
& \min \frac{1-\frac{1}{M+1}\left(\sum_{m=1}^{M} \frac{s_m^-}{x_{mo}} + \frac{s_e^-}{e_o}\right)}{1-\frac{1}{R}\sum_{r=1}^{R} \frac{s_r^+}{y_{ro}}} \\
& s.t. \sum_{n=1}^{N} \lambda_n e_n = e_o - s_e^- \\
& \sum_{n=1}^{N} \lambda_n x_{mn} = x_{mo} - s_m^-, \quad m = 1,2,\ldots,M \\
& \sum_{n=1}^{N} \lambda_n y_{rn} = y_{ro} + s_r^+, \quad r = 1,2,\ldots,R \\
& \lambda_n \geq 0, s_e^- \geq 0, s_m^- \geq 0, s_r^+ \geq 0, n = 1,2,\ldots,N
\end{aligned}
\tag{7}
$$

4.3. *Energy Efficiency Evaluation Model Based on Extended DEA*

4.3.1. Evaluation Model Considering the Impact of Carbon Emissions

In recent years, the use of clean energy has gradually increased, but the energy consumption structure has not changed in the short term, with fossil energy as the mainstay. A large number of carbon emissions, wastewater and gas caused by the use of traditional energy sources have a serious impact on the environment. Therefore, it is necessary to consider the impact of carbon emissions and wastewater emissions when evaluating energy efficiency [12]. For example, He, et al. [27] found that undesirable output has a great impact on energy efficiency when studying the energy efficiency of OECD countries. Disregarding undesirable output often leads to an overestimation of energy efficiency.

Regarding the setting of the undesirable output in the DEA method, it can be divided into strong disposability and weak disposability [28]. The first is strong disposability, which can reduce undesirable output without reducing expected output [29]. In the case of strong disposability assumptions, the undesirable output can be treated as input [30,31] or transform the data of the undesirable output, including linear transformation, inverse transformation and exponential transformation [32–34]. The second is the assumption of weak disposability of undesirable output, that is, to reduce undesirable output requires additional input or reduce expected output [35,36]. It implies that the reduction in undesirable output comes at the cost of expected output. There is also a null joint hypothesis for undesirable output, which indicates that as long as there is expected output in production activities, it must accompany undesirable output. The only solution to avoid undesirable output is to stop production.

In the existing literature, the assumption of weak disposability is widely adopted. As the setting of null-jointness is more in line with reality, that is to say, the input of energy, labor and capital will produce greenhouse gas emissions while generating economic benefits. According to Apergis, et al. [37], in the SBM-DEA model considering undesirable output, it is assumed that there are N DMUs, and the production process of each DMU is through the input of energy and other M production factors. This results in R expected outputs and K undesirable outputs. u_{qo} represents the q-th undesirable output of the o-th DMU, and the remaining variables are consistent with model (7). The SBM-DEA model based on weak disposability is shown in model (8). The constraint of model (8) on the undesirable output, $(1+\frac{1}{R}\sum\limits_{r=1}^{R}\frac{s_r^+}{y_{ro}})$ indicates that the expected output and the undesirable output have the same proportion of change.

$$\begin{aligned}
&\min\frac{1-\frac{1}{M+1}\left(\sum\limits_{m=1}^{M}\frac{s_m^-}{x_{mo}}+\frac{s_e^-}{e_o}\right)}{1+\frac{1}{R+Q}\left(\sum\limits_{r=1}^{R}\frac{s_r^+}{y_{ro}}+\sum\limits_{q=1}^{Q}\frac{s_q^-}{u_{qo}}\right)}\\
&s.t.\sum_{n=1}^{N}\lambda_n e_n=e_o-s_e^-\\
&\sum_{n=1}^{N}\lambda_n x_{mn}=x_{mo}-s_m^-,\quad m=1,2,\ldots,M\\
&\sum_{n=1}^{N}\lambda_n y_{rn}=y_{ro}+s_r^+,\quad r=1,2,\ldots,R\\
&\sum_{n=1}^{N}\lambda_n u_{qn}=\left(1+\frac{1}{R}\sum_{r=1}^{R}\frac{s_r^+}{y_{ro}}\right)u_{qo}-s_q^-,\quad q=1,2,\ldots,Q\\
&\lambda_n\geq 0,s_e^-\geq 0,s_m^-\geq 0,s_r^+\geq 0,s_q^-\geq 0,n=1,2,\ldots,N
\end{aligned}\tag{8}$$

4.3.2. Evaluation Model Considering the Network Structure

The basic DEA method regards the production system as a "black box", which means that only the initial input and final output of the system are considered to evaluate the efficiency of the decision unit [22]. However, the energy system is regarded as a complex process, in other words, the "black box" assumption ignores the process of energy conversion or transmission. For example, the conversion process of primary energy (such as coal, oil, and natural gas) to secondary energy (such as coke, liquefied petroleum gas, and thermal power) in industrial systems, as well as the process of power generation, power distribution transmission, and end-user energy consumption by power generation companies. The energy efficiency obtained under the "black box" assumption cannot provide specific process guidance in the energy production sector and the energy utilization sector to improve energy efficiency separately [38].

In order to open the "black box" to evaluate the process efficiency of DMUs, Färe and Grosskopf [22] first proposed a network DEA model considering intermediate output. Since Seiford and Zhu [39] first applied two-stage DEA to US commercial banks, network DEA has been widely used for efficiency in various industrial and commercial sectors. Since then, Tone and Tsutsui [40] proposed the SBM-Network DEA model, and Fukuyama and Weber [41] extended the SBM network DEA model to evaluate the efficiency of systems with undesirable outputs.

In the two-stage DEA structure, the DMU is divided into two sub-DMUs, where each sub-DMU consumes input to generate output, where the input of the second stage is the output of the first stage. When evaluating energy efficiency, Liu and Wang [38] divided the energy efficiency of the industrial sector into two phases of energy production and energy consumption and used a two-stage network DEA to evaluate energy efficiency, as shown in model (9).

In the network structure of model (9), the first stage is the energy production sector, where the input is primary energy and the output is secondary energy. The second stage is the energy consumption sector, whose input is the output produced in the first stage. λ^1 and λ^2 represent the linear combination coefficients of the input and output in the first and second phases, respectively; e_o is the primary energy input in the first phase; x_{mo} is the m-th input index of the o-th DMU in the first phase; z_{to} is the t-th secondary energy of the o-th DMU generated in the first stage, and it is also the input index of the second stage; y_{ro} is the r-th output index of the o-th DMU in the second stage.

$$
\begin{aligned}
&\max \tfrac{1}{2} \cdot \tfrac{1}{M+1}\Big(\sum_{m=1}^{M} \tfrac{s_m^-}{x_{mo}} + \tfrac{s_e^-}{e_o} \Big) + \tfrac{1}{2} \cdot \tfrac{1}{R} \sum_{r=1}^{R} \tfrac{s_r^+}{y_{ro}} \\
&s.t. \\
&\left.\begin{aligned}
&\sum_{n=1}^{N} \lambda_n^1 e_n = e_o - s_e^- \\
&\sum_{n=1}^{N} \lambda_n^1 x_{mn} = x_{mo} - s_m^-, m = 1,2,\ldots,M \\
&\sum_{n=1}^{N} \lambda_n^1 z_{tn} = z_{to} + s_t^{1+}, t = 1,2,\ldots,K
\end{aligned}\right\} stage1 \\
&\left.\begin{aligned}
&\sum_{n=1}^{N} \lambda_n^2 z_{tn} = z_{to} - s_t^{2-}, t = 1,2,\ldots,K \\
&\sum_{n=1}^{N} \lambda_n^2 y_{rn} \geq y_{ro} + s_r^+, r = 1,2,\ldots,R
\end{aligned}\right\} stage2 \\
&\lambda_n^1 \geq 0, \lambda_n^2 \geq 0, s_e^- \geq 0, s_m^- \geq 0, s_t^{1+} \geq 0, \\
&s_t^{2-} \geq 0, s_r^+ \geq 0, n = 1,2,\ldots,N \\
&Network - DEA - based \quad TFEE = 1 - \tfrac{s_e^-}{e_o}
\end{aligned}
\tag{9}
$$

4.3.3. Evaluation Model Considering the Dynamic Process

When the data of the evaluated DMU are panel data of multiple periods, the assumption that the production technology of the basic DEA method is the same all the time does not meet the actual situation. Therefore, Charnes and Cooper [42] introduced DEA window analysis as an extension of the traditional DEA method. DEA window analysis can analyze section data and panel data to analyze the dynamic efficiency of the evaluated object. As a commonly used dynamic efficiency analysis method, window DEA analysis is based on the moving average principle. It treats each DMU in different periods as a separate unit for efficiency measurement. When evaluating energy efficiency under the window analysis framework, the energy efficiency of a region in one period can be compared with the efficiency of other regions and its efficiency in other periods [43]. In addition, window analysis can also explore the energy efficiency of different regions in different years through a series of overlapping windows [44].

Besides, the non-parametric Malmquist productivity index (MPI), as a time series analysis technique, has also been extended to DEA models to characterize dynamic changes in efficiency. Färe, et al. [45] introduced the Malmquist index into the DEA model to evaluate the efficiency changes of different DMUs and the dynamic changes of production technology in two periods. In the field of energy efficiency evaluation, the Malmquist model is also widely used [17,46].

4.3.4. Evaluation Model Considering Game Relations

The basic DEA method maximizes the efficiency of the evaluation object by selecting a set of optimal weights, which usually leads to an overestimation of the evaluation object's efficiency value [47]. Therefore, Sexton, et al. [48] proposed cross-efficiency DEA, in which the mutual evaluation weights were added between DMUs to improve objectivity. In reality, there is often a direct or indirect competitive relationship between different DMUs. Especially in the evaluation of energy efficiency, competition may be more intense due to the scarcity of energy. Considering that when there is a competitive relationship, the cross-efficiency DEA evaluation method will not be able to evaluate the efficiency value of the DMU correctly. To this end, Liang, et al. [49] consider the game relationship between different DMUs and extend the DEA cross-efficiency to the DEA game cross-efficiency, by maximizing the cross-efficiency of each DMU without reducing the efficiency of other DMUs.

In the field of energy efficiency evaluation, DEA efficiency is widely used in game cross-efficiency. Chen, et al. [50] introduced the game cross-efficiency DEA for the first time to measure the energy efficiency of the power industry in China's provincial regions under environmental constraints. Studies have shown that the energy efficiency of eastern China is much higher than that of central and western China. Xie, et al. [51] evaluated the environmental efficiency of China's power generation industry through the game cross-efficiency DEA, and the results showed that there was a significant efficiency gap between regions. Yang and Wei [52] also used this method to analyze the energy efficiency of 26 prefecture-level cities in China. Considering that the cross-game DEA model requires multiple steps, this article does not describe the model in detail here. The detailed model can be understood by reading the relevant literature mentioned in this section.

4.3.5. Evaluation Model Considering Technical Heterogeneity

Based on a unified reference technology assumption, the basic DEA method assumes that all DMUs participate in the evaluation with the same technical benchmark. It means that the heterogeneity of different evaluation objects is ignored in the efficiency evaluation. In fact, there are huge differences in natural resources, economic foundations, urbanization levels, and industrial structures in various regions of China. For example, some provinces are dominated by agriculture and services, while others are dominated by manufacturing. In this case, each region of China may have its own characteristics of typical energy use, and the energy structure and technological level among the provinces are quite different [53,54].

In order to solve the problem of bias in efficiency evaluation results due to heterogeneity between DMUs, Battese, et al. [55] introduced a meta-frontier model for different groups with different technologies. It can calculate comparable technical efficiency for DMUs under different technologies. Since then, DEA and meta-frontier models have been widely used in energy efficiency evaluation. Yu, et al. [56] combined the meta-frontier method with super-SBM to study the energy efficiency of various regions in China during 2006–2016. Yu, You, Zhang and Ma [55] analyze the energy efficiency of 277 cities in China from 2007 to 2014, taking into account the effects of technological heterogeneity. Wang, et al. [57] classified Guangdong enterprises based on geographic boundaries and industry classification systems and used group frontier and meta-frontier direction distance functions to analyze their energy efficiency. The technical explanation and detailed settings of the meta-frontier model can be obtained in the related literature mentioned in this section.

5. Application of DEA Model in Energy Efficiency Evaluation

As DEA has become an important and commonly used analysis tool and method in the field of energy efficiency assessment, a large amount of the literature evaluates energy efficiency based on data from countries, regions, industries and companies. This section will introduce the application of DEA in energy efficiency evaluation.

5.1. Energy Efficiency Evaluation of Regions

After Hu and Wang [4] first proposed the total factor energy efficiency framework and evaluated the energy efficiency of various regions in China, the DEA method was widely used in national and regional energy efficiency evaluation. This section reviews the studies that have evaluated energy efficiency in different regions using DEA from 2015 to 2019 and the results are shown in Table 2.

Jebali, et al. [58] analyzed the energy efficiency and influencing factors of Mediterranean countries during 2009–2012. The results of the study indicate that the energy efficiency levels in Mediterranean countries are high and decline over time. Gross national income per capita, population density and the use of renewable energy can affect energy efficiency. Zhao, et al. [59] measured the energy efficiency of 35 Belt and Road countries in 2015 based on a three-stage DEA model. The results show that South Korea, Singapore, Israel, and Turkey have a TFEE of 1. Uzbekistan, Ukraine, South Africa and Bulgaria are less efficient. He, Sun, Shen, Jian and Yu [27] established an DEA-based energy efficiency evaluation model for measuring the energy efficiency of 32 OECD countries from 1995 to 2016. Additionally, the effects of environmental factors on energy efficiency assessment were compared through efficiency analysis and predicted value analysis. Wang, et al. [60] use the DEA-Malmquist method to measure the energy efficiency of 25 countries; the results of this study show that by using the same inputs as developing countries, the developed countries' balance between GDP growth and carbon dioxide emissions is more balanced. In addition, India and China increased their energy intensity during 2010–2017.

Table 2. Energy efficiency evaluation of regions.

Author	Subject of Evaluation	Model
He, Sun, Shen, Jian and Yu [27]	32 OECD countries	CCR DEA
Bampatsou, et al. [61]	15 EU countries	CCR-DEA
Zhang, Cheng, Yuan and Gao [11]	23 developing countries	BCC DEA
Zhang and Choi [14]	30 provinces in China	SBM DEA
Guo, et al. [62]	Western of China	SBM DEA
Apergis, Aye, Barros, Gupta and Wanke [37]	20 OECD countries	SBM-Undesirable DEA
Zhao, Zhang, Zeng, Li, Liu, Qin and Yuan [59]	35 Belt and Road countries	Network DEA
Jebali, Essid and Khraief [58]	24 Mediterranean countries	Network DEA
Wu, Yin, Sun, Chu and Liang [12]	30 provinces in China	Network DEA
Wu, et al. [63]	30 provinces in China	Dynamic DEA
Wang, Yu and Zhang [43]	29 provinces in China	Dynamic DEA
Guo, et al. [64]	27 countries	Dynamic DEA
Wang, Le and Nguyen [60]	25 countries	Dynamic DEA
Amowine, et al. [65]	25 African countries	Dynamic DEA
Wang, Deng, Zhang and Zhang [57]	Guangdong province	Meta-Frontier DEA
Zhang, Sun and Huang [5]	16 CDM countries	Meta-Frontier DEA
Li and Lin [13]	30 provinces in China	Meta-Frontier DEA
Yu, Zhou and Yang [53]	30 provinces in China	Meta-Frontier DEA
Sun, et al. [66]	211 cities in China	Meta-Frontier DEA
Yu, You, Zhang and Ma [56]	277 cities in China	Meta-Frontier SBM
Yang and Wei [52]	26 cities in China	Game Cross-Efficiency DEA

In addition to evaluating a country's energy efficiency, the energy efficiency of provinces and cities has also attracted the attention of many researchers, especially in China's provinces and cities. Wang, Yu and Zhang [43], Li and Lin [13], and Wu, Zhu, Yin and Song [63] adopt the DEA method to evaluate the energy efficiency of 30 provinces in China, and research shows that most provinces are less energy efficient. Eastern China has the highest energy efficiency, while western China has the worst energy efficiency. Efficiency has improved in most regions during 2006–2010. Yu, You, Zhang and Ma [56] proposed an energy efficiency evaluation model that takes into account regional technological heterogeneity and carbon emissions. By evaluating the energy efficiency of 277 cities in China between 2007 and 2014, the study found that there are large differences in the energy efficiency of Chinese cities. Sun, Wang and Li [66] considered the heterogeneity and technology gap of energy management in different regions and measured the energy efficiency of 211 cities in the country. The results show that the overall efficiency of Chinese cities is low, while that of central China is the lowest, and there is a huge technological gap between regions. Yang and Wei [52] used the game cross-efficiency DEA method to analyze the urban total factor energy efficiency of 26 prefecture-level cities in China from 2005 to 2015. The results show that the energy efficiency of cities considering competition is lower than traditionally calculated energy efficiency. During the study period, the study concluded that urban energy efficiency did not improve. There are also studies evaluating regional energy efficiency in other countries. For example, Honma and Hu [10] used the DEA method to analyze total factor energy efficiency based on data from 47 cities and counties in Japan.

5.2. Energy Efficiency Evaluation of Industries and Companies

DEA is also widely used in assessing industry energy efficiency. Through a search of the related literature, it can be found that research on industry energy efficiency is mainly concentrated in high energy consuming industries such as electricity, construction, and transportation. Table 3 shows the energy efficiency evaluation of industries.

Makridou, Andriosopoulos, Doumpos and Zopounidis [16] used the DEA method to assess the energy efficiency of five energy-intensive industries (building, power, manufacturing, mining, and transportation sectors) in 23 EU countries between 2000 and 2009. The study found that overall efficiency has improved across all sectors during this period. Lee and Choi [67] evaluated the energy and environmental efficiency of seven manufacturing sectors in South Korea from 2011 to 2017, and the results showed that energy efficiency improved by an average of 0.3% during the study period. Zhou, Xu, Wang and Wu [34] conducted an empirical study on the energy efficiency of China's industrial sector from 2010 to 2014, and the results showed that most sectors of Chinese industry performed poorly, especially those related to energy extraction. Lei, et al. [68] evaluated the energy efficiency of 30 provincial transport departments in China. The results show that the energy efficiency of the provincial transport departments in China varies widely; efficiency is better than in the midwest of China. Djordjevic and Krmac [69] uses a non-radial DEA to evaluate the energy efficiency of the transportation industry (road, railway and aviation sectors) in Europe. Studies indicate that the energy efficiency of the road sector is improving, while the energy efficiency of the railway transport sector in many assessed countries is low.

Table 3. Energy efficiency evaluation of industries.

Author & Year	Subject of Evaluation	Model
Zhou, Xu, Wang and Wu [34]	38 Chinese industrial sectors	BCC DEA
Wang, Zeng, Wei and Zhang [15]	30 Chinese provincial industrial sectors	BCC DEA
Lei, Li, Zhang, Dai and Fu [68]	30 Chinese interprovincial transport sectors	SBM-DEA
Liu and Wang [38]	30 Chinese provincial industrial sectors	Network DEA
Wu, et al. [70]	30 Chinese provincial industrial sectors	Network DEA
Makridou, Andriosopoulos, Doumpos and Zopounidis [16]	23 Energy-intensive industries in EU countries	Dynamic DEA
Lee and Choi [67]	7 Korean manufacturing sectors	Dynamic DEA
Perez, Gonzalez-Araya and Iriarte [46]	20 Chilean manufacturing industry	Dynamic DEA
Fei and Lin [71]	30 Chinese provincial agricultural sectors	Meta-Frontier Malmquist DEA
Feng and Wang [17]	30 Chinese provincial industrial sectors	Meta-Frontier Malmquist DEA
Han, et al. [72]	42 Chinese industrial sectors	Game Cross-Efficiency DEA
Xie, Gao, Zhang, Pang and Zhang [51]	30 Chinese provincial generation sectors	Game Cross-Efficiency DEA

Compared to the regional and industry levels, energy efficiency at the enterprise level is relatively low. In the existing research, Cui and Li [73] used DEA to analyze the energy efficiency of 11 airlines from 2008 to 2012. The results show that capital efficiency is an important factor to promote energy efficiency. The US financial crisis had a significant impact on energy efficiency. Zhang and Choi [14] carried out an empirical analysis of the energy efficiency of fossil fuel power generation in Korea by using the DEA method. The results show that coal-fired power plants have higher total energy efficiency than oil-fired power plants, and the technology gap of coal-fired power plants is smaller than that of oil-fired power plants. Studies show that the Korean government should promote technological innovation to reduce the technology gap in coal-fired power plants. Bi, et al. [74] analyzed the energy efficiency of Chinese fossil fuel power generation enterprises. They pointed out that the energy and environmental efficiency of the enterprises are low, and there are large differences between provinces. In addition to power generation companies, Zhang, et al. [75] also analyzed the energy efficiency of 62 power generation equipment.

6. Findings and Future Research Discussions

6.1. Main Findings

By analyzing the literature on energy efficiency evaluation using the DEA method, it can be found that a large number of studies are conducted from the perspective of theory and application based on the data of countries, regions, industries and enterprises. The research has attracted more researchers' attention and the number of publications has gradually increased since 2011. From a methodological perspective, the DEA-based energy efficiency evaluation models are more consistent with the actual situation, such as extending from a single output model to an evaluation model that considers pollution emissions. The analysis of the research stage also ranges from a single stage to a multi-stage energy conversion issue. In addition, a dynamic analysis of multi-year efficiency is also the focus of one study. In other words, the construction of the energy efficiency evaluation model based on DEA has evolved from a static structure of a simple structure to a dynamic model of a complex network structure, and the accuracy of the efficiency evaluation has also been continuously improved.

Based on the above analysis of the related research on energy efficiency using DEA, this article discusses the overall situation of existing research and existing research deficiencies as follows:

(1) From the perspective of research objects, a large number of documents use data from countries, regions, industries and companies. Many research results have been obtained. Especially as China is a large country of energy consumption and carbon emissions, a large number of studies have been conducted on energy efficiency in China. Aiming at the technical heterogeneity of energy efficiency and competitive cooperation between different research objects, existing research proposes corresponding expansion models for different scenarios to improve the accuracy of efficiency assessment. It is not difficult to find that most of the existing energy efficiency is analyzed at the regional level. Although the energy efficiency at the company level has also attracted the attention of many scholars, compared with the regional and industry sectors, the energy efficiency analysis for enterprises is relatively small.

(2) From a method point of view, a large number of scholars have improved the model from different perspectives, and the accuracy of energy efficiency assessment has also continuously improved. With the expansion of research, the level of agreement between the construction of the DEA-based energy efficiency evaluation model and the actual situation continues to increase. However, as a data-oriented efficiency assessment method, DEA is mainly based on analysis with structured and clear data. Model studies that can deal with energy efficiency issues in complex data environments such as heterogeneity, uncertainty, or big data are still lacking. As the complexity of products and services continues to increase, the depth of energy efficiency assessment objects, especially at the microdata level, such as enterprise-level data and production line data, is often unstructured, and different data structures affect DEA. The accuracy of the assessment will also have an impact, resulting in increased errors in the efficiency assessment. Therefore, with the increasing complexity of the energy

system, building a DEA model in a complex data environment will enable a more effective evaluation of energy efficiency.

6.2. Future Research Discussions

In order to inspire subsequent research on energy efficiency assessment using DEA, this paper proposes possible future studies from the perspective of application areas and models.

6.2.1. Further Research on Energy Efficiency Issues in Enterprises

This paper believes that research on the energy efficiency of enterprises will help to further improve energy efficiency if data are available. Specifically, the analysis of corporate data helps reveal the state of corporate energy-saving technologies. Besides, with the continuous improvement of carbon trading markets and policies, analyzing the energy efficiency level of enterprises will help companies to manage carbon emission quotas and improve their competitiveness.

6.2.2. Further Research on Energy Efficiency Based on Complex Data Environment

For energy efficiency assessment models based on complex data environment, as the complexity of energy systems continues to increase, it is particularly important to build evaluation models that can analyze complex data. In this article, complex data may include inaccurate or ambiguous observations of input and output data, large datasets for analysis, and heterogeneous data due to differences in input or output structure.

(1) The DEA energy efficiency evaluation model in the heterogeneous data environment.

Despite the continuous development of current information technology and the continuous improvement of data retrieval and analysis capabilities, there will still be data heterogeneity in the evaluation. Unlike the problem of data loss caused by data retrieval and data storage, the data heterogeneity discussed here is due to differences in the input or output variables caused by the complexity of the production system. For instance, Cook, et al. [76] pointed out that steel plants will produce different types of steel even if they invest the same resource structure. When the traditional DEA method is used for evaluation, the efficiency will be biased. In fact, researchers have begun to consider the heterogeneity of output indicators. Wu, et al. [77] have started to discuss the use of improved DEA analysis to evaluate the efficiency of DMUs with different input and output indicators. It is not difficult to find that under different energy consumption scenarios, especially at the microdata level, it is particularly important to expand the efficiency assessment method in the case of heterogeneous input–output variables.

(2) The DEA energy efficiency evaluation model in the uncertain data environment.

In the reality of energy efficiency assessments, the observations of input and output data may be inaccurate or ambiguous [78]. The efficiency evaluation in the uncertain environment has attracted the attention of many researchers. Among them, the fuzzy set theory proposed by Zadeh, et al. [79] and others has been widely adopted. Based on fuzzy theory, some researchers have suggested the Fuzzy DEA model [80,81]. In the field of energy efficiency assessment, the expansion and application of the Fuzzy DEA model will help to improve the accuracy of energy efficiency assessment.

(3) The DEA energy efficiency evaluation model in the big data environment.

In the big data environment, the dataset used for analysis is usually very large, which causes the traditional DEA calculation process to take a long time. Therefore, analyzing big data makes researchers face many difficulties [82]. Recently, scholars have begun to evaluate energy efficiency based on a large number of data environments. For example, Zhu, et al. [83] proposed a DEA-based method for the allocation and utilization of natural resources in China, using big data technology to characterize the production technology in each region. Li, et al. [84] uses big data theory to analyze and evaluate the efficiency of China's forest resources, taking into account many evaluation indicators and large amounts of data in the big data environment. With the continuous improvement of information and information technology and data retrieval capabilities in the future, how to make full use of the big data environment in the energy field and expand DEA models and algorithms will help further enhance the application space of DEA.

Author Contributions: Conceptualization, T.X. and J.Y.; data curation, L.S.; funding acquisition, J.Y.; methodology, T.X.; resources, H.L.; visualization, L.S.; writing—original draft, T.X.; writing—review and editing, L.S. All authors have read and agreed to the published version of the manuscript.

Funding: This research was funded by the National Natural Science Foundation of China grant number 71671125.

Conflicts of Interest: The authors declare no conflict of interest.

Nomenclatures

DEA	Data Envelopment Analysis
DMU	Decision-Making Unit
CRS	Constant Returns to Scale
TFEE	Total Factor Energy Efficiency
GDP	Gross Domestic Product
CCR	Authors' initials (Charnes, Cooper and Rhodes)
BCC	Authors' initials (Banker, Charnes and Cooper)
CDM	Clean Development Mechanism
SBM	Slack-based Model
OECD	Organization for Economic Co-operation and Development
MPI	Malmquist productivity index
EU	European Union
SFA	Stochastic Frontier Analysis

References

1. Li, M.J.; He, Y.L.; Tao, W.Q. Modeling a hybrid methodology for evaluating and forecasting regional energy efficiency in China. *Appl. Energy* **2017**, *185*, 1769–1777. [CrossRef]
2. Song, M.L.; Zhang, J.; Wang, S.H. Review of the network environmental efficiencies of listed petroleum enterprises in China. *Renew. Sustain. Energy Rev.* **2015**, *43*, 65–71. [CrossRef]
3. Patterson, M.G. What is energy efficiency? Concepts, indicators and methodological issues. *Energy Policy* **1996**, *24*, 377–390. [CrossRef]
4. Hu, J.-L.; Wang, S.-C. Total-factor energy efficiency of regions in China. *Energy Policy* **2006**, *34*, 3206–3217. [CrossRef]
5. Zhang, Y.J.; Sun, Y.F.; Huang, J. Energy efficiency, carbon emission performance, and technology gaps: Evidence from CDM project investment. *Energy Policy* **2018**, *115*, 119–130. [CrossRef]
6. Pulina, M.; Detotto, C.; Paba, A. An investigation into the relationship between size and efficiency of the Italian hospitality sector: A window DEA approach. *Eur. J. Oper. Res.* **2010**, *204*, 613–620. [CrossRef]
7. Vishwakarma, A.; Kulshrestha, M.; Kulshreshtha, M. Efficiency evaluation of municipal solid waste management utilities in the urban cities of the state of Madhya Pradesh, India, using stochastic frontier analysis. *Benchmarking* **2012**, *19*, 340–357. [CrossRef]
8. Charnes, A.; Cooper, W.W.; Rhodes, E. Measuring the efficiency of decision making units. *Eur. J. Oper. Res.* **1978**, *2*, 429–444. [CrossRef]
9. Zhu, J. *Data Envelopment Analysis: A Handbook of Models and Methods*; Springer: Boston, MA, USA, 2015; Volume 221.
10. Honma, S.; Hu, J.L. Total-factor energy efficiency of regions in Japan. *Energy Policy* **2008**, *36*, 821–833. [CrossRef]
11. Zhang, X.P.; Cheng, X.M.; Yuan, J.H.; Gao, X.J. Total-factor energy efficiency in developing countries. *Energy Policy* **2011**, *39*, 644–650. [CrossRef]
12. Wu, J.; Yin, P.Z.; Sun, J.S.; Chu, J.F.; Liang, L. Evaluating the environmental efficiency of a two-stage system with undesired outputs by a DEA approach: An interest preference perspective. *Eur. J. Oper. Res.* **2016**, *254*, 1047–1062. [CrossRef]

13. Li, K.; Lin, B.Q. Metafroniter energy efficiency with CO_2 emissions and its convergence analysis for China. *Energy Econ.* **2015**, *48*, 230–241. [CrossRef]
14. Zhang, N.; Choi, Y. Environmental energy efficiency of China's regional economies: A non-oriented slacks-based measure analysis. *Soc. Sci. J.* **2013**, *50*, 225–234. [CrossRef]
15. Wang, Z.H.; Zeng, H.L.; Wei, Y.M.; Zhang, Y.X. Regional total factor energy efficiency: An empirical analysis of industrial sector in China. *Appl. Energy* **2012**, *97*, 115–123. [CrossRef]
16. Makridou, G.; Andriosopoulos, K.; Doumpos, M.; Zopounidis, C. Measuring the efficiency of energy-intensive industries across European countries. *Energy Policy* **2016**, *88*, 573–583. [CrossRef]
17. Feng, C.; Wang, M. Analysis of energy efficiency and energy savings potential in China's provincial industrial sectors. *J. Clean. Prod.* **2017**, *164*, 1531–1541. [CrossRef]
18. Wu, J.; Zhu, Q.Y.; Chu, J.F.; Liu, H.W.; Liang, L. Measuring energy and environmental efficiency of transportation systems in China based on a parallel DEA approach. *Transp. Res. Part D* **2016**, *48*, 460–472. [CrossRef]
19. Cooper, W.W.; Seiford, L.M.; Zhu, J. *Handbook on Data Envelopment Analysis*; Springer Science & Business Media: Berlin, Germany, 2011; Volume 164.
20. Li, F.; Emrouznejad, A.; Yang, G.L.; Li, Y.J. Carbon emission abatement quota allocation in Chinese manufacturing industries: An integrated cooperative game data envelopment analysis approach. *J. Oper. Res. Soc.* **2019**. [CrossRef]
21. Banker, R.D.; Charnes, A.; Cooper, W.W. Some Models for Estimating Technical and Scale Inefficiencies in Data Envelopment Analysis. *Manag. Sci.* **1984**, *30*, 1078–1092. [CrossRef]
22. Färe, R.; Grosskopf, S. Productivity and intermediate products: A frontier approach. *Econ. Lett.* **1996**, *50*, 65–70. [CrossRef]
23. Chung, Y.H.; Färe, R.; Grosskopf, S. Productivity and Undesirable Outputs: A Directional Distance Function Approach. *J. Environ. Manag.* **1997**, *51*, 229–240. [CrossRef]
24. Tone, K. A slacks-based measure of efficiency in data envelopment analysis. *Eur. J. Oper. Res.* **2001**, *130*, 498–509. [CrossRef]
25. Ali, A.I.; Seiford, L.M. Computational Accuracy and Infinitesimals In Data Envelopment Analysis. *Inf. Syst. Oper. Res.* **1993**, *31*, 290–297. [CrossRef]
26. Tone, K. A slacks-based measure of super-efficiency in data envelopment analysis. *Eur. J. Oper. Res.* **2002**, *143*, 32–41. [CrossRef]
27. He, P.; Sun, Y.; Shen, H.; Jian, J.; Yu, Z. Does Environmental Tax Affect Energy Efficiency? An Empirical Study of Energy Efficiency in OECD Countries Based on DEA and Logit Model. *Sustainability* **2019**, *11*, 3792. [CrossRef]
28. Halkos, G.; Petrou, K.N. Treating undesirable outputs in DEA: A critical review. *Econ. Anal. Policy* **2019**, *62*, 97–104. [CrossRef]
29. Fare, R.; Grosskopf, S.; Pasurka, C. Effects on Relative Efficiency in Electric-Power Generation Due to Environmental Controls. *Resour. Energy* **1986**, *8*, 167–184. [CrossRef]
30. Tyteca, D. Linear programming models for the measurement of environmental performance of firms—Concepts and empirical results. *J. Prod. Anal.* **1997**, *8*, 183–197. [CrossRef]
31. Macpherson, A.J.; Principe, P.P.; Shao, Y. Controlling for exogenous environmental variables when using data envelopment analysis for regional environmental assessments. *J. Environ. Manag.* **2013**, *119*, 220–229. [CrossRef]
32. Scheel, H. Undesirable outputs in efficiency valuations. *Eur. J. Oper. Res.* **2001**, *132*, 400–410. [CrossRef]
33. Seiford, L.M.; Zhu, J. Modeling undesirable factors in efficiency evaluation. *Eur. J. Oper. Res.* **2002**, *142*, 16–20. [CrossRef]
34. Zhou, Z.X.; Xu, G.C.; Wang, C.; Wu, J. Modeling undesirable output with a DEA approach based on an exponential transformation: An application to measure the energy efficiency of Chinese industry. *J. Clean. Prod.* **2019**, *236*, 11. [CrossRef]
35. Fare, R.; Grosskopf, S.; Hernandez-Sancho, F. Environmental performance: An index number approach. *Resour. Energy Econ.* **2004**, *26*, 343–352. [CrossRef]
36. Yang, H.L.; Pollitt, M. The necessity of distinguishing weak and strong disposability among undesirable outputs in DEA: Environmental performance of Chinese coal-fired power plants. *Energy Policy* **2010**, *38*, 4440–4444. [CrossRef]

37. Apergis, N.; Aye, G.C.; Barros, C.P.; Gupta, R.; Wanke, P. Energy efficiency of selected OECD countries: A slacks based model with undesirable outputs. *Energy Econ.* **2015**, *51*, 45–53. [CrossRef]
38. Liu, Y.N.; Wang, K. Energy efficiency of China's industry sector: An adjusted network DEA (data envelopment analysis)-based decomposition analysis. *Energy* **2015**, *93*, 1328–1337. [CrossRef]
39. Seiford, L.M.; Zhu, J. Profitability and marketability of the top 55 US commercial banks. *Manag. Sci.* **1999**, *45*, 1270–1288. [CrossRef]
40. Tone, K.; Tsutsui, M. Network DEA: A slacks-based measure approach. *Eur. J. Oper. Res.* **2009**, *197*, 243–252. [CrossRef]
41. Fukuyama, H.; Weber, W.L. A directional slacks-based measure of technical inefficiency. *Socio-Econ. Plan. Sci.* **2009**, *43*, 274–287. [CrossRef]
42. Charnes, A.; Cooper, W.W. Preface to topics in data envelopment analysis. *Ann. Oper. Res.* **1984**, *2*, 59–94. [CrossRef]
43. Wang, K.; Yu, S.W.; Zhang, W. China's regional energy and environmental efficiency: A DEA window analysis based dynamic evaluation. *Math. Comput. Model.* **2013**, *58*, 1117–1127. [CrossRef]
44. Meng, F.Y.; Zhou, P.; Zhou, D.Q.; Bai, Y. Inefficiency and Congestion Assessment of Mix Energy Consumption in 16 APEC Countries by using DEA Window Analysis. *Enrgy Procedia* **2014**, *61*, 2518–2523. [CrossRef]
45. Färe, R.; Grosskopf, S.; Lindgren, B.; Roos, P. Productivity developments in Swedish hospitals: A Malmquist output index approach. In *Data Envelopment Analysis: Theory, Methodology, and Applications*; Springer: Berlin, Germany, 1994.
46. Perez, K.; Gonzalez-Araya, M.C.; Iriarte, A. Energy and GHG emission efficiency in the Chilean manufacturing industry: Sectoral and regional analysis by DEA and Malmquist indexes. *Energy Econ.* **2017**, *66*, 290–302. [CrossRef]
47. Wu, J.; Liang, L. A multiple criteria ranking method based on game cross-evaluation approach. *Ann. Oper. Res.* **2012**, *197*, 191–200. [CrossRef]
48. Sexton, T.R.; Silkman, R.H.; Hogan, A.J. Data envelopment analysis: Critique and extensions. *New Dir. Program Eval.* **1986**, *1986*, 73–105. [CrossRef]
49. Liang, L.; Wu, J.; Cook, W.D.; Zhu, J. The DEA Game Cross-Efficiency Model and Its Nash Equilibrium. *Oper. Res.* **2008**, *56*, 1278–1288. [CrossRef]
50. Chen, W.; Zhou, K.L.; Yang, S.L. Evaluation of China's electric energy efficiency under environmental constraints: A DEA cross efficiency model based on game relationship. *J. Clean. Prod.* **2017**, *164*, 38–44. [CrossRef]
51. Xie, B.C.; Gao, J.; Zhang, S.; Pang, R.Z.; Zhang, Z.X. The environmental efficiency analysis of China's power generation sector based on game cross-efficiency approach. *Struct. Chang. Econ. Dyn.* **2018**, *46*, 126–135. [CrossRef]
52. Yang, Z.S.; Wei, X.X. The measurement and influences of China's urban total factor energy efficiency under environmental pollution: Based on the game cross-efficiency DEA. *J. Clean. Prod.* **2019**, *209*, 439–450. [CrossRef]
53. Yu, J.; Zhou, K.; Yang, S. Regional heterogeneity of China's energy efficiency in "new normal": A meta-frontier Super-SBM analysis. *Energy Policy* **2019**, *134*, 110941. [CrossRef]
54. Broadstock, D.C.; Li, J.; Zhang, D. Efficiency snakes and energy ladders: A (meta-)frontier demand analysis of electricity consumption efficiency in Chinese households. *Energy Policy* **2016**, *91*, 383–396. [CrossRef]
55. Battese, G.E.; Prasada Rao, D.S.; O'Donnell, C.J. A metafrontier production function for estimation of technical efficiencies and technology gaps for firms operating under different technologies. *J. Prod. Anal.* **2004**, *21*, 91–103. [CrossRef]
56. Yu, A.Y.; You, J.X.; Zhang, H.; Ma, J.J. Estimation of industrial energy efficiency and corresponding spatial clustering in urban China by a meta-frontier model. *Sustain. Cities Soc.* **2018**, *43*, 290–304. [CrossRef]
57. Wang, P.; Deng, X.; Zhang, N.; Zhang, X. Energy efficiency and technology gap of enterprises in Guangdong province: A meta-frontier directional distance function analysis. *J. Clean. Prod.* **2019**, *212*, 1446–1453. [CrossRef]
58. Jebali, E.; Essid, H.; Khraief, N. The analysis of energy efficiency of the Mediterranean countries: A two-stage double bootstrap DEA approach. *Energy* **2017**, *134*, 991–1000. [CrossRef]
59. Zhao, C.H.; Zhang, H.N.; Zeng, Y.R.; Li, F.Y.; Liu, Y.X.; Qin, C.J.; Yuan, J.H. Total-Factor Energy Efficiency in BRI Countries: An Estimation Based on Three-Stage DEA Model. *Sustainability* **2018**, *10*, 278. [CrossRef]

60. Wang, L.W.; Le, K.D.; Nguyen, T.D. Assessment of the Energy Efficiency Improvement of Twenty-Five Countries: A DEA Approach. *Energies* **2019**, *12*, 1535. [CrossRef]
61. Bampatsou, C.; Papadopoulos, S.; Zervas, E. Technical efficiency of economic systems of EU-15 countries based on energy consumption. *Energy Policy* **2013**, *55*, 426–434. [CrossRef]
62. Guo, S.D.; Li, H.; Zhao, R.; Zhou, X. Industrial environmental efficiency assessment for China's western regions by using a SBM-based DEA. *Environ. Sci. Pollut. Res.* **2019**, *26*, 27542–27550. [CrossRef]
63. Wu, J.; Zhu, Q.Y.; Yin, P.Z.; Song, M.L. Measuring energy and environmental performance for regions in China by using DEA-based Malmquist indices. *Oper. Res.* **2017**, *17*, 715–735. [CrossRef]
64. Guo, X.; Lu, C.-C.; Lee, J.-H.; Chiu, Y.-H. Applying the dynamic DEA model to evaluate the energy efficiency of OECD countries and China. *Energy* **2017**, *134*, 392–399. [CrossRef]
65. Amowine, N.; Ma, Z.Q.; Li, M.X.; Zhou, Z.X.; Asunka, B.A.; Amowine, J. Energy Efficiency Improvement Assessment in Africa: An Integrated Dynamic DEA Approach. *Energies* **2019**, *12*, 17. [CrossRef]
66. Sun, J.S.; Wang, Z.H.; Li, G. Measuring emission-reduction and energy-conservation efficiency of Chinese cities considering management and technology heterogeneity. *J. Clean. Prod.* **2018**, *175*, 561–571. [CrossRef]
67. Lee, H.S.; Choi, Y. Environmental Performance Evaluation of the Korean Manufacturing Industry Based on Sequential DEA. *Sustainability* **2019**, *11*, 14. [CrossRef]
68. Lei, X.Y.; Li, L.; Zhang, X.F.; Dai, Q.Z.; Fu, Y.L. A Novel Ratio-Based Parallel DEA Approach for Evaluating the Energy and Environmental Performance of Chinese Transportation Sectors. *J. Syst. Sci. Syst. Eng.* **2019**, *28*, 621–635. [CrossRef]
69. Djordjevic, B.; Krmac, E. Evaluation of Energy-Environment Efficiency of European Transport Sectors: Non-Radial DEA and TOPSIS Approach. *Energies* **2019**, *12*, 27. [CrossRef]
70. Wu, J.; Xiong, B.B.; An, Q.X.; Sun, J.S.; Wu, H.Q. Total-factor energy efficiency evaluation of Chinese industry by using two-stage DEA model with shared inputs. *Ann. Oper. Res.* **2017**, *255*, 257–276. [CrossRef]
71. Fei, R.; Lin, B. Energy efficiency and production technology heterogeneity in China's agricultural sector: A meta-frontier approach. *Technol. Forecast. Soc. Chang.* **2016**, *109*, 25–34. [CrossRef]
72. Han, Y.M.; Long, C.; Geng, Z.Q.; Zhang, K.Y. Carbon emission analysis and evaluation of industrial departments in China: An improved environmental DEA cross model based on information entropy. *J. Environ. Manag.* **2018**, *205*, 298–307. [CrossRef]
73. Cui, Q.; Li, Y. Evaluating energy efficiency for airlines: An application of VFB-DEA. *J. Air Transp. Manag.* **2015**, *44*, 34–41. [CrossRef]
74. Bi, G.-B.; Song, W.; Zhou, P.; Liang, L. Does environmental regulation affect energy efficiency in China's thermal power generation? Empirical evidence from a slacks-based DEA model. *Energy Policy* **2014**, *66*, 537–546. [CrossRef]
75. Zhang, N.; Kong, F.B.; Choi, Y.; Zhou, P. The effect of size-control policy on unified energy and carbon efficiency for Chinese fossil fuel power plants. *Energy Policy* **2014**, *70*, 193–200. [CrossRef]
76. Cook, W.D.; Harrison, J.; Imanirad, R.; Rouse, P.; Zhu, J. Data Envelopment Analysis with Nonhomogeneous DMUs. *Oper. Res.* **2013**, *61*, 666–676. [CrossRef]
77. Wu, J.; Li, M.J.; Zhu, Q.Y.; Zhou, Z.X.; Liang, L. Energy and environmental efficiency measurement of China's industrial sectors: A DEA model with non-homogeneous inputs and outputs. *Energy Econ.* **2019**, *78*, 468–480. [CrossRef]
78. Han, Y.M.; Geng, Z.Q.; Zhu, Q.X.; Qu, Y.X. Energy efficiency analysis method based on fuzzy DEA cross-model for ethylene production systems in chemical industry. *Energy* **2015**, *83*, 685–695. [CrossRef]
79. Zadeh, L.A.; Fu, K.-S.; Tanaka, K. Fuzzy sets and their applications to cognitive and decision processes. In Proceedings of the US–Japan Seminar on Fuzzy Sets and Their Applications, Berkeley, CA, USA, 1–4 July 1974; Academic press, 2014.
80. Lertworasirikul, S.; Fang, S.-C.; Joines, J.A.; Nuttle, H.L. Fuzzy data envelopment analysis (DEA): A possibility approach. *Fuzzy Sets Syst.* **2003**, *139*, 379–394. [CrossRef]
81. Wen, M.L.; Li, H.S. Fuzzy data envelopment analysis (DEA): Model and ranking method. *J. Comput. Appl. Math.* **2009**, *223*, 872–878. [CrossRef]
82. Chu, J.F.; Wu, J.; Song, M.L. An SBM-DEA model with parallel computing design for environmental efficiency evaluation in the big data context: A transportation system application. *Ann. Oper. Res.* **2018**, *270*, 105–124. [CrossRef]

83. Zhu, Q.Y.; Wu, J.; Li, X.C.; Xiong, B.B. China's regional natural resource allocation and utilization: A DEA-based approach in a big data environment. *J. Clean. Prod.* **2017**, *142*, 809–818. [CrossRef]
84. Li, L.; Hao, T.T.; Chi, T. Evaluation on China's forestry resources efficiency based on big data. *J. Clean. Prod.* **2017**, *142*, 513–523. [CrossRef]

energies

MDPI

Article

Impact of Actual Weather Datasets for Calibrating White-Box Building Energy Models Base on Monitored Data

Vicente Gutiérrez González , Germán Ramos Ruiz and Carlos Fernández Bandera *

School of Architecture, University of Navarra, 31009 Pamplona, Spain; vgutierrez@unav.es (V.G.G.); gramrui@unav.es (G.R.R.)
* Correspondence: cfbandera@unav.es; Tel.: +34-948-425-600 (ext. 803189)

Abstract: The need to reduce energy consumption in buildings is an urgent task. Increasing the use of calibrated building energy models (BEM) could accelerate this need. The calibration process of these models is a highly under-determined problem that normally yields multiple solutions. Among the uncertainties of calibration, the weather file has a primary position. The objective of this paper is to provide a methodology for selecting the optimal weather file when an on-site weather station with local sensors is available and what is the alternative option when it is not and a mathematically evaluation has to be done with sensors from nearby stations (third-party providers). We provide a quality assessment of models based on the Coefficient of Variation of the Root Mean Square Error (CV(RMSE)) and the Square Pearson Correlation Coefficient (R^2). The research was developed on a control experiment conducted by Annex 58 and a previous calibration study. This is based on the results obtained with the study case based on the data provided by their N2 house.

Keywords: weather data; calibration; sensors; energy simulation; sensors saving; methodology; Building Energy Models (BEMs)

Citation: Gutiérrez González, V.; Ramos Ruiz, G.; Fernández Bandera, C. Impact of Actual Weather Datasets for Calibrating White-Box Building Energy Models Base on Monitored Data. *Energies* **2021**, *14*, 1187. https://doi.org/10.3390/en14041187

Academic Editor: Roberto Alonso González Lezcano

Received: 23 December 2020
Accepted: 6 February 2021
Published: 23 February 2021

1. Introduction

The building energy model (BEM) is a key element when speaking about building analytics and control applications, such as model predictive control (MPC) [1] and fault detection diagnosis (FDD) [2]. The smart grids to be built in the future will use high-quality BEMs as an important element. The European Union has funded SABINA [3], which is an innovation and research project that seeks to generate financial models and create new technologies to actively manage, connect and control storage and generation to exploit the connections between the thermal inertia of buildings and electrical flexibility. The European electricity system has the capacity to introduce an increasing amount of energy generation from renewable sources into its system. SABINA echoes this demand, as it focuses on one of the cheapest sources of green energy: thermal inertia inside buildings, also achieving the coupling between heat and power grids. Using thermal inertia as an energy store is referred to as a "power to heat" (P2H) solution [4–6].

Energy prediction relies entirely on models, and therefore one of the main pillars of SABINA is the production of high-quality models (calibrated) that can give reliability to P2H technology. These models are constructed on the basis on an initial methodology developed by Ramos et al. [7] and Bandera et al. [8], which has been recently improved and empirically validated by Gutiérrez et al. [9] based on the work carried out by Annex 58 of the Committee of the International Energy Agency Energy in Buildings and Communities program (IEA-EBC) approved in 2011 and completed in 2016. The main objectives of Annex 58 were: to develop common quality procedures for dynamic full-scale testing to come to a better performance analysis and develop models to characterize and predict the effective thermal performance of building components and whole buildings [10].

Whole building energy simulation tools allow the detailed calculation to specify building performance criteria, such as the space temperature and electric energy consumption, under the influence of external inputs such as weather [11], occupancy, ground [12] and infiltration. These calculations are carried out at time series data and EnergyPlus [13] is among the main tools that perform these calculations based on what is called a white-box model. These models are founded on physical parameters rather than mathematical or statistical formulation. The main challenge of these models is how to reduce the gap between the measured and simulated data as these models are over-parameterized and under-determined. Despite the potential benefits and the software continuous progress, a number of problems retract from a more widespread use. The gap undermines the confidence in the model prediction and curtails the adoption of BEM tools during the design, commissioning, and operation.

It is necessary that BEM closely represents the actual behavior of the real building. The calibration process can achieve this. However, the calibration of white-box modeling by aligning the measured data to a simulation is a highly under-determined, which normally yields multiple solutions [14]. This under-determined problem carries us to the uncertainty problem that should be analyzed properly [15,16]. De Wit [17] classified the various sources of uncertainty as follows:

- Specification uncertainty: Arising from incomplete or inaccurate specification of the building or systems modeled. This may include any exposed model parameters, such as the geometry, material properties, Heating Ventilation Air Conditioning (HVAC) specifications, plant and system schedules, etc.
- Modeling uncertainty: When executing highly complex physical processes, simplifications or assumptions can be made to obtain results more easily. These simplifications can be taken by the modeler (stochastic process scheduling and zoning) or internal to the calculation program (calculation algorithms).
- Numerical uncertainty: Errors introduced in the simulation and discretization of the model.
- Scenario uncertainties: Uncertainties that can be produced by external conditioning factors such as weather data, occupant behavior, etc.

In this paper, the main focus is on the uncertainty of the scenario based on outdoor weather conditions. As pointed out by some authors [18–20], one of the key parameters to produce a calibrated BEM is the weather file. According to Bhandari et al. [21] there are three existing weather files that can be fed into the energy model: future [22–24], typical and actual [25]. Typical weather files generate a weather file for a set of years, usually covering the last 20–30 years. These types of files are used to understand the building under standard conditions. For this reason, these weather files together with the so-called future weather files are not suitable for building calibration. For the study presented in this paper, the focus has been on actual weather files that are constructed from a specific location and time. The data for the generation of the weather file can be obtained from those generated by an on-site weather station [26] or by processing data from several nearby stations [25,27,28]. The latter option is often used by external or third-party data providers [29].

The calibration of a BEM normally entails the installation of an on-site weather station. This installation normally implies an extra cost for the project because, on top of the expense of the station, data handling could be an extra issue. This is one of the main problems that restrains energy service companies (ESCOs) in promoting option D (calibrated BEM for measuring energy conservation measurements) of international measurement verification protocols (IPMVP) [30]. For this reason, the goal of this study is to tackle the following research questions: First, is it possible to obtain a better weather file than that provided by the on-site weather station? Secondly, when not having the option of installing a dedicated weather station or retrieving data from a nearby one, is it still possible to obtain a calibrated BEM? The research presented in the following sections answers these two questions positively and provides a methodology for any modeler to quantify the impact of these decisions on their work.

To answer these questions, the paper is structured as follows. In Section 2, design of the method, where is explained the study used for weather file selection base on two different techniques. First, the weather file selection is performed with a base model to produce a rank of the best weather files and a selection of the best results is chosen based on two criteria: the best results at the level of uncertainty indexes and the more cost-effective solution, which does not reduce the model quality. Secondly a calibration process is performed based on the four weather files selected. This process will confirm the rank of the weather files and the initial guess that is not necessary to conduct the calibration, which is time consuming, in order to choose the most adequate weather file. In Section 3, we present our analysis of the results and explain how a better model than the on-site was achieved and a cost-effective solution was provided without greatly reducing the model quality. The paper finishes with our conclusions, which are discussed in Section 4.

2. Design of the Method

To apply the methodology proposed in this paper, the data provided in annex 58 IEA-EBC [10,31] have been used both for the generation of the energy model and for obtaining the calibrated model.

We focused on the N2 house (N2 is the name of the house) in the German town of Holzkirchen near Munich. The house is situated on a flat area and there are no buildings in the vicinity to provide shade in the summer season when the annex was tested. The house has three floors (basement, ground floor, and attic) with a free height of 2.50 m. The test proposed by the annex focuses on the ground floor spaces: the two bedrooms, the living room, the entrance, the bathroom, the corridor and the kitchen (Figure 1). This housing is optimal for the realization of the proposed study because it provides all the necessary data for the accomplishment of the work. At the same time, these data are of a high quality, which reduces the uncertainty that they may cause in the final results of the research.

We focus on house N2 (N2 is the name of the house) in the German town of Holzkirchen, about 35 km south of Munich (47.874 N, 11.728 E). It is a house situated on a flat area without any buildings that could cast shadows on it in the summer period, which was the period of analysis. It shares the typical climate of Central Europe: oceanic. The house has three floors (basement, ground floor and attic) with a clear height of 2.50 m. The study focused on the ground floor. The study focused on the ground floor, which included a living room, a kitchen, an entrance, a bathroom, a corridor and two bedrooms (Figure 1). This dwelling is optimal for the proposed study because it provides all the data necessary to carry out the work. At the same time, this data is of high quality, which reduces the uncertainty it may cause in the final results of the research.

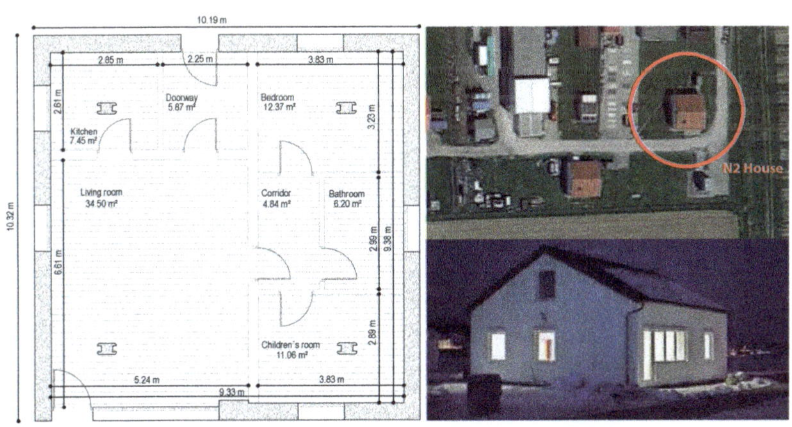

Figure 1. Plan and external views of the N2 house. Holzkirchen, Germany.

The exercise developed in annex 58 consisted of five periods of energization of the housing, each period with certain characteristics. The energy model then responds to reality, both in the energy consumed and in the temperature reached. The offered periods began with the first (initialization) duration of three days, where a constant interior temperature was maintained in the house at 30 °C. In period number 2 (Period 2, set point 30 °C), seven days in length, a constant temperature was still maintained at 30 °C inside the house, and the calibrated energy model provided the real energy consumed. In the 3rd period that lasted 14 days (Period 3, ROLBS), energy was introduced through the living room radiator in aleatory periods with a randomly ordered logarithmic binary sequence (ROLBS).

Within the EC COMPASS project [32], this sequence was developed. Its objective is to ensure that all relevant frequencies that can occur in a given time have the same weight. To achieve this, the on and off periods are chosen in logarithmically equal intervals and mixed in a quasi-random order. With this sequence, it can be ensured that there is no connection between solar gains and heat input through the HVAC systems. The energy model must be able to reproduce the internal temperature that this energy produces. In period 4 (Period 4, set point 25 °C), a constant temperature was again introduced into the house, but this time at 25 °C and the energy model reproduced the energy produced by that indoor temperature. The last period is the 5th (Period 5, free oscillation), where the house was left in free oscillation, i.e., without any energy contribution. The calibrated model, then, reproduced the interior temperatures.

For the weather files used in this exercise, a climate file offered by third parties was selected, located about 440 m in a straight line from the selected house (47.87 N, 11.73 E) and together with the data from the weather station placed on the site, a composition of the sensors was made (Table 1). Using the sensors of the on-site station as a basis, the sensors will be replaced by those of third parties, generating a total of 64 climate files. The sensors used were the outside temperature (T), global horizontal irradiation (GHI), diffuse horizontal irradiation (DHI), wind speed (WS), wind direction (WD), and relative humidity (RH).

Once the energy model of the house was created (baseline model), and having already generated the weather files that will be used in the test, we proceeded to simulate the energy model in all periods and with all the climate files, to obtain a list with the behavior of the weather files and, thereby, obtain which one is the best suited to reality.

The next step was to check that the simulations performed with all the weather files and the base model conformed to a fitting process, i.e., to subject the model to a calibration process. This process was conducted with 4 of the 64 climate archives generated. The four that were considered to be the most relevant were selected. These were:

- The weather file produced by the weather station placed on site (on-site weather file). This climate file was chosen because it allegedly best represents the weather of the area.
- Third-party weather file (third-party weather file). This file was selected as the alternative to the site's weather station.
- The weather file composed of the third-party, but replacing the outside temperature sensor with that of the site's weather station (third-party weather file + on-site temperature sensor). This climate file was selected for being one of the most cost-effective.
- The weather file that provided the best results in the simulations with the base model (weather file combination). This was chosen for being the one with the best results of adjustment with reality.

Table 1. All possible combinations of weather files. The sensors used were the outside temperature (T), global horizontal irradiation (GHI), diffuse horizontal irradiation (DHI), wind speed (WS), wind direction (WD), and relative humidity (RH).

Weather File Name	Weather Combination On-Site Weather Sensors	+	Third-Party Weather Sensors	Weather File Name	Weather Combination On-Site Weather Sensors	+	Third-Party Weather Sensors
Third-party	—	+	WS; WD; RH; GHI; DHI; T	*Weather 33*	WS	+	WD; RH; GHI; DHI; T
Weather 02	DHI	+	WS; WD; RH; GHI; T	*Weather 34*	WS; DHI	+	WD; RH; GHI; T
Weather 03	DHI; T	+	WS; WD; RH; GHI	*Weather 35*	WS; DHI; T	+	WD; RH; GHI
Weather 04	GHI	+	WS; WD; RH; DHI; T	*Weather 36*	WS; GHI	+	WD; RH; DHI; T
Weather 05	GHI; DHI	+	WS; WD; RH; T	*Weather 37*	WS; GHI; DHI	+	WD; RH; T
Weather 06	GHI; DHI; T	+	WS; WD; RH	**Combination**	WS; GHI; DHI; T	+	WD; RH
Weather 07	GHI; T	+	WS; WD; RH; DHI	*Weather 39*	WS; GHI; T	+	WD; RH; DHI
Weather 08	RH	+	WS; WD; GHI; DHI; T	*Weather 40*	WS; RH	+	WD; GHI; DHI; T
Weather 09	RH; DHI	+	WS; WD; GHI; T	*Weather 41*	WS; RH; DHI	+	WD; GHI; T
Weather 10	RH; DHI; T	+	WS; WD; GHI	*Weather 42*	WS; RH; DHI; T	+	WD; GHI; T
Weather 11	RH; GHI	+	WS; WD; DHI; T	*Weather 43*	WS; RH; GHI	+	WD; DHI; T
Weather 12	RH; GHI; DHI	+	WS; WD; T	*Weather 44*	WS; RH; GHI; DHI	+	WD; T
Weather 13	RH; GHI; DHI; T	+	WS; WD	*Weather 45*	WS; RH; GHI; DHI; T	+	WD
Weather 14	RH; GHI; T	+	WS; WD; DHI	*Weather 46*	WS; RH; GHI; T	+	WD; DHI
Weather 15	RH; T	+	WS; WD; GHI; DHI	*Weather 47*	WS; RH; T	+	WD; GHI; DHI
Third-party + On-site temp. sensor	T	+	WS; WD; RH GHI; DHI	*Weather 48*	WS; T	+	WD; RH; GHI; DHI
Weather 17	WD	+	WS; RH; GHI; DHI; T	*Weather 49*	WS; WD	+	RH; GHI; DHI; T
Weather 18	WD; DHI	+	WS; RH; GHI; T	*Weather 50*	WS; WD; DHI	+	RH; GHI; T
Weather 19	WD; DHI; T	+	WS; RH; GHI	*Weather 51*	WS; WD; DHI; T	+	RH; GHI
Weather 20	WD; GHI	+	WS; RH; DHI; T	*Weather 52*	WS; WD; GHI	+	RH; DHI; T
Weather 21	WD; GHI; DHI	+	WS; RH; T	*Weather 53*	WS; WD; GHI; DHI	+	RH; T
Weather 22	WD; GHI; DHI; T	+	WS; RH	*Weather 54*	WS; WD; GHI; DHI; T	+	RH
Weather 23	WD; GHI; T	+	WS; RH; DHI	*Weather 55*	WS; WD; GHI; T	+	RH; DHI
Weather 24	WD; RH	+	WS; DHI; GHI; T	*Weather 56*	WS; WD; RH	+	GHI; DHI; T
Weather 25	WD; RH; DHI	+	WS; GHI; T	*Weather 57*	WS; WD; RH; DHI	+	GHI; T
Weather 26	WD; RH; DHI; T	+	WS; GHI	*Weather 58*	WS; WD; RH; DHI; T	+	GHI
Weather 27	WD; RH; GHI	+	WS; DHI; T	*Weather 59*	WS; WD; RH; GHI	+	DHI; T
Weather 28	WD; RH; GHI; DHI	+	WS; T	*Weather 60*	WS; WD; RH; GHI; DHI	+	T
Weather 29	WD; RH; GHI; DHI; T	+	WS	***On-site***	WS; WD; RH; GHI; DHI; T	+	—
Weather 30	WD; RH; GHI; T	+	WS; DHI	*Weather 62*	WS; WD; RH; GHI; T	+	DHI
Weather 31	WD; RH; T	+	WS; GHI; DHI	*Weather 63*	WS; WD; RH; T	+	GHI; DHI
Weather 32	WD; T	+	WS; RH; GHI; DHI	*Weather 64*	WS; WD; T	+	RH; GHI; DHI

The research method used started with the selection of the weather files. Once they had been selected, the process of calibrating the base model began [9]. The objective of this procedure was to justify the impact that the weather files had on the calibrated model, thus checking which was the energy model that together with its weather file best fit reality and to verify that the ranking generated in the simulations of the base model with all the weather files was fulfilled (Figure 2).

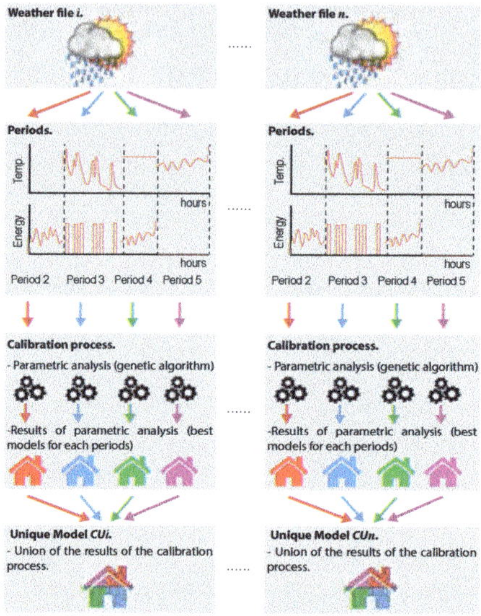

Figure 2. Generic process diagram for achieving calibration mode.

The calibration process, tested in previous studies with different buildings and with satisfactory results, will generate the model that best fits in temperature and energy to all the periods proposed by the exercise in the annex (periods 2 to 5) . To achieve this, several scripts were programmed in the EnergyPlus [33] run-time language. These commands transfer the measured temperature or energy to the model. The periods described in the previous paragraphs were also subjected to the calibration process shown in Figure 3. Although the methodology may resemble an optimization process, the objective function links the fit between the values provided by the model and the real data. In this case, the coefficient of variation of the root mean square error (CV(RMSE)) and the square Pearson correlation coefficient (R^2) were used [16]. To find the best solution, the non-dominant sorting genetic algorithm (NSGA-II) [34] was chosen as the search engine. The possible combinations of the parameters are those that will determine the search space, these are: thermal bridges, thermal mass, infiltrations and capacitances.

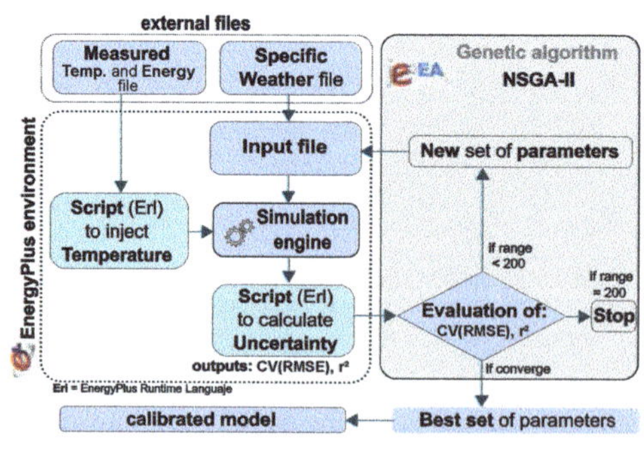

Figure 3. Calibration environment with a genetic algorithm. The coefficient of variation of the root mean square error (CV(RMSE)) and the genetic algorithm non-dominated sorting genetic algorithm (NSGA-II) [9].

Once all the periods were calibrated according to this methodology, the model that best suited all of them was obtained. This calibration operation was then repeated with the proposed weather files.

When the calibrated models (*CUi*, ..., *CUn*) were obtained, the adjustment they achieved with respect to the real temperature and energy were studied, thus discovering which was the weather file that after a calibration process of the energy model, was the closest to reality. In addition, we checked whether the adjustment classification of the base model with all the weather files had similarity with the results obtained in the calibrated models. The evaluation of the models was proposed to be performed with two types of indexes, the same ones that the genetic algorithm used in the calibration process: First, the CV(RMSE) (Equation (1)) which is the coefficient of variation of the root mean square error. The CV(RMSE) is achieved by weighting the Root Mean Square Error (RMSE) by the mean of the actual data. The measured variability is considered to be error variance by this index and therefore the American Society of Heating, Refrigerating, and Air Conditioning Engineers (ASHRAE) Guideline 14, the Federal Energy Management Program (FEMP) and the International Performance Maintenance and Verification Protocol (IPMVP) recommend its use [30,35–42]. Secondly, the coefficient of determination R^2 (Equation (2)) which is the percentage of variation of the response variable that explains its relationship with one or more predictor variables. Generally, the higher the R^2, the better the fit of the model to its data. The R^2 is always between 0 and 100%.

$$\mathrm{CV}(RMSE) = \frac{1}{\bar{y}}\left[\frac{\sum_{i=1}^{n}(y_i - \hat{y}_i)^2}{n-p}\right]^{\frac{1}{2}} \tag{1}$$

$$\mathrm{R}^2 = \left(\frac{\sum\limits_{i=1}^{n}(y_i - \bar{y})(\hat{y}_i - \bar{\hat{y}})}{\sqrt{\left(\sum\limits_{i=1}^{n}(y_i - \bar{y})^2\right)\left(\sum\limits_{i=1}^{n}(\hat{y}_i - \bar{\hat{y}})^2\right)}}\right)^2. \tag{2}$$

3. Analysis of the Results and Discussion

The first exercise that was performed for the demonstration of the methodology was the simulation of the base model with all the proposed weather file combinations. This was an attempt to discover which climate file, by simulating it with a baseline model, best fit reality. The results can be seen in Figures 4 and 5. Through the box plots, the results obtained in this phase of the methodology are shown. In both figures are highlighted the results of:

- The weather file that best fit the real data, created from the combination of sensors from the weather station placed on site and the one obtained from third parties: "Weather file combination". This configuration was: the wind speed, global, and diffuse horizontal irradiation and temperature sensors of the weather station placed on site and the wind direction and relative humidity sensors of the third-party. This is highlighted by a green circle in the figures.
- The "third-party weather file + on-site temperature sensor", created by adding, to the third-party climate file, the temperature sensor data from the site's weather station. This is highlighted in the figures by a red circle. This weather file is one of the most cost-effective.
- The weather file resulting from the sensors of the weather station placed at the site, "on-site weather file". This is highlighted with an orange circle on the figures.
- The meteorological file created with the third-party sensors "third-party weather file". This is marked in blue in the figures.

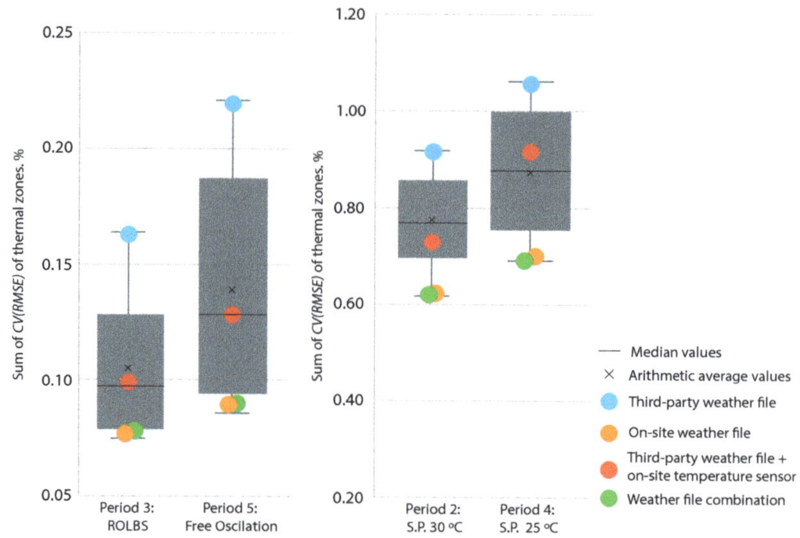

Figure 4. The simulation result of the base model with all the proposed weather files. CV(RMSE) Index.

Figure 4 shows the sum of the CV(RMSE) index of the thermal zones that come into play in the simulation (living room, bedroom, kitchen, and children's room). This index has a range from 0 to ∞, where 0 represents the model with the best adjustment. This was obtained by comparing the temperature and energy data (depending on the period) generated by the base energy model and the different weather files with those of reality. Each period has its own box plot. On the left of the figure are the periods where the model is asked for temperature, and on the right are the periods where the model is asked for energy:

The required temperature periods for the energy model:

- Period 3: During the 2 weeks that period 3 lasts, the house underwent an injection of energy through a ROLBS sequence, and the model was used to discover the temperature that this energy produces. Once the model with the different weather files was subjected to this period, we see that the weather file that behaved best was the "on-site weather file" and, almost in the same position, was the "weather file combination" both in quartile 1. The "third-party weather file + on-site temp. sensor" was at the limit of quartile 2 and 3, worsening the results with respect to period 2. Finally, the "third-party weather file" was still the worst performer at the upper limit of quartile 4.
- Period 5: In this period, the house was in free oscillation and the model represented the temperatures that were obtained when no energy was injected into the house. The behavior of the weather files in this period had a similarity with the results obtained in period 3. Quartile 1 contained both the "weather file combination" and the "on-site weather file". The latter was placed in the first position. Between quartile 2 and 3, there was the "third-party weather file + on-site temp. sensor", which was almost at the average of the results. Finally, as in all other periods, the "third-party weather file" was placed on the upper edge of quartile 4.

Required energy periods to the energy model:

- Period 2: In this period, the house was subjected to a constant temperature at 30 °C, and the energy model was required to be able to reproduce the energy needed to reach that temperature. The weather file that produced the best fit with reality was the "weather file combination" followed very closely by the on-site weather file, both located in the first quartile. Already in the second quartile, but above average, the "third-party weather file + on-site temp. sensor" was placed. The weather file that was the worst suited to reality is that of third parties, which is in the fourth quartile.

- Period 4: The house was heated to 25 °C and the energy model was able to reproduce the energy necessary to obtain that temperature. The best weather file was again the "weather combination", this time making more of a difference to the "weather on-site", although both files were in the first quartile. In the 3rd, below the average, we find the "third-party weather file + on-site temp. sensor" and, as in the other periods, located at a great distance from the rest. In the 4th quartile was the "third-party weather file".

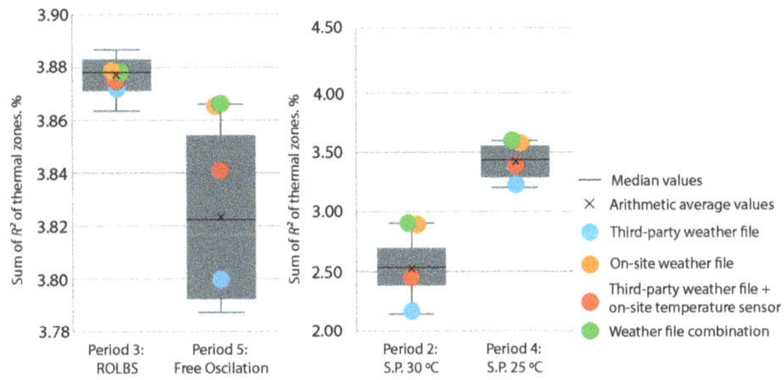

Figure 5. The simulation result of the base model with all the proposed weather files. R^2 Index.

Figure 5 shows the sum of the index R^2 of the thermal zones that enter the simulation process: living room, bedroom, kitchen, and children's room. This index measures how much the shapes of the two curves resemble each other. In this case, the temperature and energy curves produced by the simulated model and the different weather files were compared with reality. The range was from 0 to 4 (4 because this is the sum of the four thermal zones that are analyzed in the energy model) with 4 being the model with the best adjustment. Each period is shown in a box plot. As in Figure 4, on the left are the periods where the model is asked for temperature, and on the right are the periods where the model is asked for energy.

The required temperature periods for the energy model:

- Period 3: In this period, all the weather archives were much closer to each other. Even so, the one that best fit the reality was the "on-site weather file" in quartile 2, followed very closely by the "weather file combination". Both files were above average. In third position was the "third-party weather file + on-site temperature sensor" located in quartile 3 below the average. The last position was the "third-party weather file" on the border between quartile 3 and 4.
- Period 5: In this period, the difference in the results of the different weather data archives was greater. In quartile 1, the "weather file combination" was the best suited to reality. Next was the "on-site weather file". In quartile 2 and above, but at a considerable distance, was the "third-party weather file + on-site temperature sensor". In the last position, but in quartile 3, was the "third-party weather file".

The required energy periods for the energy model:

- Period 2: The "weather file combination" achieved the first position; this was the one that best fit the real data of the studied meteorological archives. This was in the first quartile followed very closely by the "on-site weather file". In the third one was the "third-party weather file + on-site temp sensor" located a little behind the average. At the bottom of the fourth quartile, we find the "third-party weather file".
- Period 4: The weather file that best matched reality was the "weather file combination" located in quartile 1, as well as the "on-site weather file" that followed it very narrowly. The "third-party weather file + on-site temperature sensor" was in the third lower quartile. Finally, at the bottom edge of quartile 4 is the "third-party weather file".

As a summary of the comments on the results of this analysis, the weather file that best fit the real data was the "weather file combination" followed at a very short distance by the "on-site weather file". Thirdly, the "third-party weather file + on-site temperature sensor" was clearly positioned most of the time in the third quartile, but above average. For the "third-party weather file", there is no doubt that it was the weather file that fit the worst to reality, placing itself, in most periods, in the lower limit of the last quartile.

Once this check was performed, the combination of sensors between those placed on site and those obtained from a third-party weather file were known to produce better adjustment results. This is the wind speed (WS), global (GHI) and diffuse (DHI) horizontal irradiation, and temperature (T) sensors of the weather station placed on the site and the wind direction (WD) and relative humidity (RH) sensors of the third-party.

The next step, as explained in the description of the methodology, is to justify the selection of the four weather files with which the calibration process was performed. They were:

- The weather file that produced the best adjustment to real data when simulated with the base energy model: "weather file combination".
- The climate file with the best ratio between cost and effectiveness: "third-party weather file + on-site temperature sensor".
- The weather file created from sensor data from the on-site weather station: "On -site weather file".
- The weather file generated from sensor data gathered from third parties: "third-party weather file".

The energy model with each type of weather file was subjected to a fitting process (explained in Section 2), with the aim of verifying whether the same results are obtained as in the classification made with the simulations of the baseline model and all the weather files (Figure 6).

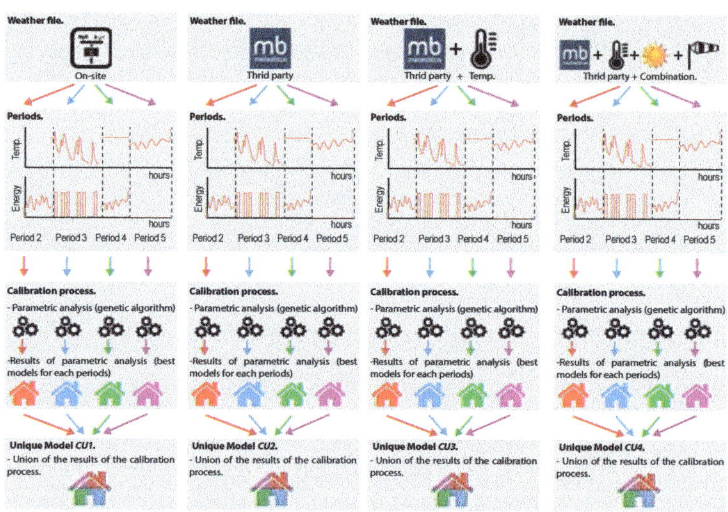

Figure 6. The process diagram for achieving the calibrated model.

The model with the four selected weather files was calibrated in each of the periods proposed in annex 58. The four calibrated models obtained for each period were joined into one, creating the model that best fits the reality in all periods. Thus, we obtained four calibrated models, one for each selected weather file:

- *CU1* model obtained with the "on -site weather file".
- *CU2* model obtained with the "third-party weather file".
- *CU3* model obtained with the "third-party weather file + on-site temperature sensor".

- *CU4* model obtained with the "weather file combination".

To present the results obtained, Table 2 was created to show the uncertainty indexes CV(RMSE) and R^2 attained by the models in the different calibration periods. For the periods where the temperature generated by the energy model was compared with the real temperature, the area-weighted average of each thermal zone analyzed was considered to produce the index result. However, in the periods where the energy consumed is the comparison, the sum of the energy spent per thermal zone was carried out to achieve the uncertainty index.

To combine the effects on the weather file in all calibration periods in a unique result, the value of the temperature and energy indices obtained (normalized indices) was normalized, so that they have the same basis to be able to add them up and thus obtain a unified value (sum indices).

Table 2. The normalized results in the checking periods with the different weather files. Randomly ordered logarithmic binary sequence (ROLBS).

Weather	Model	Checking Period	Index		Normalized Indices	Indices Sum
				Average house temperatures		
		Period 3 (ROLBS)	CV(RMSE)	1.23%	0.000	
			R^2	99.07%	0.000	
		Period 5 (Free oscillation)	CV(RMSE)	1.63%	0.109	
			R^2	98.66%	0.021	
On-site	Model *CU1*			Sum of housing energy		
		Period 2 (Set point 30 C)	CV(RMSE)	9.65%	0.049	
			R^2	91.80%	0.091	
		Period 4 (Set point 25 C)	CV(RMSE)	14.46%	0.200	
			R^2	92.77%	0.072	0.544
				Average house temperatures		
		Period 3 (ROLBS)	CV(RMSE)	4.87%	1.000	
			R^2	90.26%	0.453	
		Period 5 (Free oscillation)	CV(RMSE)	4.70%	0.953	
			R^2	79.62%	1.000	
Third-party	Model *CU2*			Sum of housing energy		
		Period 2 (Set point 30 C)	CV(RMSE)	17.84%	0.307	
			R^2	69.66%	0.532	
		Period 4 (Set point 25 C)	CV(RMSE)	39.93%	1.000	
			R^2	46.20%	1.000	6.245
				Average house temperatures		
		Period 3 (ROLBS)	CV(RMSE)	1.33%	0.026	
			R^2	98.75%	0.017	
		Period 5 (Free oscillation)	CV(RMSE)	1.45%	0.059	
			R^2	98.55%	0.027	
Third-party + Temp.	Model *CU3*			Sum of housing energy		
		Period 2 (Set point 30 C)	CV(RMSE)	12.53%	0.140	
			R^2	88.47%	0.157	
		Period 4 (Set point 25 C)	CV(RMSE)	20.60%	0.393	
			R^2	88.09%	0.165	0.985

Table 2. *Cont.*

Weather	Model	Checking Period	Index		Normalized Indices	Indices Sum
				Average house temperatures		
		Period 3 (ROLBS)	CV(RMSE)	1.37%	0.039	
			R^2	98.87%	0.011	
		Period 5 (Free oscillation)	CV(RMSE)	1.68%	0.123	
			R^2	98.21%	0.044	
Combination	Model *CU4*					
				Sum of housing energy		
		Period 2 (Set point 30 C)	CV(RMSE)	11.16%	0.097	
			R^2	91.22%	0.103	
		Period 4 (Set point 25 C)	CV(RMSE)	8.08%	0.000	
			R^2	96.37%	0.000	0.416

When studying the results obtained, the weather file that obtained the best results was the "weather file combination", with an index sum value of 0.416, exceeding, by a very small margin, the "on-site weather file" (0.544). In third position was the "third-party weather file + on-site temperature sensor" with a normalized index sum value of 0.985. In addition, the "third-party weather file" obtained a normalized index sum value of 6.245, which was well behind the other weather files. As can be seen in the table, we confirmed that the ranking the weather files produced with the base model was repeated in the calibration process.

When analyzed in more detail, we concluded that in the periods where the result sought was temperature (period 3 and 5), both the "weather file combination", "on-site weather file" and "third-party weather file + on-site temperature sensor", had very similar results, all obtaining very good values. The biggest differences occurred when we examined the energy periods (periods 2 and 4). Here, the "weather file combination" and the "on-site weather file" produced a difference compared with the other files.

In addition, in period 4, the "weather file combination" was much better adjusted to the expected result of 8.08% CV(RMSE) compared to the 14.46% obtained by the "on-site weather file".

Another analysis made to the weather archives was to determine if they complied with the objectives proposed by the international standards: The International Performance Measurement and Verification Protocol (IPMVP) determines that a model achieves best fit values when it manages to obtain a CV(RMSE) of less than ±20% on an hourly scale. On the other hand, ASHRAE and the Federal Energy Management Program (FEMP) state that a model can be considered to be fit when it obtains a CV(RMSE) index of less than ±30% on the same scale (Table 3). As for the R^2 index, ASHRAE recommends that models should not have an index of less than 75% to be considered calibrated.

Table 3. The calibration criteria of Federal Energy Management Program (FEMP), American Society of Heating, Refrigerating, and Air Conditioning Engineers (ASHRAE), and international performance measurement and verification protocol (IPMVP).

Data Type	Index	FEMP 3.0 Criteria	ASHRAE G14-2002	IPMVP
Calibration Criteria				
Monthly Criteria %	CV(RMSE)	±15	±15	-
Hourly Criteria %	CV(RMSE)	±30	±30	±20

In Table 2, the results that are not colored comply with international standards, the results that comply with ASHRAE and FEMP have been colored in orange, while the periods that do not comply with any standard have been marked in red. Both the "weather file combination" and the "on-site weather file" met the standard expectations in all the periods analyzed. The "third-party weather file + on-site temperature sensor" achieved very good rates, although, in period 4, it did not manage to comply with the IPMVP standard in the CV(RMSE) rate value where it stayed with 20.60% being the limit to be able to comply with 20.00%. The "third-party weather file" failed to respond satisfactorily in period 2, where it did not meet the value of R^2, and in period 4 where it did not reach a good value for either index.

4. Conclusions

Based on the results obtained with the study conducted in this paper, a new method is proposed to establish the degree to which the sensors that generate the climate file were affected in the process of adjusting the energy models of the building. Thus, we were able to select the best composition of sensors depending on the needs of the project.

Different types of sensors come into play in the process of creating the weather file: the wind speed and direction, global and diffuse horizontal irradiation, the outdoor temperature, and the relative humidity. Data from these sensors can be obtained from weather stations placed at the experiment site or can be acquired from third parties.

A weather station placed on the site requires a strong economic investment not only when buying or renting the group of sensors needed, as well as when processing and validating the data generated. Although weather data acquired from third parties are much cheaper, they have the disadvantage of being less accurate.

In this article, a new fast methodology was developed to determine the degree of adjustment of the different weather files composed of site data and third-party data. To this end, the sensors of the on-site weather station and those of third parties were combined with the aim of creating as many climate files as possible, resulting in a total of sixty-four files.

The results obtained in the experiment confirmed that there was a combination of sensors that offered a better degree of adjustment between the simulated temperatures and energies and reality than those offered by the site's weather station. Using these sensors to compose the weather file allowed the energy model to represent reality more accurately. This combination used the wind speed (WS), global and diffuse solar radiation (GHI), and temperature (T) sensors from the weather station placed at the site and the wind direction (WD) and relative humidity (RH) sensors from the data provided by third parties (weather file combination).

Good results were obtained by the climate file composed of the third-party data plus the temperature sensor of the site's weather station (third-party + on-site temp. sensor). With a very low economic cost, as it was simply necessary to place a temperature sensor outside the building, very positive adjustment results were achieved. The model behaved with excellent performance in all periods, complying with the parameters proposed by ASHRAE both in the CV(RMSE) and in R^2 index. This model obtained a normalized index sum of 0.985 compared to 0.544 as obtained by the model calibrated with the site's climate file. This is a difference to be considered, but is far from the one that occurs with the model adjusted with the climate file composed of third-party data, which obtained an index sum of 6.245. Depending on the degree of precision that must be obtained in the energy model, this option can be a good way to deal with the adjustment process. The weather file had the best cost-effectiveness ratio.

One of the most relevant conclusions obtained when developing the proposed methodology was the fact that it was not necessary to go through a process of adjustment of the model to be able to indicate the effect of the weather file in the final result. The results obtained when simulating all the proposed climate archives with the base model were the same as those obtained when executing an adjustment process. Even the positions in which the files were placed, as well as the distance between them, were met.

With the methodology proposed here, in a swift and simple way, we opened the option of being able to choose the composition of the meteorological file for use in the energy model adjustment process. This decision can be considered for economic, precision, and effectiveness factors, among others.

Author Contributions: V.G.G. and C.F.B. supervised the methodology used in the article, performed the simulations and the analysis, and wrote the manuscript. G.R.R. developed the EnergyPlus model and participated in the data analysis. All authors have read and agreed to the published version of the manuscript.

Funding: The work was funded by the research and innovation program Horizon 2020 of the European Union under Grant No. 731211, project SABINA.

Acknowledgments: We would like to thank the promoters of annex 58, the possibility of accessing the data of the houses. Without this data it would not have been possible to complete this work.

Conflicts of Interest: The authors declare no conflict of interest.

Abbreviations

The following abbreviations are used in this manuscript:

BEM	Building Energy Model
CV(RMSE)	Coefficient of Variation of Mean Square Error
MPC	Model Predictive Control
R^2	Square Pearson Correlation Coefficient
FDD	Fault Detection Diagnosis
SABINA	SmArt BI-directional multi eNergy gAteway
P2H	Power to Heat
IEA-EBC	International Energy Agency Energy in Buildings and Communities
HVAC	Heating Ventilation Air Conditioning
ESCOs	Energy Service Companies
IPMVP	International Measurement a Verification Protocol
EU	European Union
WD	Wind Direction
°C	Celsius degrees
WS	Wind Speed
m	Meter
N	North
E	East
T	Temperature
GHI	global radiation
WD	Wind Direction
RH	Relative Humidity
DHI	diffuse radiation
FEMP	Federal Energy Management Program
NSGA-II	Non-Dominated Sorting Genetic Algorithm
RMSE	Root Mean Square Error
ASHRAE	American Society of Heating, Refrigerating, and Air Conditioning Engineers
%	Percentage

References

1. Zhang, Z.; Chong, A.; Pan, Y.; Zhang, C.; Lam, K.P. Whole building energy model for HVAC optimal control: A practical framework based on deep reinforcement learning. *Energy Build.* **2019**, *199*, 472–490, doi:10.1016/j.enbuild.2019.07.029.
2. Hensen, J.L.; Lamberts, R. *Building Performance Simulation for Design and Operation*; Routledge: London, UK, 2012.
3. SABINA H2020 EU Program. Available online: http://sindominio.net/ash (accessed on 10 June 2020).
4. Kohlhepp, P.; Hagenmeyer, V. Technical potential of buildings in Germany as flexible power-to-heat storage for smart-grid operation. *Energy Technol.* **2017**, *5*, 1084–1104, doi:10.1002/ente.201600655.
5. Bloess, A.; Schill, W.P.; Zerrahn, A. Power-to-heat for renewable energy integration: A review of technologies, modeling approaches, and flexibility potentials. *Appl. Energy* **2018**, *212*, 1611–1626, doi:10.1016/j.apenergy.2017.12.073.
6. Fernández Bandera, C.; Pachano, J.; Salom, J.; Peppas, A.; Ramos Ruiz, G. Photovoltaic Plant Optimization to Leverage Electric Self Consumption by Harnessing Building Thermal Mass. *Sustainability* **2020**, *12*, 553, doi:10.3390/su12020553.
7. Ruiz, G.R.; Bandera, C.F.; Temes, T.G.A.; Gutierrez, A.S.O. Genetic algorithm for building envelope calibration. *Appl. Energy* **2016**, *168*, 691–705, doi:10.1016/j.apenergy.2016.01.075.
8. Fernández Bandera, C.; Ramos Ruiz, G. Towards a new generation of building envelope calibration. *Energies* **2017**, *10*, 2102, doi:10.3390/en10122102.

9. González, V.G.; Ruiz, G.R.; Bandera, C.F. Empirical and Comparative Validation for a Building Energy Model Calibration Methodology. *Sensors* **2020**, *20*, 5003, doi:10.3390/s20175003.
10. Strachan, P.; Svehla, K.; Heusler, I.; Kersken, M. Whole model empirical validation on a full-scale building. *J. Build. Perform. Simul.* **2016**, *9*, 331–350, doi:10.1080/19401493.2015.1064480.
11. Agüera-Pérez, A.; Palomares-Salas, J.C.; de la Rosa, J.J.G.; Florencias-Oliveros, O. Weather forecasts for microgrid energy management: Review, discussion and recommendations. *Appl. Energy* **2018**, *228*, 265–278, doi:10.1016/j.apenergy.2018.06.087.
12. González, V.G.; Ruiz, G.R.; Segarra, E.L.; Gordillo, G.C.; Bandera, C.F. Characterization of Building Foundation in Building Energy Models. In Proceedings of the Building Simulation, Rome, Italy, 2–4 September 2019.
13. United States Department of Energy. *EnergyPlus Engineering Reference: The Reference to EnergyPlus Calculations*; Lawrence Berkeley National Laboratory: Berkeley, CA, USA, 2009.
14. Coakley, D.; Raftery, P.; Keane, M. A review of methods to match building energy simulation models to measured data. *Renew. Sustain. Energy Rev.* **2014**, *37*, 123–141.
15. Ruiz, G.R.; Bandera, C.F. Analysis of uncertainty indices used for building envelope calibration. *Appl. Energy* **2017**, *185*, 82–94, doi:10.1016/j.apenergy.2016.10.054.
16. González, V.G.; Colmenares, L.Á.; Fidalgo, J.F.L.; Ruiz, G.R.; Bandera, C.F. Uncertainy's Indices Assessment for Calibrated Energy Models. *Energies* **2019**, *12*, 2096, doi:10.3390/en12112096.
17. De Wit, S.; Augenbroe, G. Analysis of uncertainty in building design evaluations and its implications. *Energy Build.* **2002**, *34*, 951–958, doi:10.1016/S0378-7788(02)00070-1.
18. Segarra, E.L.; Ruiz, G.R.; González, V.G.; Peppas, A.; Bandera, C.F. Impact Assessment for Building Energy Models Using Observed vs. Third-Party Weather Data Sets. *Sustainability* **2020**, *12*, 6788, doi:10.3390/su12176788.
19. Erba, S.; Causone, F.; Armani, R. The effect of weather datasets on building energy simulation outputs. *Energy Procedia* **2017**, *134*, 545–554, doi:10.1016/j.egypro.2017.09.561.
20. Voisin, J.; Darnon, M.; Jaouad, A.; Volatier, M.; Aimez, V.; Trovão, J.P. Climate impact analysis on the optimal sizing of a stand-alone hybrid building. *Energy Build.* **2020**, *210*, 109676, doi:10.1016/j.enbuild.2019.109676.
21. Bhandari, M.; Shrestha, S.; New, J. Evaluation of weather datasets for building energy simulation. *Energy Build.* **2012**, *49*, 109–118, doi:10.1016/j.enbuild.2012.01.033.
22. Sandels, C.; Widén, J.; Nordström, L.; Andersson, E. Day-ahead predictions of electricity consumption in a Swedish office building from weather, occupancy, and temporal data. *Energy Build.* **2015**, *108*, 279–290, doi:10.1016/j.enbuild.2015.08.052.
23. Du, H.; Barclay, M.; Jones, P. Generating high resolution near-future weather forecasts for urban scale building performance modelling. In Proceedings of the 15th Conference of International Building Performance Simulation Association, San Francisco, CA, USA, 7–9 August 2017.
24. Du, H.; Jones, P.; Segarra, E.L.; Bandera, C.F. Development of a REST API for obtaining site-specific historical and near-future weather data in EPW format. In Proceedings of the Building Simulation and Optimization 2018, Cambridge, UK, 11–12 September 2018.
25. Silvero, F.; Lops, C.; Montelpare, S.; Rodrigues, F. Generation and assessment of local climatic data from numerical meteorological codes for calibration of building energy models. *Energy Build.* **2019**, *188*, 25–45, doi:10.1016/j.enbuild.2019.02.001.
26. Cuerda, E.; Guerra-Santin, O.; Sendra, J.J.; Neila, F.J. Understanding the performance gap in energy retrofitting: Measured input data for adjusting building simulation models. *Energy Build.* **2020**, *209*, 109688, doi:10.1016/j.enbuild.2019.109688.
27. Song, S.; Haberl, J.S. Analysis of the impact of using synthetic data correlated with measured data on the calibrated as-built simulation of a commercial building. *Energy Build.* **2013**, *67*, 97–107, doi:10.1016/j.enbuild.2013.07.057.
28. Ciobanu, D.; Eftimie, E.; Jaliu, C. The influence of measured/simulated weather data on evaluating the energy need in buildings. *Energy Procedia* **2014**, *48*, 796–805, doi:10.1016/j.egypro.2014.02.092.
29. Afram, A.; Janabi-Sharifi, F. Theory and applications of HVAC control systems–A review of model predictive control (MPC). *Build. Environ.* **2014**, *72*, 343–355, doi:10.1016/j.buildenv.2013.11.016.
30. Cowan, J. International Performance Measurement and Verification Protocol: Concepts and Options for Determining Energy and Water Savings-Vol. I. International Performance Measurement & Verification Protocol, 1. LBNL Report #: LBNL/PUB-909. Available online: https://escholarship.org/uc/item/68b7v8rd (accessed on 10 June 2020)
31. Annex 58: Reliable Building Energy Performance Characterisation Based on Full Scale Dynamic Measurement. 2016. Available online: https://bwk.kuleuven.be/bwf/projects/annex58/index.htm (accessed on 10 June 2020).
32. van Dijk, H.; Tellez, F. Measurement and Data Analysis Procedures, Final Report of the JOULE IICOMPASS project (JOU2-CT92-0216). 1995.
33. Crawley, D.B.; Hand, J.W.; Kummert, M.; Griffith, B.T. Contrasting the capabilities of building energy performance simulation programs. *Build. Environ.* **2008**, *43*, 661–673, doi:10.1016/j.buildenv.2006.10.027.
34. Deb, K.; Pratap, A.; Agarwal, S.; Meyarivan, T. A fast and elitist multiobjective genetic algorithm: NSGA-II. *IEEE Trans. Evol. Comput.* **2002**, *6*, 182–197, doi:10.1109/4235.996017.
35. Guideline, A. *Measurement of Energy and Demand Savings*; Guideline 14-2002; American Society of Heating, Ventilating, and Air Conditioning Engineers: Atlanta, GA, USA, 2002.
36. Lucas Segarra, E.; Du, H.; Ramos Ruiz, G.; Fernández Bandera, C. Methodology for the quantification of the impact of weather forecasts in predictive simulation models. *Energies* **2019**, *12*, 1309, doi:10.3390/en12071309.

37. Ruiz, G.R.; Bandera, C.F. Importancia del clima en la simulación energética de edificios. In *Jornadas Internacionales de Investigación en Construcción: Vivienda: Pasado, Presente y Futuro: Resúmenes y Actas*; Instituto Eduardo Torroja: Madrid, Spain, 2013.
38. Du, H.; Bandera, C.F.; Chen, L. Nowcasting methods for optimising building performance. In Proceedings of the Building Simulation 2019: 16th Conference of International Building Performance Simulation Association, Rome, Italy, 2–4 September 2019.
39. Li, H.; Hong, T.; Sofos, M. An inverse approach to solving zone air infiltration rate and people count using indoor environmental sensor data. *Energy Build.* **2019**, *198*, 228–242, doi:10.1016/j.enbuild.2019.06.008.
40. Martínez, S.; Eguía, P.; Granada, E.; Moazami, A.; Hamdy, M. A performance comparison of multi-objective optimization-based approaches for calibrating white-box building energy models. *Energy Build.* **2020**, *216*, 109942, doi:10.1016/j.enbuild.2020.109942.
41. Reddy, T.A.; Maor, I.; Panjapornpon, C. Calibrating Detailed Building Energy Simulation Programs with Measured Data—Part I: General Methodology (RP-1051). *HVAC&R Res.* **2007**, *13*, 221–241, doi:10.1080/10789669.2007.10390952.
42. Carroll, W.; Hitchcock, R. Tuning simulated building descriptions to match actual utility data: Methods and implementation. *ASHRAE Trans. Am. Soc. Heat. Refrig. Aircond. Eng.* **1993**, *99*, 928–934.

energies

MDPI

Article

Life Cycle Assessment of Dynamic Water Flow Glazing Envelopes: A Case Study with Real Test Facilities

Belen Moreno Santamaria [1], Fernando del Ama Gonzalo [2,*], Matthew Griffin [2], Benito Lauret Aguirregabiria [1] and Juan A. Hernandez Ramos [3]

1 Department of Construction and Architectural Technology, Technical School of Architecture of Madrid, Technical University of Madrid (UPM), Av. Juan de Herrera 4, 28040 Madrid, Spain; belen.moreno@upm.es (B.M.S.); benito.lauret@upm.es (B.L.A.)
2 Keene State College, 229 Main St, Keene, NH 03435, USA; Matthew.Griffin@ksc.keene.edu
3 Escuela Department of Applied Mathematics, School of Aeronautical and Space Engineering, Technical University of Madrid (UPM), Plaza Cardenal Cisneros 3, 28040 Madrid, Spain; juanantonio.hernandez@upm.es
* Correspondence: fernando.delama@keene.edu

Abstract: High initial costs hinder innovative technologies for building envelopes. Life Cycle Assessment (LCA) should consider energy savings to show relevant economic benefits and potential to reduce energy consumption and CO_2 emissions. Life Cycle Cost (LCC) and Life Cycle Energy (LCE) should focus on investment, operation, maintenance, dismantling, disposal, and/or recycling for the building. This study compares the LCC and LCE analysis of Water Flow Glazing (WFG) envelopes with traditional double and triple glazing facades. The assessment considers initial, operational, and disposal costs and energy consumption as well as different energy systems for heating and cooling. Real prototypes have been built in two different locations to record real-world data of yearly operational energy. WFG systems consistently showed a higher initial investment than traditional glazing. The final Life Cycle Cost analysis demonstrates that WFG systems are better over the operation phase only when it is compared with a traditional double-glazing. However, a Life Cycle Energy assessment over 50 years concluded that energy savings between 36% and 66% and CO_2 emissions reduction between 30% and 70% could be achieved.

Keywords: water flow glazing; dynamic building envelope; life cycle assessment

Citation: Santamaria, B.M.; Gonzalo, F.d.A.; Griffin, M.; Aguirregabiria, B.L.; Hernandez Ramos, J.A. Life Cycle Assessment of Dynamic Water Flow Glazing Envelopes: A Case Study with Real Test Facilities. *Energies* **2021**, *14*, 2195. https://doi.org/10.3390/en14082195

Academic Editor: Fitsum Tariku

Received: 14 March 2021
Accepted: 8 April 2021
Published: 14 April 2021

1. Introduction

In recent years, clean energy use has steadily grown. However, energy consumption has not altered its pattern, with fossil fuels acting as the primary energy consumption and generation source. Many carbon emissions and greenhouse gasses caused by conventional energy sources have severe consequences on the environment. Therefore, building codes, regulations, and energy directives consider carbon emissions' impact when evaluating energy efficiency [1]. New technologies emerge rapidly and force building designers to act without a thorough environmental impact analysis [2]. To control the construction sector's environmental impact is the most critical challenge that the architecture, engineering, and construction (AEC) industry must face soon [3,4]. Life Cycle Assessment (LCA) is a comprehensive and internationally standardized method. It quantifies all relevant emissions, resources consumed, and the related environmental aspects associated with any goods or services [5,6].

1.1. Literature Review

Extensive investigation has been conducted to study the environmental impact of building materials [7]. A Life Cycle Assessment can significantly help improve a sustainable building design by presenting how different materials or processes contribute to the building's overall environmental impact [8,9]. According to some authors, Life Cycle Assessment considers four phases, the extraction of resources, the construction phase, the building operation, and the building's demolition [10]. The European standards (EN) standards for Building Life Cycle Assessment maintain a list of 24 total categories that describe potential environmental impact [11–13]. A broadly accepted standard of the environmental impact of buildings is energy consumption [14]. The Life Cycle Energy analysis (LCE) only accounts for energy inputs at different life cycle stages, including the operational energy and embodied energy in buildings over their lifetime. Life Cycle Energy analysis evaluates the embodied energy of products, design modifications, and strategies used to optimize operational energy. The most common period for major renovations in the residential sector in practice is, according to some authors, 30–40 years [15]. Other authors estimate the lifetime of buildings as 50 years in their LCC approach [16]. The optimization of building envelopes should examine both the energy-saving and the Life Cycle Cost goals [17]. Different authors have proposed mathematical models to calculate each envelope material's Life Cycle Cost and heating and cooling system [18]. Construction costs, return on investment, increased market value, and maintenance and operation costs are the factors that determine consumers' response to sustainable buildings [19]. Since buildings have long lifespans, the design decisions have long-term consequences, considering that upfront costs amount to less than 30% of the total Life Cycle Costs [20,21].

Global warming potential (GWP) is a measure of the amount of heat trapped in the atmosphere. This variable is measured in carbon dioxide equivalents ($kgCO_{2eq}$). This equivalency means that the total greenhouse potential of a specific emission type is given concerning CO_2 [22]. Since the calculation for global warming potential includes the residence time of gases in the atmosphere, a total time range for an assessment can be defined at 100 years [23]. Emissions are substances that are released into the environment, which includes the air, water, and soil. This, in turn, negatively impacts human and environmental health. Emissions typically enter into the environment as a waste product from different industrial processes. The most common (and therefore well-known) emissions are called greenhouse gas (GHG) emissions [24]. Some articles have studied building envelope design from its energy-saving potential, environmental consequences, and social impacts. It is of the utmost importance to optimize the balance between the cost and energy savings [25,26].

Dynamic or active buildings adapt their thermal performance according to different inputs, such as outdoor and indoor conditions, and can produce part of the building's energy over its operational life. Technical research and numerical simulation tools on active building envelopes have increased over the last decade. However, the ratio of dynamic facades in the building industry remains stable [27]. Some dynamic envelopes change their opacity or vary their transmission or reflection properties. Electrochromic glass, Polymer dispersed liquid crystal, and Suspended Particle Devices are hindered by their high cost and non-standardized manufacturing processes [28,29]. Active envelopes can also produce renewable energy on-site. Photovoltaic panels (PV) are considered the most reliable on-site renewable energy generation technology due to the wide range of electricity use and cables' flexibility that transport the energy [30]. Solar thermal collectors are considered a renewable and CO_2 free energy source. However, the pipes' stiffness transporting warm water has limitations in their applicability as a part of the building envelope [31].

This paper will examine the Life Cycle Assessment (LCA), Life Cycle Energy (LCE), and Life Cycle Cost (LCC) calculations for dynamic Water Flow Glazing (WFG) envelopes. Coupled with a plug and play piping system, it would produce a high-performance building envelope and innovative heating and cooling system [32,33]. Water Flow Glazing is a technology that can be integrated into transparent building envelopes, either in new buildings or as a retrofit for traditional glazing [34]. Flowing water through WFG panels captures an extensive percentage of the solar infrared radiation and keeps the glazing transparency [35]. WFG can absorb solar energy to provide domestic hot water to plumbing fixtures when the solar irradiance is high enough [36,37].

1.2. Objectives and Innovation

Water Flow Glazing has proven its potential for energy savings and for increasing comfort of occupants in previous studies. The main contribution of this article has been to carry out a thorough analysis of a 50-year life cycle from the energy and cost perspectives. To accomplish this task, the authors have employed a tested methodology used in previous scientific articles. This study considered two prototypes in different locations, glazing compositions and energy systems. Real-world data were collected from these prototypes and analyzed to determine the actual energy performance of WFG systems. This paper is structured into four separate parts. Section 2 of this article provides background information on Water Flow Glazing technology, a description of the WFG test facilities, and finally the methodology of Life Cycle Energy analysis (LCE) and Life Cycle Cost (LCC) calculations. Section 3 provides an analysis of the year-long data collection that occurred at the WFG facilities. This section also contains information pertaining to the embodied energy, the operational energy, and the renewable energy production of the test facilities. Section 4 is a discussion on LCE and LCC and their impact on global warming potential based on a multi-index evaluation. In addition to this, this section has a discussion on the limitations of the methodology. The fifth and final section is the conclusions section.

2. Materials and Methods

Water flow glazing can be used as both a high-performance envelope as well as an element of the heating and cooling system. The WFG module presented in this paper is made up of three components: an extruded aluminum frame, the glazing, and a circulating device. The glazing is a compound of several layers of laminated glass and coatings, with thermal and spectral properties provided by the glass manufacturers. It combines coatings and Polyvinyl butyral layers with a variable water mass flow rate to absorb or reject incoming infrared solar radiation. The circulating device is defined by a water pump, an exchanger that can regulate heat, and different sensors (such as the water flow and water temperature) to regulate the different fluid variables involved. Finally, the aluminum assembly provides the frame with structure. Although WFG can absorb and transport thermal energy, alternative heating and cooling sources might be added to compensate for the heat losses and gains and maintain comfort conditions. The initial cost of the glazing exceeds the cost of a traditional double or triple glazing panel. However, its performance has to be evaluated over its life cycle to consider potential energy savings. In addition, the water flow captures the solar infrared radiation and increases its temperature through the window. This water is transported and eventually releases the energy in buffer tanks so that thermal energy can be used in hydronic heating systems. The renewable energy integrated into the building envelope might not be enough to meet the energy needs, so it is necessary to study different energy sources and systems to compare their final energy consumption and greenhouse gas emission potential. The Life Cycle Assessment of Water Flow Glazing includes four phases. Phase 1 is the extraction, production of construction materials, and transportation of materials from the extraction point to the construction site. Phase 2 is the construction phase, including required energy to run construction machinery, any additional materials for construction, and any waste disposal. Phase 3 includes the energy required for the building's actual operation (including all energy used during the building occupation over the total lifespan of the structure), general maintenance, repairs, and finally, any required material replacement for the building. Finally, phase 4 involves demolition and transport of waste to recycling plants or landfills.

2.1. Water Flow Glazing Thermal Properties

Heat flux through any glazing depends on the difference between the indoor and outdoor temperatures $(\theta_e - \theta_i)$ and the direct and diffuse solar radiation, i_0. Equation (1) shows the heat flow, q, in glazing panels with gas chambers depending on the thermal transmittance, U, and g-factor. Equation (2) illustrates that the variable fluid's temperature and mass flow rate impact the heat flow through the glazing. Equations (1) and (2) are valid assuming steady conditions, constant values for convective and radiative coefficients, negligible thermal resistance and thermal mass of the glass panes and the water chamber, and the uniform flow inside the water chamber.

$$q = U(\theta_e - \theta_i) + g i_0, \tag{1}$$

$$q = U(\theta_e - \theta_i) + U_w(\theta_{IN} - \theta_i) + g i_0, \tag{2}$$

where U_w is the thermal transmittance between the water chamber and the interior, U is the glazing thermal transmittance, θ_e is the outdoor temperature, θ_i is the indoor temperature, and θ_{IN} is the inlet temperature of the fluid into the WFG system. Equations (3)–(5) are taken from a previous article [38]. They show the thermal transmittances of WFG, along with the g-factor. All the parameters depend on the mass flow rate, which is estimated uniform inside the glass pane.

$$U = \frac{U_i\, U_e}{\dot{m}c + U_e + U_i}, \tag{3}$$

$$U_w = \frac{U_i\, \dot{m}c}{\dot{m}c + U_e + U_i}, \tag{4}$$

$$g = \left(\frac{U_i}{\dot{m}c + U_e + U_i}\right)\left(\left(A_1\left(\frac{U_e}{h_e}\right) + A_2\left(\frac{1}{h_g} + \frac{1}{h_e}\right)U_e + A_3\left(\frac{U_e}{h_i}\right) + A_w\right)\right) + A_i + T. \tag{5}$$

Water flow glazing can change the thermal performance by varying the mass flow rate per unit of surface, $\dot{m}$. U and U_w are two additional thermal transmittances that depend on the mass flow rate. The product $\dot{m}c$ denotes the potential of the water to transfer energy. U_i and U_e thermal transmittances were obtained utilizing the convective heat coefficients, h_e, h_i, h_g, h_w. The thermal resistance, $(1/U_e)$ is the sum of the thermal resistances from the water chamber to the outdoors. Similarly, the thermal resistance, $(1/U_i)$, is the sum of the thermal resistances from the water chamber to indoors. Hence, U_e represents the thermal transmittance between the water chamber and outdoors, and U_i, the thermal transmittance between that water chamber and indoors. A_w is the water absorptance, whereas A_1, A_2, A_3 are the glass panes absorptances.

This study included two separate prototypes. The first one was placed in Peralveche, Spain (latitude 40°36′42″ N, longitude 2°26′57″ W, altitude 1111 MAMSL). The second one was built and tested in Sofia, Bulgaria (42°39′1″ N, 23°23′26″ E, Elevation: 590 m a.s.l.). The Peralveche WFG cabin was made of double glazing with a water chamber. Figure 1 shows the WFG cabin and energy system schematics.

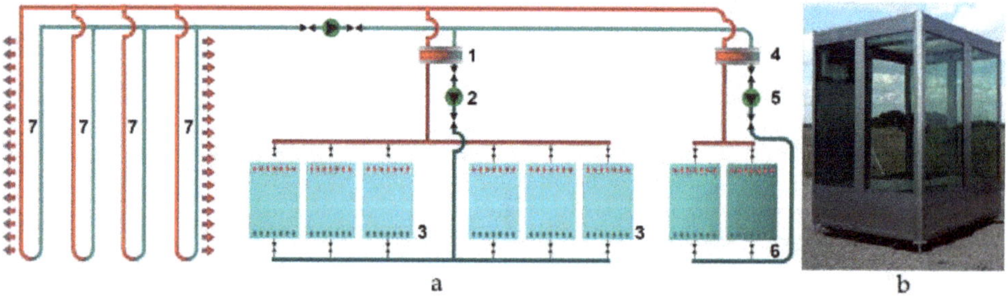

Figure 1. Peralveche prototype: (**a**) schematics of energy system: 1—plate heat exchanger, 2—water circulation pump and precision flow meters, 3—wall water flow glazing panels, 4—plate heat exchanger, 5—water circulation pump and precision flow meters, 6—roof water flow glazing panels, 7—borehole heat exchangers; (**b**) Water Flow Glazing (WFG) cabin.

The glass panes utilized to build this glazing assembly were Planiclear 6 + 6 mm with Poly-Vinyl Butyral layers (1 × 0.38 mm), a water chamber measuring 16 mm, and Planiclear 8 + 8 mm with Poly-Vinyl Butyral layers (1 × 0.38 mm). The cabin had six sides and each side was a square of 2 m × 2 m. The WFG panels were connected to four borehole heat exchangers buried 50 m underneath the cabin. The WFG cabin managed two different closed loop circulating systems. The first loop consisted of pipes that distributed refrigerant fluid from the borehole heat exchangers to the circulation pump. The WFG cabin included a second loop that circulated water through vertical facades and the horizontal roof and the mass flow rate was set to 0.9 $L \cdot min^{-1} \cdot m^{-2}$. Table 1 shows the glazing's thermal and spectral values from previous articles [38,39]. This WFG cabin was compared with another, referred to as Reference prototype, which had double glazing with an air cavity.

Table 1. Thermal properties of Peralveche glazing.

			$\dot{m}$ = 0 $L \cdot min^{-1} \cdot m^{-2}$			$\dot{m}$ = 0.9 $L \cdot min^{-1} \cdot m^{-2}$		
	Orientation	**Area (m^2)**	U ($W \cdot m^{-2} \cdot K^{-1}$)	U_w ($W \cdot m^{-2} \cdot K^{-1}$)	g	U ($W \cdot m^{-2} \cdot K^{-1}$)	U_w ($W \cdot m^{-2} \cdot K^{-1}$)	g
WFG [1]	S + E + W + roof	15.8	4.797	0.0	0.396	0.762	5.802	0.27
Reference [1]	S + E + W + roof	15.8	2.6	-	0.67	-	-	-

[1] Values taken from [39].

As per the Sofia prototype, the square plan dimensions were 7 m × 7 m. The walls are parallelograms of 7 m by 3.4 m comprised of five WFG modules facing east, west, and south, respectively. Figure 2 shows the unitized facade module. The circulating system incorporated a solar water pump and a plate heat exchanger connected to the water distribution pipes. The unitized WFG panels measured 3000 mm high and 1300 mm width, and the mass flow rate was set at 2 $L \cdot min^{-1} \cdot m^{-2}$. In the Southern glazing, the following layers were employed: a single 10 mm diamant glass pane, a 16 mm argon chamber, a low-emissivity coating Planitherm XN, Planiclear (8 mm), 2 Saflex R solar (SG41), Planiclear (8 mm), water chamber (24 mm), Planiclear (8 mm), 4 Saflex R standard clear (RB11), Planiclear (8 mm). Eastern and western glazing composition was meant to reject energy, so a highly reflective coating was included, instead of the low emissivity coating.

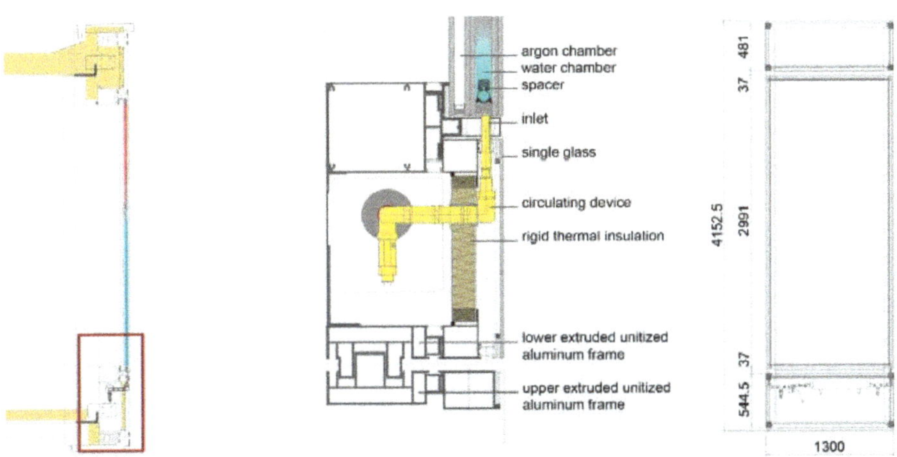

(a) (b) (c)

Figure 2. Sofia prototype description: (a) section; (b) detail of the unitized facade and circulating device; (c) elevation of the unitized module.

Table 2 shows the thermal properties of the glazing taken from a previous article [40]. The total glass area was 60 m^2. This WFG cabin was compared with another one, referred to as Reference prototype, with triple glazing with an air cavity and an argon cavity. The glass layers and coatings were the same as the WFG.

Table 2. Thermal properties of Sofia glazing.

			$\dot{m}$ = 0 L·min^{-1}·m^{-2}			$\dot{m}$ = 2 L·min^{-1}·m^{-2}		
	Area (m^2)	Orientation	U (W·m^{-2}·K^{-1})	U_w (W·m^{-2}·K^{-1})	g	U (W·m^{-2}·K^{-1})	U_w (W·m^{-2}·K^{-1})	g
WFG [1]	19.2	S Wall	1.041	0.0	0.59	0.066	6.459	0.24
WFG [1]	38.4	E–W Wall	0.995	0.0	0.27	0.063	6.462	0.22
Reference	19.2	S Wall	1.0	-	0.57	-	-	-
Reference	38.4	E–W Wall	1.0	-	0.30	-	-	-

[1] Values taken from [40].

2.2. *Life Cycle Energy (LCE) Analysis*

The total embodied energy of building elements involves the energy consumed directly at the primary material extraction, manufacturing, and assembly. These amounts of energy constitute the element's initial embodied energy, and the operational energy includes the heating and cooling energy consumption. Primary energy also considers the energy required to produce the final energy consumed in the building, and it varies according to fuel type and transportation losses. Primary energy is proportional to energy-related CO_2 emissions [41,42]. Life Cycle Energy comprises the building's operational energy, initial and recurrent embodied energy over its lifetime, and, finally, the energy for demolition and disposal. Equation (6) was used to calculate the Life Cycle Energy.

$$E = E_i + E_{err} + E_o(n) + E_d, \quad (6)$$

where E is the total energy of the building element, E_i is the initial embodied energy, E_{err} is the recurrent embodied energy for future maintenance and refurbishment (5% of the initial embodied energy), E_o is the total annual operational energy, n is the lifetime of the element in years, and E_d is the embodied energy required for demolition and disposal (3% of the initial embodied energy). Some LCE analysis studies include energy requirements such as lighting, cooking, hot water, and appliances. However, this study includes only the heating and cooling energy through the envelope. For the purposes of this paper, the energy absorbed by the WFG panels was considered as renewable energy production over the operational time in the final energy balance. Therefore, when calculating the total annual operational energy variable, this will be determined by taking the total energy consumption value and subtracting the amount of energy provided by the WFG panels in the structure.

2.3. *Life Cycle Cost (LCC) Analysis*

In addition to improving the thermal performance and reducing the environmental impact, the design of an efficient building envelope needs to pay attention to reducing the economic costs. Building owners demand the selection of cost-effective elements of the building envelope in a sustainable building design. Therefore, in terms of effective decision-making, it is essential to have a complete insight into the construction and running costs throughout the building's lifespan. The Life Cycle Cost (LCC) approach is based on optimizing design solutions and minimizing the sum of construction and operating expenses over the building lifetime. The building envelope's Life Cycle Cost has included its construction cost, C_1, operation cost, C_2, and demolition cost, C_3. The present value interest factor of the annuity (PVIFA) was used to calculate the operation cost's current value. The present value interest factor (PVIF) was used to estimate the demolition and disposal cost's current value at the end of the envelope life cycle. C represents the building envelope's total Life Cycle Cost and the heating and cooling system; r represents the interest rate; n represents the design operating life in years. Therefore, the proposed Life Cycle Cost model for building facades is presented in Equation (7).

$$C = C_1 + C_2\left(\frac{1}{r} - \frac{1}{r(1+r)^n}\right) + C_3\left(\frac{1}{(1+r)^n}\right) \quad (7)$$

3. Results

This section started by defining the parameters to quantify the environmental performance: primary energy, equivalent CO_2 emissions, and cost. The building reference time was defined as 50 years.

Depending on the local context, the embodied impact may surpass the operational impact. The indicators that measure energy performance can be split into economic factors and physical energy factors.

3.1. Embodied Energy Calculation

The building elements' embodied energy has been taken from different sources, whereas the operational energy was calculated with experimental data from the prototypes. It was assumed that the structures were not renovated during this time and had no change in their usage mode throughout their useable life. Energy use, in the viewpoint of the various stages of a building's life cycle, in cold-weather temperatures, the operational stage can reach upwards of 80%. In contrast, the building materials and in-situ construction account for 10–20% [43]. For simplification purposes, the energy for demolition and disposal is a percentage of the total primary energy. Embedded Energy (EE) is the total energy needed to produce goods and services, including processing, mining and extraction, manufacturing, and transport of products. Table 3 shows the emissions associated with the EE are the result of energy and emissions quantification based on the ITeC database [18]. The following assumptions have been considered: The embodied energy associated with the replacement, refurbishment, and substitution of materials and products is assumed to be 5% every ten years [44]. The embodied energy associated with demolition and disposal was assumed to be 3% of the total building Life Cycle Energy [45–47]. The considered energy for replacement, demolition, and transportation to landfill does not exceed 10 kWh.m^{-2}.

Table 3. Embodied Energy (EE) and Embodied Carbon (EC) of materials.

Material	EE (MJ·kg^{-1})	EC (kgCO$_{2eq}$·kg^{-1})	Weight (kg·m^{-2})	EE (MJ·m^{-2})	EC (kgCO$_{2eq}$·m^{-2})	Cost (€·m^{-2})
Double Reference	12.95	1.59	69.00	1500.00	110.01	120.50
Triple Reference	15.30	1.70	104.50	2002.30	180.05	220.50
Double WFG	13.25	1.63	70.00	1666.00	114.10	239.10
Triple WFG	15.90	1.75	105.00	2299.50	183.75	370.22
Extr. Al. Stick	220.00	14.76	25.8	5676.00	380.81	110.85
Extr. Al. Stick (R)	14.60	1.02	25.8	376.68	26.32	110.85
Extr. Al. Unitized	220.00	14.76	84.00	18480.00	1239.84	415.20
Extr. Al. Unitized (R)	14.60	1.02	84.00	1226.40	85.68	415.20
Circulating device [1]	-	-	-	237.20	79.68	200.00

[1] A circulating device can serve 4 m^2 of WFG.

The energy systems used for this study were borehole heat exchangers coupled with a ground-source heat pump for the WFG Peralveche prototype, Air-to-Water heat pump for WFG Sofia prototype, and Air-to-Air heat pump for both reference prototypes. All the heat pumps are considered to be 7 to 20 kW. The chosen system boundaries include the production of the component, starting from raw materials, the use stage, and the dismantling stage of both systems, along a temporal horizon of 50 years. Closed-loop borehole heat exchangers were made of high-density polyethylene vertical pipes with water (78%) and ethylene glycol (22%) mixture flowing. The installation process included drilling boreholes and trenches, inserting a vertical loop, and grouting operations of the borehole with concrete and bentonite mixture [48–50]. The dismantling process included the disposal of the glycol, the sealing of the borehole, and the disposal at the heat pump's end of life. Table 4 shows the embodied energy, emissions, and cost associated with the energy systems used to operate the different prototypes.

Table 4. Embodied Energy (EE) and Embodied Carbon (EC) of energy systems.

Material	EE (MJ)	EC ($kgCO_{2eq}$)	Cost (€)
Borehole heat exchangers [1]	9975.50	1661.98	5695.30
Ground-source heat pump	2193.60	390.30	7396.21
Air-to-Water heat pump	2210.50	350.70	8078.65
Air-to-Air heat pump	2393.70	380.50	3700.70

[1] Four 50 m borehole heat exchangers, diameter 118 mm.

The compressor and structure were made of reinforced steel and the evaporator and condenser from low alloyed steel. The pipes, cables, and expansion valves were made of copper. Pipes were insulated with a polymer and the cables were insulated with PVC. The refrigerant was assumed to be the same (tetrafluoroethane) for all the heat pumps. The heat pumps were considered maintenance-free, and their lifetime, 25 years.

3.2. Renewable Energy Production

Renewable primary energy (RPE in $kWh{\cdot}m^{-2}$) was produced as the water flowed through the glazing. The data were measured over a year. The RPE can cover part of the winter's heating needs and was subtracted from the non-renewable primary energy needed to maintain the prototype's indoor temperature within a comfort range. The heat absorption rate in the southern facades is closely related to the glazing composition and the orientation. When the solar radiation hits in the glazing units, the water chamber absorbs part of that energy. Heating was assumed to be delivered by a hydronic system using the energy absorbed by the WFG envelope. When needed, a heat pump operated to deliver the energy to keep comfort indoor conditions. Figure 3 shows the water heat gain of the prototypes' envelope. The horizontal panels in Peralveche show the largest water heat gain in July with 81.44 $kWh{\cdot}m^{-2}$. When it comes to the southern facades, the Sofia prototype shows the largest heat gain. The mass flow rate was set higher than any other prototypes (2 $L{\cdot}min^{-1}{\cdot}m^{-2}$), and the southern glazing properties demonstrate the highest absorptance. The peak solar heat gain in the Sofia prototype was 97.39 $kWh{\cdot}m^{-2}$ in July.

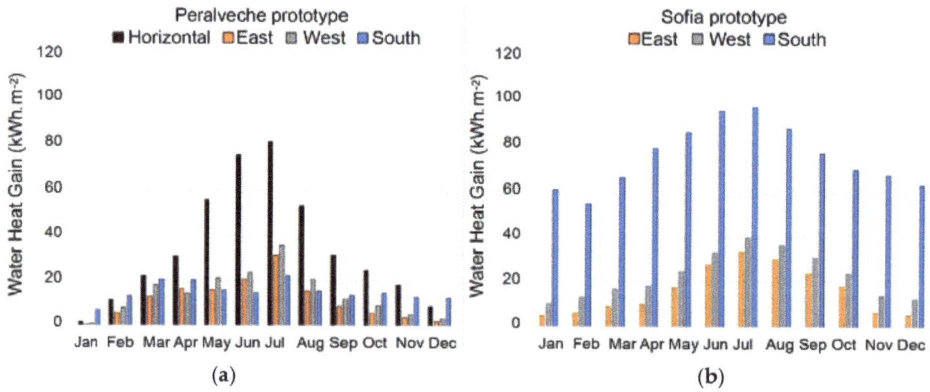

Figure 3. Experimental data. Water heat gain in WFG envelopes; (**a**) Peralveche prototype WFG; (**b**) Sofia WFG prototype in $kWh{\cdot}m^{-2}$.

3.3. Operational Energy

Heating and cooling energy loads (in kWh·m^{-2}) were calculated by using Equation (1) for the reference glazing and Equation (2) for WFG. The inputs were measured over a year time. Real data, obtained from the prototype throughout the year, allowed the correlation between the WFG cabin and the Reference cabin. Figure 4 shows the heating and cooling energy in kWh·m^{-2} to keep the inside within comfort temperature over the year. The figure considers only the energy that goes through the transparent envelope. The cooling load of the Peralveche Reference cabin showed by far the highest energy consumption with a peak of 118 kWh·m^{-2} in July.

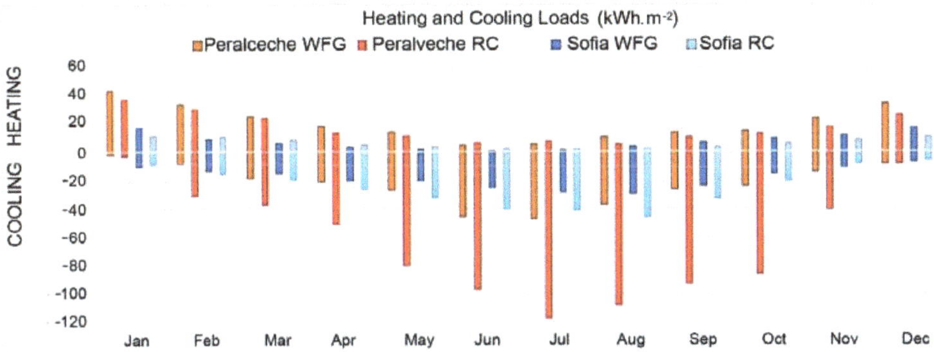

Figure 4. Experimental data. Heating and cooling energy through the transparent building envelopes in kWh·m^{-2}.

Table 5 shows conversion factors between non-renewable primary energy, NRPE, and final energy, FE (kWh$_{NRPE}$/kWh$_{FE}$), as well as CO_2 emissions conversion factors provided by the Spanish Regulation of Thermal Installations in Buildings [42]. The price per kWh of different energy carriers or fuels was taken from Eurostat documents [43].

Table 5. Conversion factor (f) from final energy (FE) to non-renewable primary energy (NRPE), CO_2 emissions, and price of energy carriers.

	f_{NRPE} (kWh$_{NRPE}$/kWh$_{FE}$)	f_{TPE} (kWh$_{TPE}$/kWh$_{FE}$)	f_{CO2} (kgCO_2/kWh$_{FE}$)	Price [1] (€/kWh)
Mainland electricity	1.954	2.368	0.331	0.239
Fuel	1.179	1.182	0.331	0.0713
Natural gas	1.190	1.195	0.252	0.102
Biomass (pellets)	0.085	1.113	0.018	0.0462

[1] values taken from [43].

The non-renewable primary energy consumption of the existing building per unit of envelope area and year, *NRPEC*, in kWh·m^{-2} per year, was calculated using Equation (8).

$$NRPE = f_{NRPE}FE_{heat} + f_{NRPE}FE_{cool}, \quad (8)$$

The equivalent CO_2 emissions of the existing building per unit of envelope area and year, in kgCO_2.m^{-2} per year, were calculated with Equation (9).

$$CO_{2eq} = f_{CO2}FE_{heat} + f_{CO2}FE_{cool}, \quad (9)$$

Tables 6 and 7 illustrate the non-renewable primary energy consumption (NRPE) with different energy generators in Peralveche and Sofia, respectively. The conversion factor between final energy and non-renewable primary energy (kWh NRPE/kWh FE) and the factor of emitted CO_2 for electricity were taken from Table 3 for mainland electricity: 1.954

for the conversion factor between final energy and non-renewable primary energy (kWh NRPE/kWh FE) and 0.331 for CO_2 emissions for electricity. In the final energy balance, the renewable energy production was subtracted from the WFG prototype heating loads. The ground-source heat pumps' performance depends on the source inlet temperature in the heat pump, and the inlet temperature in the WFG (θ_{IN}). A typical value of source inlet temperature ranges from 20 °C in ground-source heat pumps (GSHP) to 35 °C in other Water-to-Water (W-W) heat pumps. The parameters that influence Air-to-Air (A-A) heat pumps' performance are the dry bulb exterior air temperature and the dry bulb interior return air temperature [51].

Table 6. Peralveche prototype. Final energy, non-renewable primary energy, and CO_2 emissions.

	Heating		Cooling	
Peralveche	**WFG(W-W)**	**RC(A-A)**	**WFG(W-W)**	**RC(A-A)**
Energy (kWh·m^{-2})	42.12	206.95	279.70	755.05
COP [1]	6.60	4.20	5.90	3.30
FE (kWh·m^{-2})	6.38	49.27	47.41	228.80
NRPE (kWh·m^{-2})	12.47	96.28	92.63	447.08
CO_{2eq} (kgCO_2.m^{-2})	2.11	16.31	15.69	75.73
OE cost (€·m^{-2} per year)	1.53	11.78	11.33	54.68

[1] COP values are taken from [51].

Table 7. Sofia prototype. Final energy, non-renewable primary energy, and CO_2 emissions.

	Heating		Cooling	
Sofia	**WFG(W-W)**	**RC(A-A)**	**WFG(W-W)**	**RC(A-A)**
Energy (kWh·m^{-2})	0.00	76.58	219.10	298.07
COP [1]	5.30	4.20	4.60	3.30
FE (kWh·m^{-2})	0.00	18.23	47.63	90.32
NRPE (kWh·m^{-2})	0.00	35.63	93.07	176.49
CO_{2eq} (kgCO_2.m^{-2})	0.00	6.04	15.77	29.90
OE cost (€·m^{-2} per year)	0.00	4.36	11.38	21.59

[1] COP values are taken from [51].

The yearly operating energy was assumed to remain steady during the entire building operation. The resource mix supplying electricity to the buildings is assumed to be unvarying. The HVAC systems' efficiency and the operation schedule were assumed to remain unchanged during the Life Cycle Assessment.

4. Discussion

WFG can be a part of hydronic heating and cooling systems, and it is compatible with ground-source heat pumps and boilers. In this study, the authors have considered only mainland electricity as the energy source, ground source heat pumps for the Peralveche WFG prototype, Air-to-Water heat pumps for the Sofia WFG prototype, and Air-to-Air heat pumps for the reference prototypes.

4.1. Life Cycle Cost Evaluation

The method to evaluate the Life Cycle Cost (LCC) has been shown in Section 2. The parameters to calculate the envelope's Life Cycle Cost and the energy system initial cost have been discussed in Section 3. The operation cost was calculated with data provided by Table 5 and energy prices for mainland electricity. Maintenance costs for the envelope over 50 years have been calculated as a percentage of the initial cost (1% for the reference glass and 5% for WFG), whereas the heat pump's lifetime was 25 years. The cost of the studied WFG unitized facade was 985 €·m^{-2}, whereas the price for the circulating water pump was 20 €·m^{-2}. Replacing the water pumps after a 10-year cycle is included in the maintenance cost. Finally, the demolition and disposal costs were calculated as a percentage of the initial cost (3%). Therefore, the total proposed Life Cycle Cost for the building envelope and energy system was calculated according to Equation (7). The total construction costs of the envelopes and the energy systems were calculated by multiplying each envelope component's quantity by the total unit price. The operation cost was converted to the present value based on annual heating and cooling loads through the envelope. The considered demolition and disposal values were 3% of the total construction costs and converted to the present value. This article considered a discount rate of 2% calculated as an average of the harmonized consumer price index in Spain over the last 20 years. The price of energy was taken from Eurostat reports [43]. The material prices of other components come from the ITeC database [18]. Table 8 shows the cost analysis parameters for all cases.

Table 8. Cost analysis parameters.

	C_1	C_2	C_3	Total (50 Years)
Peralveche WFG	17,529	640	7921	40,611
Peralveche RC	7373	1239	3921	47,767
Sofia WFG	65,626	1320	10,046	110,865
Sofia RC	40,795	1923	4923	103,066

Water flow glazing envelopes showed a higher Coefficient C_1. The initial cost reflected the investment in borehole heat exchangers and circulating devices. However, the coefficient C_2 reflected that the reference yearly operation cost is twice as much as the WFG cost in Peralveche and 1.5 more in Sofia. The total cost considered a 50-year life cycle by converting the operation and disposal costs to the present value. In Peralveche, the reference prototype cost surpassed the WFG one. In Sofia, the WFG is slightly higher than the reference one.

4.2. Life Cycle Energy Evaluation

Life Cycle Energy included the materials initial embodied energy, the operational energy for heating and cooling over 50 years, the recurrent embodied energy over its lifetime, and, finally, the energy for demolition and disposal. Equation (6) was used to calculate the Life Cycle Energy. Table 9 shows the initial embodied energy E_i, the recurrent embodied energy for future maintenance E_{err}, calculated as a percentage of the initial embodied energy, the total annual operational energy, and the embodied energy required for demolition and disposal E_d.

Table 9. Life Cycle Energy parameters in GJ.

	E_1	E_{err}	E_o	E_d	Total (50 Years)
Peralveche WFG	193	19	300	58	571
Peralveche RC	116	12	1553	35	1716
Sofia WFG	1216	122	978	61	2377
Sofia RC	1198	120	2230	60	3608

The most considerable initial embodied energy *Ei* was shown in both WFG prototypes. It was more significant in the Peralveche prototype than in Sofia because of the high embodied energy in the borehole heat exchangers. However, the lower operational energy over 50 years compensated for the higher initial energy consumption. Table 10 illustrates the CO_2 emissions, which is the parameter used to assess the global warming potential (GWP) in $kgCO_{2eq}$. The life cycle emissions included the same phases: manufacture and transport of construction materials, maintenance, operation of the building, and waste disposal.

Table 10. Life cycle Embodied Carbon in $kgCO_{2eq}$.

	EC_1	EC_{err}	EC_o	EC_d	Total (50 Years)
Peralveche WFG	10,107	707	14,133	505	25,452
Peralveche RC	8145	3654	73,080	407	85,286
Sofia WFG	83,713	2302	46,048	4186	136,250
Sofia RC	83,269	5247	104,945	4163	197,625

Improving the thermal performance of the building envelope with WFG caused an increase in the initial cost. However, the amount of energy and CO_2 emissions declined in both cases over the considered life cycle.

4.3. Multi-Index Evaluation Model

Usually, the optimum building envelope's thermal performance is often accompanied by an increase in the cost and environmental load. This study established a multi-index evaluation model, including non-renewable primary energy consumption NRPE, global warming potential GWP, and cost, as indicators of the environment's aspects to evaluate a building envelope. Figure 5 shows that the initial cost of the WFG prototype in Peralveche doubled the reference one's cost because of the investment in borehole heat exchangers. Over a 50-year lifetime cycle, the reference prototype's accumulative cost surpasses the WFG prototype, due to the operational cost difference. When it comes to the Life Cycle Energy and the global warming potential the WFG showed a better performance. The reference prototype consumed three times as much energy as the WFG. The accumulative CO_2 emissions for the WFG envelope were 25,247 $kgCO_{2eq}$, whereas the reference glass envelope was responsible for 85,286 $kgCO_{2eq}$. When the cost is increased by 100%, the total Life Cycle Energy decreased by 1145 GJ. It has been estimated that the initial investment for the WFG system would cost €17,529, while the Reference system would require an initial investment of €7373. The WFG system would require €640.88 for maintenance per year, with an end-of-product, total demolition, and removal cost of €7921. Summing together these values and the initial investment for the system, the final total Life Cycle Cost of the Peralveche WFG system is €40,611. Meanwhile, for Peralveche RC, this system would cost €1239 in maintenance costs per year, with an end-of-product, total demolition, and removal cost of €3921. Combining these values with the initial investment cost shows that the final total Life Cycle Cost of the Peralveche RC system would be €47,767. Therefore, in the Peralveche case, it is apparent that the successful construction and implementation of the WFG system would save the owner a total of €7156 over the total building life cycle period. When it comes to Life Cycle Energy assessment and global warming potential, the WFG system in the Peralveche prototype used 571 GJ of energy during its lifetime. It was also determined that the system would contain 25,452 $kgCO_{2eq}$ during its 50 years of use. Meanwhile, for the Peralveche RC prototype, the system was calculated to use 1716 GJ of energy, while containing 85,286 $kgCO_{2eq}$ during its 50 years of use. Therefore, the Peralveche WFG system, when compared to the RC, will save 1145 GJ of energy as well as 59,834 $kgCO_{2eq}$ of Embodied Carbon over its entire lifetime.

Figure 6 shows that the HVAC operation cost over 50 years did not compensate for the higher initial cost of the WFG prototype. Over a 50-year lifetime cycle, the WFG envelope showed a better performance in the Life Cycle Energy and the global warming potential. The accumulative Life Cycle Energy for the WFG envelope was 2377 GJ, whereas the Life Cycle Energy of the reference glass was 3608 GJ. A model integrating Life Cycle Cost with Life Cycle Energy is used to assess the envelope schemes and select the optimal one in the decision-making process. The Sofia WFG system is estimated to initially cost €65,626, while the Sofia Reference system would cost €40,795.

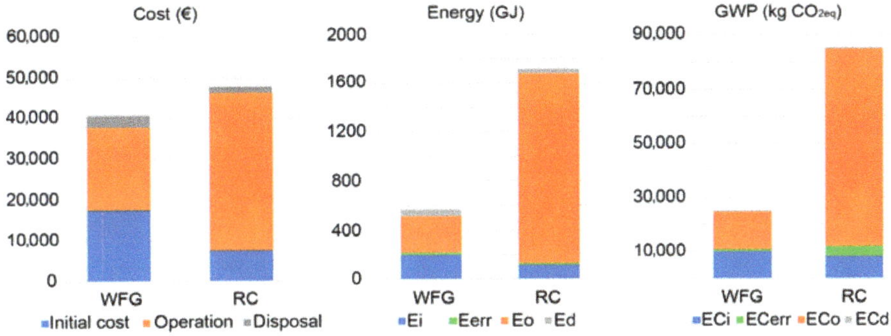

Figure 5. Life Cycle Assessment outcomes of the Peralveche prototype.

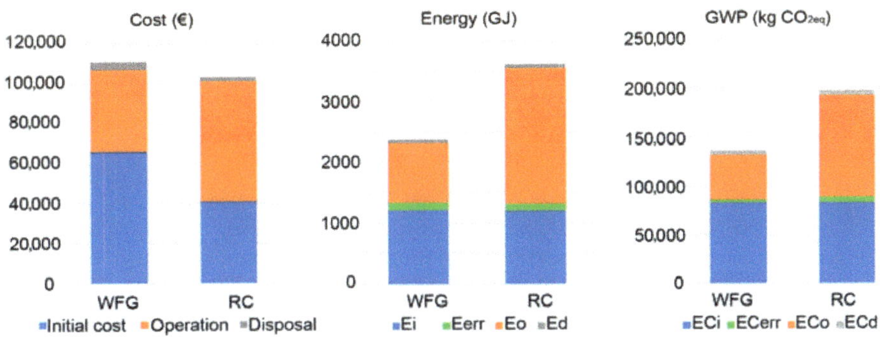

Figure 6. Life Cycle Assessment outcomes of the Sofia prototype.

For Sofia prototypes, the WFG system would require €1320 for maintenance per year, with an end-of-product, total demolition, and removal cost of €10,046. Summing together these values, as well as the initial investment for the system, it can be seen that the final total Life Cycle Cost of the Sofia WFG system is €110,865. Meanwhile, for Sofia RC, this system would cost €1923 in maintenance costs per year, with an end-of-product, total demolition and removal cost of €4923. Combining these values with the initial investment cost, the final total Life Cycle Cost of the Sofia RC system would be €103,066. The Sofia WFG system would cost an additional €33,184 compared to Sofia RC over the structure's lifespan.

The WFG system in the Sofia prototype used 2377 GJ of energy, so the system would contain 136,250 kgCO_{2eq} during its 50 years of use. Meanwhile, for the Sofia RC prototype, the system used 3608 GJ of energy while containing 197,625 kgCO_{2eq} during its 50 years of service. In this case, the WFG system will save 1231 GJ of energy while also containing 6137 kgCO_{2eq} less of Embodied Carbon over its entire lifetime as compared to the Reference system.

5. Conclusions and Limitations

This paper aimed to develop a conceptual framework to assess the building envelope energy consumption throughout their entire life cycle. By analyzing two case studies, the results can assist building designers during the decision-making process at early stages and consider water flow glazing as an option to reduce energy consumption and CO_2 emissions.

As has been demonstrated, WFG glazing technology typically retains a higher price point for initial construction than the reference prototypes, requiring a substantial investment early on. This is because of the additional equipment needed for the successful operation of WFG panels. The Peralveche reference prototype's initial cost is 42% of the

Peralveche WFG cost, whereas the Sofia reference prototype initial cost is 62% of the WFG initial cost. The steeper initial investment for WFG technology can serve as a deterrent for the technology in the eyes of building design professionals. However, when the total Life Cycle Costs of the WFG and RC are also taken into account, WFG can potentially be a much more economical option.

It is not until we consider each system's overall Life Cycle Costs that the economic benefits of WFG systems become apparent. For this study, the building lifespan was determined to be 50 years. The total Life Cycle Cost (LCC) of the Peralveche WFG is 85% of the total Reference system LCC. Meanwhile, the Sofia Reference LCC is 92% of the total WFG LCC. The conclusion derived from these findings is that selecting high-performance triple glazing is better than a WFG in Life Cycle Cost. A WFG system is better over the operation phase only when it is compared with a traditional double-glazing system, as has been demonstrated in Peralveche.

Another important factor that should be taken into account in the analysis between a much more traditional glazing system versus a WFG system is the Life Cycle Energy (LCE) and global warming potential (GWP) variables. The Peralveche WFG system, as compared to the Peralveche reference prototype, has demonstrated a savings of 66% in LCE with a 70% reduction of CO_2 emission. In Sofia, there are similar results. Sofia WFG demonstrated a 36% savings in LCE with a 30% reduction in CO_2 emissions. This analysis shows that a high-performance triple glazing system improves the Reference prototype performance, but WFG performs better in LCE and GWP in both cases.

The WFG system, however, does have several limitations. Firstly, there is an apparent lack of interoperability with the rest of the building systems present in modern structures, especially concerning the ventilation system. In addition to this, it is not always possible to retrieve all the detailed information needed as input for Water Flow Glazing operation. The maintenance of the building systems operation and the control of the building's indoor environmental conditions according to its user's comfort is of the utmost importance, and smart meters can assist in this. However, these devices pose considerable limitations concerning the quality, frequency, and accuracy of data. Therefore, taking these limitations into account, several future steps of research should be undertaken. Firstly, a development of a testing method to evaluate the performance of the unitized module components should be explored. In addition to this, more case studies in several different climate regions should be analyzed. Thirdly, the development of a management system to control the water pump in the circulating device should be realized. The life cycle of the water pump, which is another point of future research, depends on the operating hours and the on-off cycle. Finally, an integration of a whole evaluation protocol, including maintenance, environmental, and economic aspects, should be explored. This could be used by stakeholders involved in the design, maintenance, and monitoring process in future, potential projects.

After monitoring the WFG systems for a year, several uncertainties, misfunctions, and system issues must be addressed. WFG systems are limited by a high initial investment cost coupled with the need for an energy management system integrated with the other required equipment, especially if the system is coupled with borehole heat exchangers combined with a ground source heat pump. The heating and cooling devices must be adequately dimensioned to avoid misfunctions, especially the Air-to-Water heat pump. Further research must include monitoring energy performance much more accurately by attaching sensors to monitor the amount of electricity powering the heat pump to compare the actual thermal and electricity consumption. In addition to this, further standardization of the manufacturing and deployment process is required to bring down upfront investment costs and payback periods. Finally, another potential further research component would be to control indoor relative humidity, which would be achieved by integrating WFG with efficient ventilation systems.

Author Contributions: Conceptualization, B.M.S., F.d.A.G., J.A.H.R.; methodology, B.M.S., F.d.A.G.; software, J.A.H.R.; formal analysis, M.G., F.d.A.G.; data curation, F.d.A.G., J.A.H.R.; writing—original draft preparation, B.M.S., F.d.A.G., J.A.H.R.; writing—review and editing, M.G., B.M.S.; visualization, M.G., B.M.S., B.L.A.; supervision, J.A.H.R., B.L.A.; project administration, F.d.A.G.; funding acquisition, F.d.A.G. All authors have read and agreed to the published version of the manuscript.

Funding: This article has been funded by KSC Faculty Development Grant (Keene State College, NH, USA).

Acknowledgments: This work was supported by program Horizon 2020-EU.3.3.1: Reducing energy consumption and carbon footprint by smart and sustainable use, project Ref. 680441 InDeWaG: Industrialized Development of Water Flow Glazing Systems. The authors wish to thank the municipality of Peralveche, Spain, for its generous support.

Conflicts of Interest: The authors declare that they have no conflict of interest.

Abbreviations

Symbol	Meaning
AEC	Architecture, engineering, and construction.
EE	Embedded energy.
FE	Final energy.
GHG	Greenhouse gas.
GSHP	Ground source heat pump.
GWP	Global warming potential.
LCA	Life Cycle Assessment.
LCC	Life Cycle Cost.
LCE	Life Cycle Energy.
NRPE	Non-renewable primary energy.
PV	Photovoltaic
PVIF	Present value interest factor.
PVIFA	Present value interest factor of the annuity.
RC	Reference cabin.
RPE	Renewable primary energy.
WFG	Water Flow Glazing.
A_j	Absorptance of glass layers.
C_1	Construction cost.
C_2	Operation cost.
C_3	Demolition cost.
E	Total energy of the building element.
E_i	Initial embodied energy.
E_{err}	Recurrent embodied energy for future maintenance and refurbishment.
E_o	Total annual operational energy.
E_d	embodied energy required for demolition and disposal.
h_j	Convective heat coefficients.
i_0	Solar irradiance ($W{\cdot}m^{-2}{\cdot}K^{-1}$).
$\dot{m}$	Mass flow rate per unit of surface ($Liter{\cdot}min^{-1}{\cdot}m^{-2}$).
n	Lifetime of the element in years.
P	Heat gain in the water chamber (W).
q	Heat flow ($W{\cdot}m^{-2}{\cdot}K^{-1}$).
r	Interest rate.
θ_i	Interior temperature (K).
θ_e	Exterior temperature (K).
θ_{IN}	Inlet temperature of the water chamber (K).
U	Thermal transmittance ($W{\cdot}m^{-2}{\cdot}K^{-1}$).
U_i	Interior thermal transmittance ($W{\cdot}m^{-2}{\cdot}K^{-1}$).
U_e	Exterior thermal transmittance ($W{\cdot}m^{-2}{\cdot}K^{-1}$).
U_w	Thermal transmittance (water chamber–interior) (K).

References

1. Wu, J.; Yin, P.Z.; Sun, J.S.; Chu, J.F.; Liang, L. Evaluating the environmental efficiency of a two-stage system with undesired outputs by a DEA approach: An interest preference perspective. *Eur. J. Oper. Res.* **2016**, *254*, 1047–1062. [CrossRef]
2. Webler, T.; Holewinski, M.; Orrick, B.; Kaur, R. Toward a method for the rapid collection of public concerns and benefits of emerging energy technologies. *J. Risk Res.* **2020**, *23*, 35–46. [CrossRef]
3. Manfren, M.; Sibilla, M.; Tronchin, L. Energy Modelling and Analytics in the Built Environment—A Review of Their Role for Energy Transitions in the Construction Sector. *Energies* **2021**, *14*, 679. [CrossRef]
4. Jezierski, W.; Sadowska, B.; Pawłowski, K. Impact of Changes in the Required Thermal Insulation of Building Envelope on Energy Demand, Heating Costs, Emissions, and Temperature in Buildings. *Energies* **2021**, *14*, 56. [CrossRef]
5. Oshiro, K.; Fujimori, S. Stranded investment associated with rapid energy system changes under the mid-century strategy in Japan. *Sustain. Sci.* **2021**, *16*, 477–487. [CrossRef]
6. Wang, H.; Chen, W.; Zhang, H.; Li, N. Modeling of power sector decarbonization in China: Comparisons of early and delayed mitigation towards 2-degree target. *Clim. Chang.* **2020**, *62*, 1843–1856. [CrossRef]
7. Allan, K.; Phillips, A.R. Comparative Cradle-to-Grave Life Cycle Assessment of Low and Mid-Rise Mass Timber Buildings with Equivalent Structural Steel Alternatives. *Sustainability* **2021**, *13*, 3401. [CrossRef]
8. Omrany, H.; Soebarto, V.; Sharifi, E.; Soltani, A. Application of Life Cycle Energy Assessment in Residential Buildings: A Critical Review of Recent Trends. *Sustainability* **2020**, *12*, 351. [CrossRef]
9. Zabalza-Bribián, I.; Aranda-Usón, A.; Scarpellini, S. Life cycle assessment in buildings: State-of-the-art and simplified LCA methodology as a complement for building certification. *Build. Environ.* **2009**, *44*, 2510–2520. [CrossRef]
10. Bruce-Hyrkäs, T.; Pasanen, P.; Castro, R. Overview of Whole Building Life-Cycle Assessment for Green Building Certification and Ecodesign through Industry Surveys and Interviews. *Procedia CIRP* **2018**, *69*, 178–183. [CrossRef]
11. European Commission, Joint Research Centre, Institute for Environment and Sustainability. *International Reference Life Cycle Data System (ILCD) Handbook—General Guide for Life Cycle Assessment—Detailed Guidance*, 1st ed.; EUR 24708 EN; Publications Office of the European Union: Luxembourg, 2010.
12. ISO 14044:2006. *Environmental Management—Life Cycle Assessment—Requirements and Guidelines*; International Organization for Standardization: Geneva, Switzerland, 2006.
13. EN 15978 (2012). *Sustainability of Construction Works—Assessment of Environmental Performance of Buildings—Calculation Method. European Committee for Standardisation*; International Organization for Standardization: Geneva, Switzerland, 2012.
14. Chau, C.K.; Leung, T.M.; Ng, W.Y. A review on life-cycle assessment, life-cycle energy assessment and life-cycle carbon emissions assessment on buildings. *Appl. Energy* **2015**, *143*, 395–413. [CrossRef]
15. Jausovec, M.; Sitar, M. Comparative Evaluation Model Framework for Cost-Optimal Evaluation of Prefabricated Lightweight System Envelopes in the Early Design Phase. *Sustainability* **2019**, *11*, 5106. [CrossRef]
16. Gluch, P.; Baumann, H. The Life Cycle Costing (LCC) Approach: A Conceptual Discussion of Its Usefulness for Environmental Decision-Making. *Build. Environ.* **2004**, *39*, 571–580. [CrossRef]
17. Hu, M. A Building Life-Cycle Embodied Performance Index—The Relationship between Embodied Energy, Embodied Carbon and Environmental Impact. *Energies* **2020**, *13*, 1905. [CrossRef]
18. ITeC (Instituto de la Tecnología de la Construcción de Cataluña). Available online: https://itec.es/servicios/bedec/ (accessed on 3 February 2021).
19. Oyedele, L.; Tham, K.; Fadeyi, M.; Jaiyeoba, B. Total Building Performance Approach in Building Evaluation: Case Study of an Office Building in Singapore. *J. Energy Eng.* **2012**, *138*, 25–30. [CrossRef]
20. Sharma, A.; Saxena, A.; Sethi, M.; Varun, V.S. Life cycle assessment of buildings: A review. *Renew. Sustain. Energy Rev.* **2011**, *15*, 871–875. [CrossRef]
21. Ferrara, M.; Fabrizio, E.; Virgone, J.; Filippi, M. Energy Systems in Cost-Optimized Design of Nearly Zero-Energy Buildings. *Autom. Constr.* **2016**, *70*, 109. [CrossRef]
22. EN-European Standard. *EN15978-Sustainability of Construction Works-Assessment of Environmental Performance of Buildings-Calculation Method*; No. EN 15978:2011; European Committee for Standarization. s.l.: Brussels, Belgium, 2011.
23. Connolly, R.; Connolly, M.; Carter, R.M.; Soon, W. How Much Human-Caused Global Warming Should We Expect with Business-As-Usual (BAU) Climate Policies? A Semi-Empirical Assessment. *Energies* **2020**, *13*, 1365. [CrossRef]
24. Wang, X.; Jiang, D.; Lang, X. Climate Change of 4 °C Global Warming above Pre-industrial Levels. *Adv. Atmos. Sci.* **2018**, *35*, 757–770. [CrossRef]
25. Lu, S.L.; Wang, Z.C.; Zhang, T.S. Quantitative analysis and multi-index evaluation of the green building envelope performance in the cold area of China. *Sustainability* **2020**, *12*, 437. [CrossRef]
26. Chiradeja, P.; Ngaopitakkul, A. Energy and economic analysis of tropical building envelope material in compliance with thailand's building energy code. *Sustainability* **2019**, *11*, 6872. [CrossRef]
27. Luo, Y.; Zhang, L.; Bozlar, M.; Liu, Z.; Guo, H.; Meggers, F. Active building envelope systems toward renewable and sustainable energy. *Renew. Sustain. Energy Rev.* **2019**, *104*, 470–491. [CrossRef]
28. Abdulmohsin, H.; Aritra, G.; Senthilarasu, S.; Tapas, K.M. Evaluation of thermal performance for a smart switchable adaptive polymer dispersed liquid crystal (PDLC) glazing. *Sol. Energy* **2020**, *195*, 185–193.

29. Ghosh, A.; Norton, B.; Duffy, A. Behaviour of a SPD switchable glazing in an outdoor test cell with heat removal under varying weather conditions. *Appl. Energy* **2016**, *180*, 695–706. [CrossRef]
30. Frattolillo, A.; Canale, L.; Ficco, G.; Mastino, C.C.; Dell'Isola, M. Potential for Building Façade-Integrated Solar Thermal Collectors in a Highly Urbanized Context. *Energies* **2020**, *13*, 5801. [CrossRef]
31. Arpino, F.; Cortellessa, G.; Frattolillo, A. Experimental and numerical assessment of photovoltaic collectors performance dependence on frame size and installation technique. *Sol. Energy* **2015**, *118*, 7–19. [CrossRef]
32. Gutai, M.; Kheybari, A.G. Energy consumption of water-filled glass (WFG) hybrid building envelope. *Energy Build.* **2020**, *218*, 110050. [CrossRef]
33. Gil-Lopez, T.; Gimenez-Molina, C. Influence of double glazing with a circulating water chamber on the thermal energy savings in buildings. *Energy Build.* **2013**, *56*, 56–65. [CrossRef]
34. Moreno Santamaria, B.; Ama Gonzalo, F.; Lauret Aguirregabiria, B.; Hernandez Ramos, J.A. Evaluation of Thermal Comfort and Energy Consumption of Water Flow Glazing as a Radiant Heating and Cooling System: A Case Study of an Office Space. *Sustainability* **2020**, *12*, 7596. [CrossRef]
35. Gutai, M.; Kheybari, A.G. Energy consumption of hybrid smart water-filled glass (SWFG) building envelope. *Energy Build.* **2021**, *230*, 110508. [CrossRef]
36. Lyu, Y.L.; Chow, T.T.; Wang, J.L. Numerical prediction of thermal performance of liquid-flow window in different climates with anti-freeze. *Energy* **2018**, *157*, 412–423. [CrossRef]
37. Chow, T.T.; Li, C. Liquid-filled solar glazing design for buoyant water-flow. *Build. Environ.* **2013**, *60*, 45–55. [CrossRef]
38. Sierra, P.; Hernandez, J.A. Solar heat gain coefficient of water flow glazing. *Energy Build.* **2017**, *139*, 133–145. [CrossRef]
39. Moreno Santamaria, B.; del Ama Gonzalo, F.; Pinette, D.; Gonzalez-Lezcano, R.-A.; Lauret Aguirregabiria, B.; Hernandez Ramos, J.A. Application and Validation of a Dynamic Energy Simulation Tool: A Case Study with Water Flow Glazing Envelope. *Energies* **2020**, *13*, 3203. [CrossRef]
40. Moreno Santamaria, B.; del Ama Gonzalo, F.; Pinette, D.; Lauret Aguirregabiria, B.; Hernandez Ramos, J.A. Industrialization and Thermal Performance of a New Unitized Water Flow Glazing Facade. *Sustainability* **2020**, *12*, 7564. [CrossRef]
41. López-Ochoa, L.M.; Las-Heras-Casas, J.; López-González, L.M.; García-Lozano, C. Energy Renovation of Residential Buildings in Cold Mediterranean Zones Using Optimized Thermal Envelope Insulation Thicknesses: The Case of Spain. *Sustainability* **2020**, *12*, 2287. [CrossRef]
42. Spanish Regulation of Thermal Installations in Buildings RITE. *Factores de Emisión de CO2 y Coeficientes de paso a Energía Primaria de Diferentes Fuentes de Energía Final Consumidas en el Sector de Edificios en España*; Instituto para la Diversificación y Ahorro de la Energía (IDAE): Madrid, Spain, 2016.
43. Energy Prices in 2019 Household Energy Prices in the EU Increased Compared with 2018. Available online: https://ec.europa.eu/eurostat/documents/2995521/10826603/8-07052020-AP-EN.pdf/2c418ef5-7307-5217-43a6-4bd063bf7f44 (accessed on 3 February 2021).
44. Yuan, Z.; Zhou, J.; Qiao, Y.; Zhang, Y.; Liu, D.; Zhu, H. BIM-VE-Based Optimization of Green Building Envelope from the Perspective of both Energy Saving and Life Cycle Cost. *Sustainability* **2020**, *12*, 7862. [CrossRef]
45. Copiello, S. Economic implications of the energy issue: Evidence for a positive non-linear relation between embodied energy and construction cost. *Energy Build.* **2016**, *123*, 59–70. [CrossRef]
46. Crawford, R.H. Post-occupancy life cycle energy assessment of a residential building in Australia. *Archit. Sci. Rev.* **2014**, *57*, 114–124. [CrossRef]
47. Crawford, R.H.; Bontinck, P.A.; Stephan, A.; Wiedmann, T.; Yu, M. Hybrid life cycle inventory methods—A review. *J. Clean. Prod.* **2018**, *172*, 1273–1288. [CrossRef]
48. Bonamente, E.; Aquino, A. Life-Cycle Assessment of an Innovative Ground-Source Heat Pump System with Upstream Thermal Storage. *Energies* **2017**, *10*, 1854. [CrossRef]
49. Bonamente, E.; Aquino, A. Environmental Performance of Innovative Ground-Source Heat Pumps with PCM Energy Storage. *Energies* **2020**, *13*, 117. [CrossRef]
50. Saner, D.; Juraske, R.; Kübert, M.; Blum, P.; Hellweg, S.; Bayer, P. Is it only CO_2 that matters? A life cycle perspective on shallow geothermal systems. *Renew. Sustain. Energy Rev.* **2010**, *14*, 1798–1813. [CrossRef]
51. Priarone, A.; Silenzi, F.; Fossa, M. Modelling Heat Pumps with Variable EER and COP in EnergyPlus: A Case Study Applied to Ground Source and Heat Recovery Heat Pump Systems. *Energies* **2020**, *13*, 794. [CrossRef]

energies

MDPI

Article

Zero Energy Building Economic and Energetic Assessment with Simulated and Real Data Using Photovoltaics and Water Flow Glazing

Fernando del Ama Gonzalo [1,*], Belen Moreno Santamaria [2], José Antonio Ferrándiz Gea [3], Matthew Griffin [1] and Juan A. Hernandez Ramos [4]

1 Department of Sustainable Product Design and Architecture, School of Sciences, Sustainability and Health, Keene State College, 229 Main St, Keene, NH 03435, USA; Matthew.Griffin@ksc.keene.edu
2 Department of Construction and Architectural Technology, Technical School of Architecture of Madrid, Technical University of Madrid (UPM), Av. Juan de Herrera, 4, 28040 Madrid, Spain; belen.moreno@upm.es
3 Department of Technology, Norfolk State University, 700 Park Avenue, Norfolk, VA 23504, USA; jaferrandiz@nsu.edu
4 Escuela Department of Applied Mathematics, School of Aeronautical and Space Engineering, Technical University of Madrid (UPM), Plaza Cardenal Cisneros 3, 28040 Madrid, Spain; juanantonio.hernandez@upm.es
* Correspondence: fernando.delama@keene.edu

Citation: del Ama Gonzalo, F.; Moreno Santamaria, B.; Ferrándiz Gea, J.A.; Griffin, M.; Hernandez Ramos, J.A. Zero Energy Building Economic and Energetic Assessment with Simulated and Real Data Using Photovoltaics and Water Flow Glazing. *Energies* **2021**, *14*, 3272. https://doi.org/10.3390/en14113272

Academic Editor: Francesco Asdrubali

Received: 5 May 2021
Accepted: 28 May 2021
Published: 3 June 2021

Abstract: The new paradigm of Net Zero Energy buildings is a challenge for architects and engineers, especially in buildings with large glazing areas. Water Flow Glazing (WFG) is a dynamic façade technology shown to reduce heating and cooling loads for buildings significantly. Photovoltaic panels placed on building roofs can generate enough electricity from solar energy without generating greenhouse gases in operation or taking up other building footprints. This paper investigates the techno-economic viability of a grid-connected solar photovoltaic system combined with water flow glazing. An accurate assessment of the economic and energetic feasibility is carried out through simulation software and on-site tests on an actual prototype. The assessment also includes the analysis of global warming potential reduction. A prototype with WFG envelope has been tested. The WFG prototype actual data reported primary energy savings of 62% and 60% CO_2 equivalent emission reduction when comparing WFG to a reference triple glazing. Finally, an economic report of the Photovoltaic array showed the Yield Factor and the Levelized Cost of Energy of the system. Savings over the operating lifetime can compensate for the high initial investment that these two technologies require.

Keywords: water flow glazing; building integrated PV panels; levelized cost of energy

1. Introduction

As the world economy continues to develop, the building industry continues to maintain a great expense of energy utilization and carbon dioxide emission. Consumers are central participants in this specific circumstance, as adaptability in demand is necessarily required to embrace the discontinuous nature of most renewable energy sources [1]. The poor performance of existing buildings offers essential possibilities for energy retrofit. Integrating renewable solar systems into the building envelope and storage of thermal and electric energy can be the solution to this challenge. [2]. The Energy Performance Buildings Directive (EPBD) promotes policies that will produce highly energy-efficient and decarbonized structures by 2050 [3]. Starting 31 December 2020, all new buildings will have to be Nearly Zero Energy Buildings [4]. Zero-energy buildings are required to produce equivalent energy from renewable technologies to balance their energy consumption [5]. Designing passive measures and implementing highly efficient building systems are the first steps to accomplish Net Zero Energy objectives [6]. The second step includes the integration of active surfaces for energy generation and management. Solar thermal collectors and photovoltaic panels are the most reliable ways to provide electrical and thermal energy from renewable sources [7]. Therefore, the current and future architectural design philosophies must consider the floor, roof, wall, and any other surface space available to be able to place energy generation devices. The building's energy footprint is a new concept to be considered at the design stage [8]. In relation to the architectural building design, the possible renewable supply options are in the building's physical footprint [9,10]. Architects and building designers have to face the challenges of designing buildings and their energy footprint and considering how this energy footprint will influence the habitats, cities, and landscapes of tomorrow [11].

Photovoltaic panels placed on building roofs can generate enough electricity without causing greenhouse gases or occupying other building footprints. The integration of Photovoltaic (PV) systems in the building envelope can reduce the primary energy consumption and decrease the global warming potential by cutting down on CO_2 emissions [12,13]. Energy generated by PV panels can be used where it is consumed and can be used to power any building request, from lighting to electrical heating and cooling generators. The system can allow construction costs reduction if used to substitute traditional building materials [14]. Some authors regard Building-integrated PV panels as an optimal solution for zero energy building design due to the ability to generate energy [15,16]. PV modules are available at the price of about 1.5 €/W. However, the price is lower in the European market, with the lowest price of about 0.75 €/W [17]. A global annual growth rate of 40% is expected in the BIPV installed capacity from 2017 to 2021 [18,19]. Although policies for promoting BIPV deployment have begun over the last few years, there exist barriers to the massive use of this technology in the building industry [20]. The lack of national programs or building codes covering the technical specifications of the devices and long pay-back periods for solar installations hinder the massive deployment of this technology [21]. Financial boundaries continue to be the primary barrier to a renewably-powered community [22]. Battery packages for energy storage are required in areas without access to electric grids, although integrating batteries shows a high energy product cost [23,24]. Extensive research has been conducted to assess the PV cells' efficiency. The reported efficiency for the polycrystalline silicon PV and amorphous silicon PV cells usually used in BIPV is 25.7% and 10.2%, respectively [25,26].

Researchers commonly recognize the impacts of the envelope on the building energy balance. Many design features such as insulation, structural stiffness, and aesthetics pertain to fabrication, assembly, and deployment [27]. The use of curtain wall systems enables effective daylighting in high-grade, institutional and commercial buildings [28]. A stick Curtain Wall System comprises mullions, transoms, and glass panels placed on-site. [29]. Unitized Curtain Walls are produced, constructed, and glazed in an off-site facility; these unitized units are transported to the construction site and attached to the building structure.

There are many current trends in energy efficiency for transparent building envelopes. These include, but are not limited to: addressing solar heat gain coefficient (SHGC) with coatings on glazing [30], dynamic glazing solutions such as electrochromic glazing (EC) [31], suspended particle device glazing (SPD) [32], and polymer dispersed liquid crystal (PDLC) [33]. In addition to this, one can improve the thermal resistance of the building envelope by using multi-layered glazing and reduce cooling demand with shading devices [34,35]. A fluid medium that can cool the glass itself can preheat the air before it enters the interior space [36]. An alternative to circulating air is using a circulating water chamber which benefits from capturing solar energy and turning that potential energy load into a renewable energy source [37]. This idea is known as Water Flow Glazing (WFG), a disruptive active façade technology that aims to compliment the HVAC system in a structure. Water-flow glazing (WFG) has been studied as a curtain wall component. Energy savings depend on the water layer's absorption for energy management of the building envelope [38,39]. Building envelopes with high window-to-wall ratios significantly impact the built environment's ecological footprint due to the high operational energy demand.

WFG panels absorb heat which can be transported to a thermal storage unit [40]. Some authors successfully tested this technology and showed saving potential between 52% and 72% compared with traditional double glazing and between 34% and 61% compared with triple glazing [41]. Other authors have assessed the WFG life cycle showing that its cost payback time and energy payback time is less than three years [42]. Some experimental facilities called 'water houses' have been built to evaluate their viability and compare the water-filled envelope with existing construction techniques [43]. The present article continued this line of research, presenting a new Water-flow glazing unitized façade. This facade has been tested in a real-world facility built in Sofia, Bulgaria. The objective of this paper was to assess the energy performance and the economic feasibility of a WFG façade coupled with a PV array system and heat pump to improve the energy resilience of buildings while allowing for desirable architectural features. The first section describes the proposed system and presents the data about this research. The second section explains the performance indicators used to evaluate the system operation. This study includes an analysis to study the influence of variable input on technical performance and economic feasibility. This article further presents a model for expanding the enhancement of the PV energy source. It was reported that the potential of renewable energy for buildings and reduction in CO_2 potential can be influenced through different government policies and strategy measures.

The main novelties of this work comprise several goals. Firstly, to fulfill the demand of an office building for space heating and cooling using a PV-driven heat pump. Secondly, to compare Building Information Modeling (BIM) and Energy+ simulation with measured data from a real facility. Thirdly, dynamic simulations were performed to analyze the thermo-economic performances at different peak powers of PV by means of two different software tools. Finally, technical and economical parameters (including the tilt angle, peak power of PV, electricity price, and conversion factors for primary energy) have been taken into account and presented.

This paper is organized into five different sections. Section 2 contains a description of the tested facility, weather data from the site, and a description of the water flow glazing system and PV array. Section 3 presents actual data from the tested facility, calculates the heating and cooling loads, showcases the renewable energy production of WFG, and finally demonstrates simulation data from commercial software. Section 4 takes into consideration the energy and economic analysis. Lastly, Section 5 covers conclusions and limitations.

2. Materials and Methods

In this section, the authors intend to explain the testing facility, provide a quick overview of the water flow glazing envelope, the energy system employed in the testing facility, and finally, an overview of the PV array system. The testing facility was a structure built in Sofia, Bulgaria, that collected data over the timespan of one year. This structure was built using water flow glazing panels, a façade typology that contains a triple-glazed panel. The chamber located towards the exterior contained a traditional gas-filled chamber, while the interior chamber was filled with a circulating fluid medium. The energy system used in this prototype is a water-to-water heat pump that collected heat from the water flow glazing panels and stores it in a buffer tank below ground level. When needed, this heat recirculated into the building. Finally, a photovoltaic (PV) array provided the required electricity to the building and building peripherals, including heating, lighting, and computer systems.

2.1. Description of Testing Facility and Energy Management System

WFG modules were tested in an experimental facility built in Sofia, Bulgaria (42°39′1″ N, 23°23′26″ E, Elevation: 590 m a.s.l.). The building is a stand-alone construction with a square geometry, deployed in a single plan of outer dimensions 7 × 7 m, and a ceiling height of 2.7 m, with open office design and a central service core. Three of the facades are fully glazed, from floor to ceiling, except the north façade which is 100% opaque. The glass facades are composed of unitized WFG modules with dimensions of 2.7 × 1.3 m. Eastern, western and southern facades were triple glazing with a 16 mm argon cavity and a 24 mm water cavity. The unitized WFG panels were put together in a workshop and then served for on-site application with an aluminum enclosure that held the circulating device. A water pump, a plate heat exchanger, flow meters at the inlet and outlet, and temperature probes made up this circulating device. The design flow rate was 2 L/min m^2. The southern glazing showed high near-infrared and low-infrared FIR absorptions, making it an ideal solution for large glazing facades in severely cold climates. The goal for eastern and western glazing was to prevent heat from entering the internal area, especially in summer with high front reflectance and low infrared absorptance. Besides, the water flow absorbed solar infrared radiation increasing the water temperature within the window. The energy transported by the water is released in tanks so that thermal energy can be collected and utilized in radiant heating systems and domestic hot water. This renewable energy integrated into the water flow glazing panels might be enough to meet the heating loads in winter. However, it is necessary to study other renewable energy sources to address the electric loads over the year. Figure 1 illustrates the prototype, the WFG panels and the building integrated PV array. Southern WFG showed a low Far-Infrared absorptance, high Near-Infrared absorptance, and high front Far-Infrared reflectance. Its ability to absorb heat made it the right solution for heating up water. The infrared absorptance was very low in eastern and western WFG, whereas the infrared front reflectance was high. This glazing would show the best performance for preventing heat from entering the indoor space, but it would not be relevant to heat water.

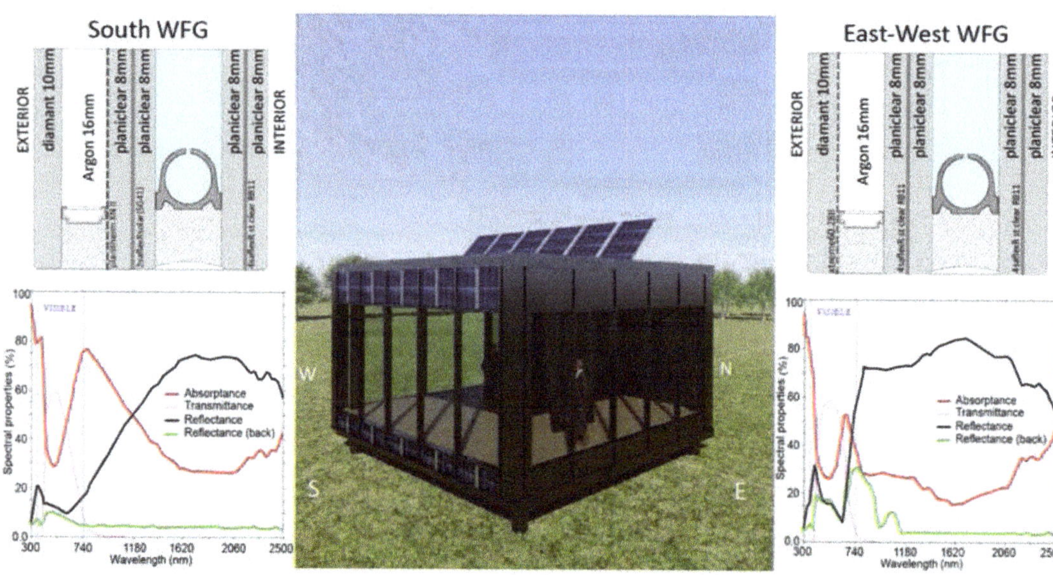

Figure 1. Description of Sophia's prototype glazing.

The space heating and cooling demand was met by the energy management system made of WFG panels, a cylindrical 370-L buffer tank with a 2 cm rigid insulation, and a water-to-water heat pump coupled to a building-integrated PV array. The tank's internal diameter was 0.6 m and the height 1.55 m. It was placed underneath the prototype, with the top accessible from a crawl space, and connected to the water-to-water heat pump., The PV array satisfied the electricity demand formed by: (i) the electrical appliances installed in the building, (ii) the lighting equipment and (iii) the space heating and cooling by electric-driven heat pumps for (iv) the hydraulic auxiliary system for WFG panels. The PV capacities are calculated to maximize the self-consumed renewable energy. The PV array is sized to fulfill the electricity demand. When the PV output is lower than the demand, the remaining amount of electricity demand is withdrawn from the power grid. Moreover, if the PV plant output exceeds the demand, the excess output is delivered to the grid. Therefore, it is assumed that there is no electricity storage system. Figure 2 shows the proposed layout that consisted of two main thermal loops. The water flow glazing loop is based on water heated by the WFG modules and supplied to the buffer tank. The second loop connects the buffer tank with the water-to-water heat pump. In summer, WFG panels absorbed solar radiation and released the heat in the buffer tank underneath the prototype. The heat pump distributes cold water through the WFG façade and interior panels. In winter, WFG façade panels absorbed solar radiation and heated up the buffer tank. Hot water is circulated through the interior WFG panels and released heat inside the building. The water-to-water heat pump worked as a backup system and was operating if the buffer tank temperature is not high enough to supply the required heating load.

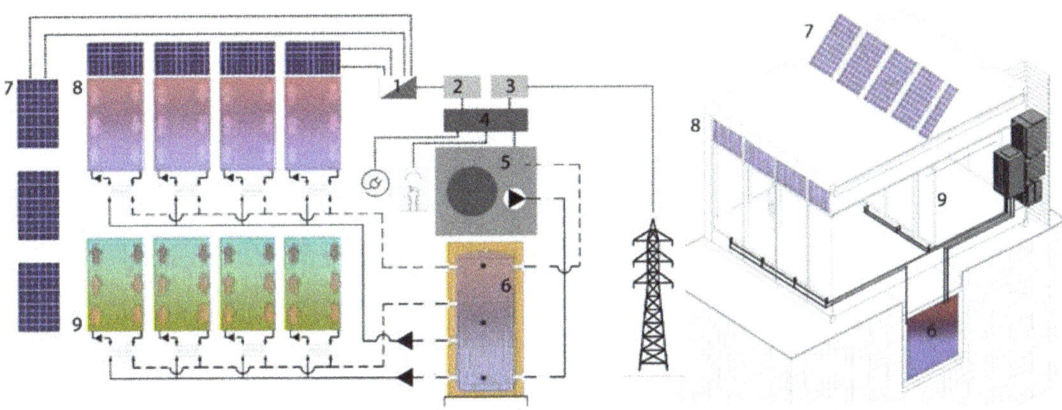

Figure 2. Description of Sophia's prototype energy management system. 1. Inverter; 2. Production meter; 3. Utility meter; 4. Customer panel; 5. Electric loads (heat pump, lighting, power); 6. Buffer tank; 7. Roof PV open rack; 8. WFG facade with building integrated PV panels; 9. WFG interior panels.

2.2. Estimation of Solar Irradiances and Outdoor Temperature

To properly calculate the total irradiance that will be received on the test prototype, there needs to exist a breakdown of variables that will present the total irradiance when summed together. Firstly, the Global Horizontal Irradiance (Ghi) maintains the total irradiance received on horizontal surfaces. This variable is calculated by summing the horizontal components of diffuse and direct (beam) irradiance types. This beam irradiance is the most relevant variable to consider when calculating for technologies that rely on solar power. Secondly, the Diffuse Horizontal Irradiance (Dhi) is the horizontal component of irradiance that is scattered due to the presence of the atmosphere. In conjunction with beam irradiance, diffuse irradiance is a crucial component to consider when calculating the performance of the solar module. Thirdly and finally, the Direct Normal Irradiance (Dni) is the amount of solar irradiance delivered directly from the sun. The authors have estimated these crucial solar irradiance components (direct beam, diffuse, and reflected radiation) to show the importance of each type.

Figure 3 demonstrates solar irradiances received in January. Direct (Beam) Horizontal Irradiance (Ebh) is the horizontal type of Direct Normal Irradiance. In this figure it is understood that days one, two, and three are sunny days, while days five and six are cloudy. During sunny days both direct normal irradiance and diffuse radiation are present;

Meanwhile, during the cloudy days only diffuse radiation is present. Outdoor temperatures during these days ranged from −5 °C to 7 °C.

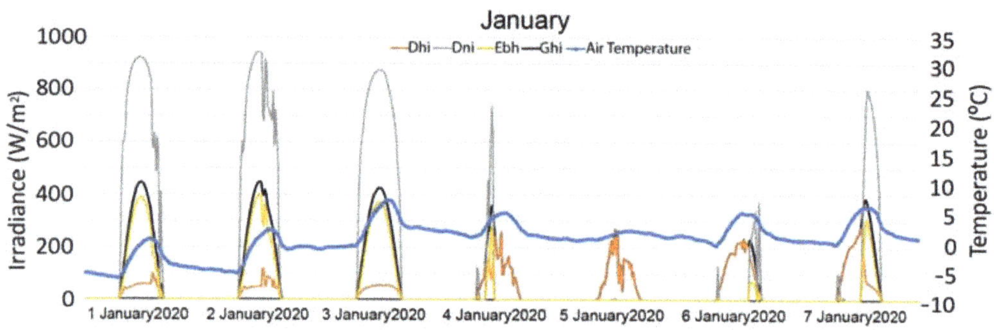

Figure 3. Solar irradiances and outdoor temperature in January.

Figure 4 demonstrates solar irradiances in the month of July, where Direct (Beam) Horizontal Irradiance (Ebh) is the horizontal component of Direct Normal Irradiance. During this month, it can be understood that the exterior temperature ranges from 15 °C to 30 °C.

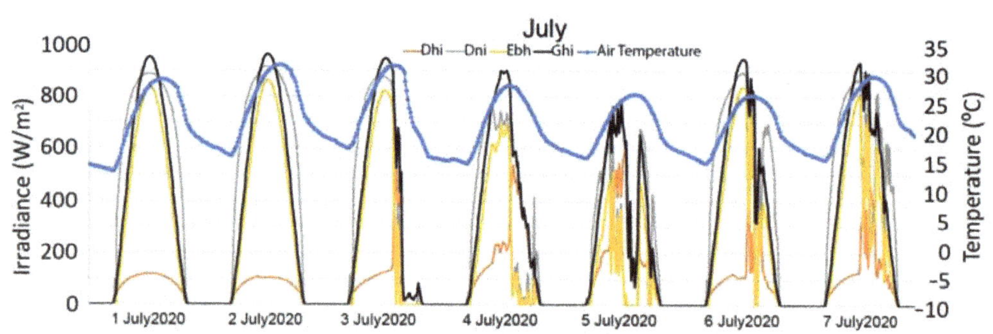

Figure 4. Solar irradiances and outdoor temperature in July.

2.3. Water Flow Glazing Thermal Properties

Previous articles have shown the equations that explain the spectral and thermal behavior of water flow glazing assuming steady conditions [44,45]. The variable $\dot{m}$ denotes the mass flow rate per unit of surface; this variable can be manipulated in order to change the thermal performance of the water flow glazing (WFG) system. c is the variable that measures the heat capacity of the fluid medium. Meanwhile, c, when multiplied by $\dot{m}$ (the mass flow rate), gives the capacity of the fluid medium to absorb heat. Equation (1) demonstrates the variable g-factor of a triple-glazed WFG panel.

$$g = \left(\frac{U_i}{\dot{m}c + U_e + U_i}\right)\left(\left(A_1\left(\frac{U_e}{h_e}\right) + A_2\left(\frac{1}{h_g} + \frac{1}{h_e}\right)U_e + A_3\left(\frac{U_e}{h_i}\right) + A_w\right)\right) + A_3\left(1 - \frac{U_i}{h_i}\right) + T, \quad (1)$$

where:
U_e = thermal transmittance between water chamber and exterior environment,
U_i = thermal transmittance between water chamber and interior,
A_w = Water absorptance,
A_1, A_2, A_3 = Glazing panel absorptances of each respective pane in window assembly,

h_e, h_i, h_g, h_w = Convective heat coefficients,
T = energy transmittance of glazing.

U_i, U_e are obtained through the manipulation of the convective heat coefficient variables, h_e, h_i, h_g, h_w. Equations (2) and (3) are used to calculate the variable thermal transmittances of WFG. U is the glazing thermal transmittance; U_w is the total thermal transmittance between the indoor environment and the water chamber.

$$U = \frac{U_i\, U_e}{\dot{m}c + U_e + U_i}, \tag{2}$$

$$U_w = \frac{U_i\, \dot{m}c}{\dot{m}c + U_e + U_i}. \tag{3}$$

Equation (4) illustrates the expression of q, the heat flow through glazing with gas cavities. The heat flow depends on the difference between the indoor and outdoor temperatures $(\theta_e - \theta_i)$, thermal transmittance, U, g-factor, and direct and diffuse solar radiation, i_0.

$$q = U(\theta_e - \theta_i) + g i_0. \tag{4}$$

Equation (5) illustrates the heat flow, q, in water flow glazing panels, where the fluid's temperature and mass flow rate are transient, which affects the heat flow. θ_{IN} is the variable that measures the inlet temperature of the fluid medium in the WFG assembly.

$$q = U(\theta_e - \theta_i) + U_w(\theta_{IN} - \theta_i) + g i_0, \tag{5}$$

where θ_{IN} is the inlet temperature of the fluid into the WFG system. Thermal transmittances and g-factors of the water flow glazing change at different flow rate conditions. When the fluid is flowing at a rate of $\dot{m} > 2\ \mathrm{L/m^2}$ min, g and U are denoted with the superscript *ON*. When there is no flow ($\dot{m} = 0$) the g-factor and U values, denoted with the superscript *OFF*, depend only on the spectral properties of the glass.

2.4. Photovoltaic

Photovoltaic (PV) systems have many potentialities, thanks to their ability to produce electricity and the enormous recent cost reduction. In this project, a grid-connected PV arrangement supplied the grid when the building demand exceeded the produced energy. The viability of the private grid-connected PV system came after a techno-economic feasibility study that included the initial costs and the energy savings over the facility's operating life cycle. The research was conducted using different PV simulation software and analyzing various techno-economic indicators. These indicators benchmarked the system production at a particular spot upon the system's production under possible operational patterns. The inputs were based on actual energy consumption results over a year. The highest daily consumption and the minimum amount of peak solar hours are required to determine a photovoltaic array. The grid-connected systems stated in this article were integrated in the WFG southern modules and installed on the rooftop of the tested facility. The estimation of rooftop areas is the first step to fitting the design specifications. In terms of roof construction, the tested facility had a flat rooftop that facilitates the installation of a PV array. This PV system eliminated the expense of storage. The system was broken down into PV arrays, inverters, and transformers, as shown in Figure 2. The excess electricity, exceeding consumption by the connected equipment, is supplied to the grid. The PV system monthly energy output has been calculated with the solar thermal power which is absorbed per unit area of a PV system, determined using Equation (6).

$$E_{PV} = I_t\ A_m\ N_m, \tag{6}$$

where I_t is monthly average daily of total absorbed solar irradiance on a tilted surface, A_m is the area of a PV panel, and N_m is the number of PV panels that made up the PV array. Equation (7) illustrates the AC energy production from a PV system.

$$E_{PV-DC} \ = E_{PV} \ \eta_e \ \eta_{inv} \ , \tag{7}$$

where η_e is the PV efficiency of the panel and η_{inv} is the inverter's efficiency. The monthly average hourly beam irradiance on a tilted surface (Wh/m^2) is given by Equation (8).

$$I_B \ = I_{Bn} \ cos\theta, \tag{8}$$

where I_{Bn} is the average beam irradiance for normal incidence (Wh/m^2), and θ is the angle of incidence. Equation (9) shows the diffuse irradiance on a tilted surface (Wh/m^2).

$$I_D \ = I_{Dn} \ \left(\frac{1+cos\beta}{2}\right), \tag{9}$$

where the monthly average hourly diffuse irradiance for normal incidence, I_{Dn} (Wh/m^2), depends on the tilt angle, β [46]. Table 1 shows data from the energy management system, including the PV array, the Water Flow Glazing thermal and spectral properties, the buffer tank, the reference glazing, and the heat pump. WFG prototype was connected to a buffer tank and a water-to-water heat pump, whereas the reference glazing prototype was simulated with an air-to-water heat pump. It has been assumed that the former had a heat pump, with a Coefficient of Performance (COP) of 5.4 and an Energy Efficiency Ratio (EER) equivalent to 6.0. The latter heat pump's coefficients were COP of 3.5 and EER of 4.0 [47].

Table 1. Description of the energy management system.

Parameter	Description	Value		Unit
P_{PV}	Maximum power at STC	192		(W/m^2)
V_{mp}	Optimum operating Voltage	33.7		(V)
I_{mp}	Optimum operating Current	9.5		(A)
η_e	Module efficiency	19.18		%
T_m	Operating module Temperature	−40 to +90		(°C)
NOCT	Nominal Operating Cell Temperature	47		(°C)
Prototype				
A_{roof}	Area of the roof	49		(m^2)
A_{wall}	Area of the opaque walls	20		(m^2)
A_{floor}	Area of the floor	49		(m^2)
U_{roof}	U value of the roof	0.2		(W/m^{2}K)
U_{wall}	U value of the opaque walls	0.3		(W/m^{2}K)
U_{floor}	U value of the floor	0.3		(W/m^{2}K)
WFG		E-W	S	
A_{WFG}	Area of WFG	38.4	19.2	(m^2)
g^{ON}	g-factor at $\dot{m}$ = 2 L/m^{2}min	0.22	0.24	
g^{OFF}	g-factor at $\dot{m}$ = 0 L/m^{2}min	0.27	0.59	
U^{ON}	U value of WFG at $\dot{m}$ = 2 L/m^{2}min	0.063	0.066	(W/m^{2}K)
U^{OFF}	U value of WFG at $\dot{m}$ = 0 L/m^{2}min	0.995	1.041	(W/m^{2}K)
Reference				
A_R	Area of reference glazing	38.4	19.2	(m^2)
g_R	g-factor	0.30	0.57	(W/m^{2}K)
U_R	U value of reference glazing	1.01	1.01	(W/m^{2}K)
Heat Pump		Air-to-Water	Water-to-Water	
COP [1]	Coefficient of performance	3.5	5.4	
EER [1]	Energy efficiency ratio	4.0	6.0	

[1] Values taken from [47].

3. Results

This section reviewed several critical energy parameters, demonstrating both initial predictions and collected real-world data from the facility. The heating and cooling loads, both with and without WFG panels, was demonstrated. The amount of PV energy produced is also listed. Finally, heat gains from the internal water chamber were presented. The tested prototype had an opening schedule from 8:00 to 20:00, (Monday to Friday). During scheduled operating hours the considered internal loads were equipment loads: 7 W/m^2, lighting loads: 10 W/m^2, occupation loads: 5 W/m^2. The heat balance was broken down in internal heat loads provided by the equipment in operation and occupancy of the building, the heat transmitted between outside and inside through the transparent and opaque envelope, solar gain through the glazing (considering the amount of direct and diffuse radiation which penetrates to the interior), and ventilation loads. Domestic hot water consumption has been considered zero.

3.1. Energy Plus Simulation Results

This section aimed to evaluate how the different technologies implemented in the facility are performing related to the construction of zero energy buildings, as set out in the 2010/31/UE directive, and using renewable solar energy. Subsequently, the electric type energy consumption is represented due to the lighting, the heat pump, and the operation of equipment, all of them monthly, in kWh. The consumption of domestic hot water has been considered zero. For performing the energy simulations, the EnergyPlus 8.4.1 (https://energyplus.net accessed on 2 February 2021) software tool has been used to calculate the building's energy consumption. For graphical interface, the Open Studio Application Suite (https://energyplus.net/extras accessed on 2 February 2021), a tool able to support the whole building energy modeling when using Energy Plus. Table 2 illustrates the estimated heating loads of the WFG prototype on a cloudy winter day. The number of occupants (n) at each hour was used to calculate ventilation loads (Vent). Values for area and thermal transmittance of the roof, opaque walls, and water flow glazing were taken from Table 1.

Table 2. Winter heating loads on 5 January 2020.

Hour	θ_i (°C)	θ_e (°C) [1]	n	$\sum UA(\theta_i - \theta_e)$ (Wh)	$\sum U_G A_G(\theta_i - \theta_e)$ (Wh)	Vent (Wh)	Total (Wh)
0–6	-	-	-	-	-	-	-
7–8	22	−4	0	793	96	0	889
8–9	22	−3.5	2	778	94	191	1063
9–10	22	−2.6	2	750	91	185	1025
10–11	22	−1.2	2	708	86	174	967
11–12	22	−0.5	4	686	83	338	1107
12–13	22	2.2	4	604	73	297	974
13–14	22	1.8	2	616	74	152	842
14–15	22	1.5	2	625	76	154	855
15–16	22	0.5	2	656	79	161	896
16–17	22	−0.6	4	689	83	339	1112
17–18	22	−0.8	4	695	84	342	1121
18–19	22	−1.3	2	711	86	175	971
19–20	22	−1.8	2	726	88	179	992
20–24	-	-	-	-	-	-	-

[1] Values are taken from Figure 3.

Table 3 shows the estimated cooling loads of the WFG prototype. Internal loads (*IL*) were determined with the metabolic rate of an office, the number of people, and 20 W/m^2 for lighting. Solar radiation (*SR*) was calculated using the g-factor from Table 1 for WFG.

Table 3. Summer cooling loads on 2 July 2020.

Hour	θ_i (°C)	θ_e (°C) [1]	n	$\sum UA(\theta_i - \theta_e)$ (Wh)	$\sum U_G A_G(\theta_i - \theta_e)$ (Wh)	Vent (Wh)	IL (Wh)	SR (Wh)	Total (Wh)
0–6	23	18	0	0	0	0	0	0	0
7–8	23	18.5	0	−137	−17	0	200	45	91
8–9	23	20.3	2	−82	−10	−81	400	80	307
9–10	23	21.5	2	−46	−6	−45	400	90	394
10–11	23	23.7	2	21	3	21	400	98	542
11–12	23	25.8	4	85	10	168	600	120	984
12–13	23	28.9	4	180	22	354	600	143	1298
13–14	23	31.2	2	250	30	246	400	170	1096
14–15	23	29.7	2	204	25	201	400	174	1004
15–16	23	27.8	2	146	18	144	400	170	878
16–17	23	25.3	4	70	8	138	600	130	947
17–18	23	25.2	4	67	8	132	600	94	901
18–19	23	23.1	2	3	0	3	400	68	474
19–20	23	20.5	2	−76	−9	−75	400	28	267
20–24				0	0	0	0	0	0

[1] Values are taken from Figure 4.

3.2. Operational Energy and Thermal Energy Production

Heating and cooling energy loads were calculated by using Equation (4) for the reference glazing and Equation (5) for WFG. The inputs were taken from real data over a year time. The thermal energy production was obtained from the prototype's real data throughout the year. Equation (10) verified the absorbed energy per unit of area per day [44]. The output of the equation is the water heat gain. The daily water heat gain efficiency is defined as the ratio of total water heat gain compared to the daily solar radiation. In this research, the daily water heat gain is accumulated if the inlet temperature is lower than the outlet temperature.

$$P = \frac{\dot{m}c}{\dot{m}c + U_e + U_i}(A_v i_0 + U_i(\theta_i - \theta_{IN}) + U_e(\theta_e - \theta_{IN})), \quad (10)$$

where:

U_e = thermal transmittance between water chamber and exterior environment,

U_i = thermal transmittance between water chamber and interior,

$A_v i_0$ = energy absorbed by the water, plus the energy transferred by convection due to the glass panels heat absorption,

θ_{IN} = inlet temperature in the water flow glazing

$\dot{m}c$ = capacity of the fluid medium to absorb heat.

Equation (11) defines A_v as the absorptance of triple glazing with a water chamber facing indoors [44].

$$A_v = A_1\left(\frac{U_e}{h_e}\right) + A_2\left(\frac{1}{h_g} + \frac{1}{h_e}\right)U_e + A_3\left(\frac{U_e}{h_i}\right) + A_w, \quad (11)$$

where:

U_e = thermal transmittance between water chamber and exterior environment,

U_i = thermal transmittance between water chamber and interior,

A_w = Water absorptance,

A_1, A_2, A_3 = Glazing panel absorptances of each respective pane in window assembly,

h_e, h_i, h_g, h_w = Convective heat coefficients.

In Figures 5 and 6, the southern facing wall was responsible for the majority of water heat gain, while the walls in the eastern and western orientations provided much less. Two causes can explain this phenomenon. The first being the simple fact that in winter, the southern direction is the orientation that receives the most amount of direct solar radiation, while the walls in the eastern and western orientations receive much less. The second reason involves the selection of the glazing types used for each direction. In the eastern and western orientations, the glazing was selected to have a high near-infrared reflectance. This high reflectance, in turn, caused a low absorption of solar heat gains in near-infrared radiation. The authors were seeking out this effect because of two reasons. Reason one was because the solar irradiance in eastern and western orientations was predicted to be negligible. Therefore, it is not warranted to invest the same in these directions as in the Southern orientation. Reason two was to prevent overheating during summertime when the solar radiation is perpendicular to eastern and western facades. Meanwhile, in the southern orientation, the glazing was chosen to have a low reflectance and a high absorptance of near-infrared radiation. This absorptance causes a high heat gain leading to increasing temperatures in the fluid medium in the WFG panels. This heated water can then be used for other purposes.

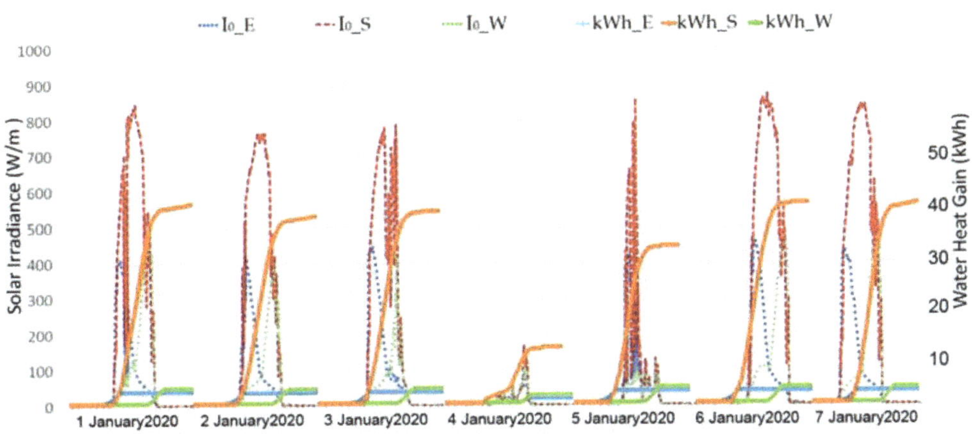

Figure 5. Solar irradiances on the façade and water heat gain in January.

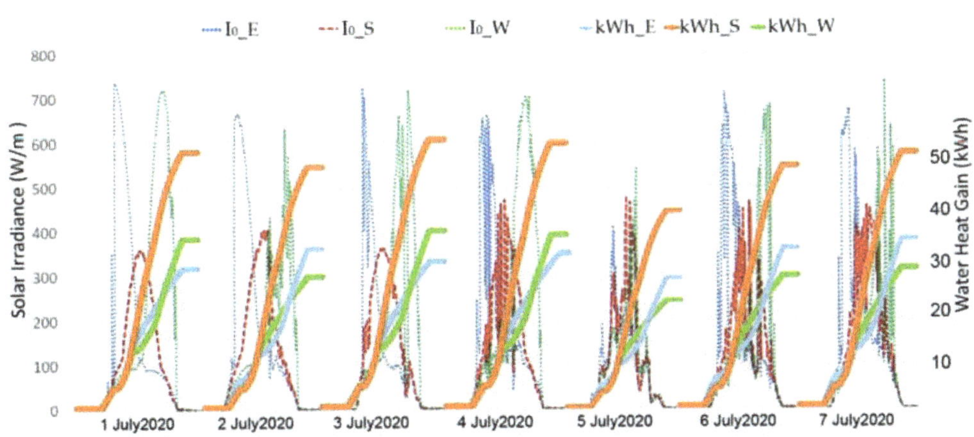

Figure 6. Solar irradiances on the façade and water heat gain in July.

Figure 5 demonstrates the measured solar irradiance and the water heat gains during the selected winter period in three different orientations. Panels oriented in a southern direction had a daily energy absorption rate of 39 kWh on sunny winter days when the peak solar irradiance on the southern façade was 800 W/m^2. On cloudy winter days, the energy absorption rate in the eastern and western orientations was negligible. The higher gain in southern glazing can also be explained by the solar angle and the hours of sun on the southern facade. The angle and sun hours are lower on eastern and western facades, which results in less heat absorption.

Figure 6 shows the measured solar irradiance and the water heat gains over the first week of July. The average water heat gain in the southern WFG panels was 47 kWh per day, whereas the average water heat gain in the eastern and western WFG panels was 33 kWh per day. As expected, most of the heat on eastern and western facades was rejected, whereas the water heat gain was high on the southern facade. Solar radiation on eastern and western facades showed low values in winter. The peak solar radiation was slightly over 400 W/m^2, whereas in summer, the peak solar radiation was over 700 W/m^2. In addition, there were more sun hours in summer than in winter. The east-west gain is different depending on the day. Figure 6 shows that on 2 July, solar heat gains are higher in the east than in the west because solar radiation showed high values. On 4 July, Figure 6 shows that the solar radiation in the western facade was higher, so was the solar heat gain over that day.

The water-energy absorption can be used as renewable primary energy production. Table 4 shows the total thermal energy production compared to the heating load needed to keep the indoor temperature at a comfortable range. A hydronic system made of indoor WFG panels delivered this energy inside the building. It can be seen in Table 4 that in December and January, the heating loads are slightly above what the water heat gains can provide. During these months, another energy source will be required to provide thermal comfort conditions. Meanwhile, during the remainder of the year, especially during the summer months, the exact opposite occurs. The WFG panels absorb an excessive amount of solar heat.

Table 4. Yearly water heat gain and heating loads for the WFG prototype.

Month	Water Heat Gain (kWh)			Heating Loads (kWh)	Total (kWh)
	East	West	South		
January	94	199	1184	1681	−204
February	117	257	1064	1040	398
March	176	328	1287	747	1044
April	199	351	1539	453	1636
May	339	480	1682	251	2250
June	538	644	1865	72	2975
July	645	771	1899	90	3225
August	585	702	1711	270	2728
September	468	597	1502	428	2139
October	351	456	1359	900	1266
November	129	269	1313	1143	568
December	102	242	1225	1589	−20

An excess of water heat gains will lead to uncomfortable interior conditions for building occupants. Therefore, the heat gains should be applied to something that will be beneficial to building occupants in a way other than heating. Possible remedies could include domestic hot water for residential buildings, and seasonal heating storage by use of a buffer tank.

3.3. PV Production and Electric Loads

This subsection aimed to demonstrate that the remaining heating loads, cooling loads, artificial lighting, and power for building equipment were covered by energy produced from the PV array. Figure 7 demonstrates the total loads per month, with lighting and power being considered the same in both prototypes. For the Water Flow Glazing structure, most of the heating loads were covered by heating the water in the WFG panels, heat storage in the buffer tank, and finally heat delivery by indoor WFG panels acting as traditional radiators. The coefficient of performance (COP) for the WFG water-to-water heat pump was 5.4, while the water-to-water heat pump's energy efficiency ratio (EER) was 6. Meanwhile, with the traditional reference cabin, energy consumption was much higher as compared to the WFG facility. In the reference cabin, the COP of air-to-water heat pump was 3.5, whereas EER of air-to-water heat pump was 4. The significant reduction in electricity consumption becomes especially apparent when viewing the total electrical loads of the reference glazing, 1000 kWh, compared to 575 kWh in the WFG prototype during August.

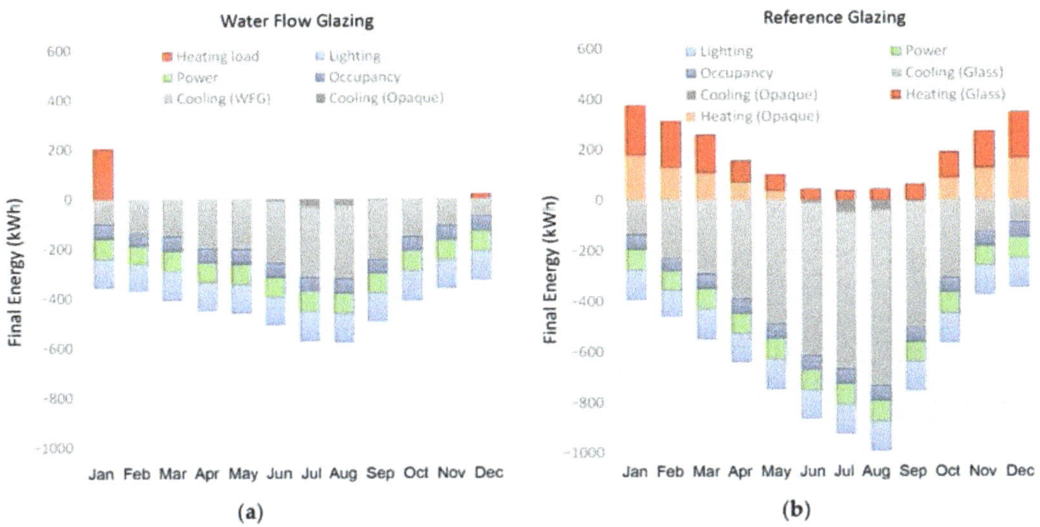

Figure 7. Estimation of final electric energy consumption of the prototypes: (**a**) Water Flow Glazing prototype with a water-to-water heat pump; (**b**) Reference glazing prototype with an air-to-air heat pump.

The National Renewable Energy Laboratory PVWatts Calculator [48] was used to estimate the power production of the building integrated PV panels installed on the roof and the south facade. The goal was to maximize Alternating Current (AC) energy production, reducing the energy supplied to and from the electrical grid. Table 5 summarizes the two options that were apparent to the authors of this text. Option 1 was 8.3 m^2 of building integrated PV panels on the south façade. Additionally, there would exist an extra 15 m^2 of open-rack PV panels on the roof that would maintain a 60° tilt. Option 2 would be 15 m^2 open-rack PV panels with a 30° tilt, with an additional 15 m^2 of panels with a tilt of 60° on the roof. The AC production of each of these options is higher than the electric demand shown in Figure 7. In January, the demand was 563 kWh, while the production was 276 kWh for option 1 and 245 kWh for option 2.

Figure 8 demonstrates the hourly energy balance during the seven days of January. The differences in solar radiation observed on different days were due to the impact of different factors on the PV system performance, such as temperature, wind, and the reduced sun hours in January. On sunny days the peak PV production surpassed 5 kW at noon, whereas the demand was almost steady at 1.3 kW every hour. Meanwhile, on cloudy days the system produced more than 1 kW every hour, with the demand continuing to be steady at 1.3 kW every hour.

Figure 9 shows the hourly energy balance during seven days in July. The PV yield is almost the same throughout the entire period. Peak production during the central hours of the day surpassed the demand.

Table 5. Yearly water heat gain and heating loads for the WFG prototype.

Month	Option 1. AC System Output (kWh)			Option 2. AC System Output (kWh)			Electric Demand [1]
	8.3 m² (90°)	15 m² (60°)	Total	15 m² (30°)	15 m² (60°)	Total	(kWh)
January	89	187	276	120	125	245	563
February	127	276	403	182	184	366	369
March	128	304	432	221	203	424	408
April	136	352	488	272	235	507	449
May	138	383	521	313	255	568	461
June	146	446	592	375	297	672	499
July	164	493	657	406	329	735	543
August	190	527	717	408	352	760	554
September	179	434	613	310	290	600	489
October	137	307	444	209	205	414	410
November	72	157	229	107	105	212	358
December	67	140	207	91	93	184	348
Total	1573	4006	5579	3014	2673	5687	5451

[1] Values are taken from Figure 7.

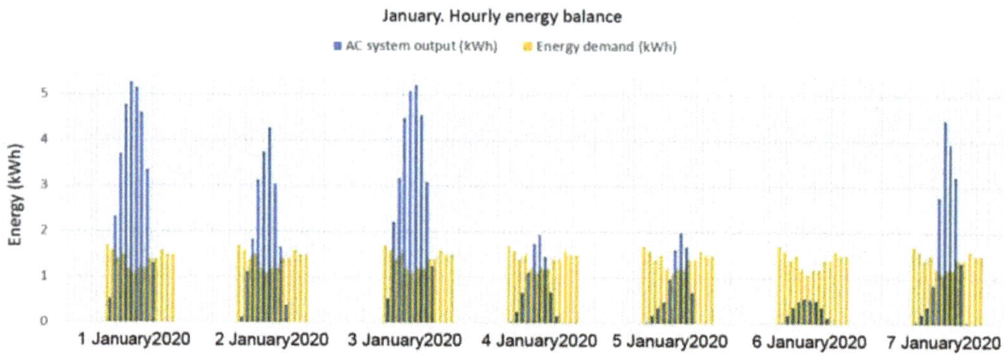

Figure 8. Hourly energy balance. January.

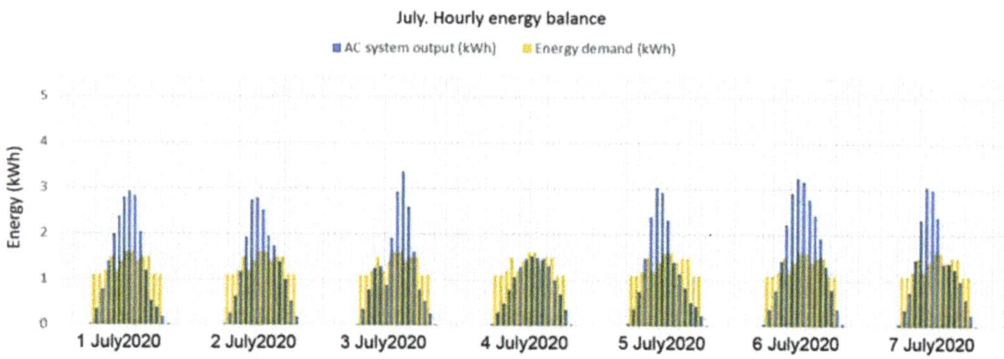

Figure 9. Hourly energy balance. July.

4. Discussion

4.1. Operational Energy Analysis and Environmental Assessment

Data from the prototype was used to calculate the yearly operational energy. In this case study, it was assumed that the building experienced no renovations and no change in the usage mode throughout its life cycle. It has been assumed that all the building's envelope elements have the same service life, and the lifetime of the PV array and heat pumps was 25 years. Non-renewable primary energy, equivalent CO_2 emissions, and cost were the parameters utilized to assess the environmental performance. The electricity conversion factor for Bulgaria was 0.47, the Conversion Factors from final energy (FE) to non-renewable primary energy (NRPE) in Bulgaria was 2.17, and the cost of electricity in Bulgaria was 0.239 €/kWh [49,50]. Table 6 shows the primary energy consumption, global warming potential, and operational energy (OE) cost of the prototypes.

Table 6. Final energy, non-renewable primary energy, and CO_2 emissions per year.

	Water Flow Glazing Prototype			Reference Glazing Prototype		
Electric Load	**Heating**	**Cooling**	**Light-Power**	**Heating**	**Cooling**	**Light-Power**
Energy (kWh/year)	225	13,212		7724	30,023	
COP/EER [1]	5.4	6		3.5	4	
FE (kWh/year)	42	2202	2345	2207	7506	2345
NRPE (kWh/year)	90	4778	5089	4789	16,287	5089
CO_{2eq} ($KgCO_2$/year)	20	1035	1102	1037	3528	1102
OE cost (€/year)	10	526	560	527	1794	560

[1] COP and EER values are taken from [47].

The total operational cost of the WFG prototype was 1096 €, whereas the reference glazing prototype was 2881 €. The total CO_{2eq} emissions were 2157 $KgCO_{2eq}$/year for the WFG prototype and 5667 $KgCO_{2eq}$/year for the reference glazing prototype for non-renewable energy sources. These results demonstrated the energy savings and reduction of global warming potential of WFG compared with a high performance triple glazing. As shown in Section 3.3, PV arrays of options 1 and 2 produce more AC energy than the WFG prototype electric demand, so it is necessary to compare the global warming potential in terms of total CO_{2eq}.

The embodied energy of materials is highly correlated with the production of technology and components [51,52]. Table 7 shows the embodied carbon and energy of each material and the costs of a PV panel.

Table 7. Embodied Energy (EE), Embodied Carbon (EC), and Cost of PV panel materials.

Material	EE (MJ/kg)	EC ($kgCO_{2eq}$/kg)	EE (MJ/m^2)	EC ($kgCO_{2eq}/m^2$)	Cost (€/m^2)
Aggregate	0.083	1.59	-	-	-
Aluminum frame	155	1.70	-	-	-
Glass	15	1.63	-	-	-
Silicon	2355	1.75	-	-	-
Lithium	853	14.76	-	-	-
Plastic	80	1.02	-	-	-
Wires	36	14.76	-	-	-
Polycrystalline PV panel	-	-	4070	208	518 [1]

[1] value taken from [52].

The global warming potential of the manufacturing process of PV panels showed that the CO_{2eq} emissions were 208 $KgCO_{2eq}/m^2$. The PV area of option 1 was 23.3 m^2. The PV area of option 2 was 30 m^2. The total CO_{2eq} emissions during the PV array manufacturing process were 4846 $KgCO_{2eq}$ for option 1 and 6240 $KgCO_{2eq}$ for option 2.

Several outcomes can be drawn from the examination of this data. First, using WFG can reduce emissions during the operating lifetime in glazed buildings. Second, by producing AC energy with PV arrays, the operating lifetime emissions are zero. Third, manufacturing PV panels also has a global warming potential due to the high emission potential of processing Silicon and Lithium. This section has shown that the minimum global warming potential of manufacturing the PV array to turn the WFG prototype into a Zero Energy

Building was 4846 $KgCO_{2eq}$. In contrast, the yearly global warming potential due to the building operation was 2157 $KgCO_{2eq}$/year, so in three years the reduction in CO_2 emissions compensated for the PV array's initial manufacturing emissions.

4.2. Economic Analysis

The highest cost of a solar PV system happens at the installation stage. Therefore, once a PV system is operating, it produces electricity for free. A solar PV array financial report determined the economic feasibility, considering initial cost requirements, operating and maintenance costs. Yield factor (YF) is the energy generated by a PV array over time, and it takes into account the module efficiencies and array designs. Equation (12) shows the expression of yield factor measured in kWh/kWp, by dividing the energy yield (E_t) by the PV array's nominal power at standard test conditions (P_{STC}).

$$YF = \frac{E_t}{P_{STC}}. \tag{12}$$

The initial cost included the PV modules, circuit, breakers, cables, initial labor, and grid connection cost. Operating costs included yearly scheduled operation costs that comprised screening, maintenance, repairs, panel cleaning, and insurance. To evaluate and compare the cost of energy production between selected case studies and each configuration of PV systems, the concept of Levelized Cost of Energy (LCOE) needs to be introduced. The LCOE is the cost at which electricity must be produced from a source to break even over the project's life [53]. The Levelized Cost of energy expression shown in Equation (13) was taken from previous articles as the ratio between the total life cycle cost and the total life cycle energy output [54].

$$LCOE = \frac{I_0 + \sum_{t=1}^{n} \frac{M_t + F_t}{(1+r)^t}}{\sum_{t=1}^{n} \frac{E_t}{(1+r)^t}}, \tag{13}$$

where I_0 is the initial investment cost, M_t represents the cost of operation and maintenance in the yeart, F_t is the fuel expense in the year t, E_t is the electrical energy generated in the year t, and r is the discount rate. In this article, fuel cost F_t was zero. It was assumed that the replacement rate for WFG would be the same as the reference triple glazing over the considered lifetime. Other components, such as the circulating water pumps, are certified to work for 25 years. The performance evaluation of the solar PV system has been carried out by using Equations (12) and (13). Table 8 presents the yearly yield factor and the levelized cost of energy of two options. The yield factor (YF), which measured the productivity of the system, was found to being 1247 kWh/kWp per year, and 987 kWh/kWp per year, for option 1 and option 2, respectively.

Table 8. Final energy, non-renewable primary energy, and CO_2 emissions per year.

	P_{STC} (kW)	I_0 (€)	M_t (€/Year)	n (Year)	E_t (kWh/Year)	YF (kWh/kWp)	r (%)	LCOE (€/kWh)
							2	4.35
Option 1	4474	12,069	121	25	5579	1247	3	6.51
							4	8.68
							2	5.49
Option 2	5760	15,540	155	25	5687	987	3	8.22
							4	10.96

Final LCOE results depended on average temperatures and solar radiation on the selected site. The relationship between the final results and the selected climate zone was addressed in the conclusions section. Another relevant factor was the discount rate. In this study, three different discount rates have been considered, being 3%, in line with historical data for the Euro Area [55]. Equipment, labor, connection charges, and taxes make up the installation costs. The replacement of the inverter in the 12th year was considered the only maintenance cost. Due to this prototype's small size, there were no operating costs, and maintenance cost was fixed at 1% yearly. The investment costs were different for each country, according to the study performed by the International Renewable Energy Agency (IRENA) [55]. An annual degradation of 0.8% has been estimated based on the specifications provided by manufacturers and thorough analysis on this subject. This degradation leads to decreasing the production, especially after 10 years. [56] The investment costs (I_0) can be divided into installation costs, soft costs, and hardware costs. Installation costs are the expenses related to the settings of the PV system, including mechanical and electrical fixtures. In contrast, soft costs include all relevant permits administrative costs connected with the system. Hardware costs comprise module, inverter, racking, and electrical wiring. The total investment costs were different for each country. Typical values for Bulgaria were taken from previous articles and databases [22]. A low discount rate increases the reduction of the overall cost of the PV system, decreasing the price of generating one unit of energy. Option 1 showed a better performance when it came to the yield factor and the levelized cost of energy at all discount rates.

5. Conclusions

This paper aimed to develop a comprehensive comparative analysis from the operating energy, global warming potential, and economic point of view, of two different technologies. This study measured real data of a small office building located in Sofia, Bulgaria to understand the thermal behavior of a structure using WFG panels. This energy system also included a PV system connected to the grid to provide electric energy for heating, cooling, lighting and power supply. In addition to the water flow glazing panels and the PV array, the proposed building-energy system layout also includes radiant panels that delivered the necessary heating and cooling loads. Electrical heat pumps are also modelled for backup heating and for balancing the system cooling demands.

In order to validate the energy savings and the reduction on global warming potential, the water flow glazing facade has been compared with a high-performance triple glazing one. This analysis showed that the selected water flow glazing system coupled with a water-to-water heat pump improves the performance of highly efficient triple glazing. Coefficient of performance and energy efficiency ratio of water-to-water heat pumps are usually higher than air-to-water or air-to-air heat pumps. WFG system, as compared to the reference prototype, has demonstrated actual savings of 60% in non-renewable primary energy consumption with a 62% reduction of CO_2 emission.

WFG panels can work as integrated solar thermal collectors. This article has shown that water heat gain throughout the year compensate for 86% of the prototype's heating loads. To maximize water heat gains in the southern orientation, the water flow glazing was chosen to have a low reflectance and a high absorptance of near-infrared radiation. This absorptance causes a high heat gain leading to increasing temperatures in the fluid medium in the WFG panels. In southern and western elevations, a high reflectance panel has been selected to avoid overheating in summer.

The goal of zero-energy building was achieved first, by reducing the heating and cooling demand and second, by implementing a PV array.

Due to the high emission potential of some PV panel components such as Silicon and Lithium, manufacturing PV panels has also a global warming potential This article has compared the CO_2 emissions of manufacturing the PV system with the emissions over the operating lifetime. It was found that in three years the reduction in CO_2 emissions compensate for the initial manufacturing emissions of the PV array.

Two factors have been used to assess the economic aspects of a PV system, the Yield factor (YF) and the Levelized Cost of Energy (LCOE). This study has not included SELL and BUY.

Option 1 included integrated PV panels on a vertical façade. This option has shown a yield factor of 1247 (kWh/kWp), which was 21% more than option 2.

Levelized Cost of Energy (LCOE) has shown to be dependent on the discount rate. Option 1 has shown a better LCOE ranging from 4.35 €/kWh to 8.68 €/kWh with discount rates of 2% and 4% respectively.

Combining water flow glazing facades and building integrated PV panels can reduce the energy demand and produce renewable energy (thermal and electrical) within the building footprint. However, the proposed system has limitations and room for improvement. The primary disadvantage is the capital cost required for initial implementation. Seasonal thermal storage systems are necessary to use the excess heat produced by the WFG façade. Office buildings have low hot water consumption, and much thermal energy is wasted in summer. Increasing the volume of buffer tanks may lead to achieving seasonal storage, although it will increase the energy system's cost. Building Integrated Photovoltaic panels (BIPV) are negatively affected by several factors in the outdoor environment, such as temperature variations and variable solar irradiation. More prototypes in various climate zones should be analyzed. Another future research line is the assessment of comfort. Traditional cooling systems focused on air movement have lacked comfort by blowing cold air instead of focusing on the mean radiant temperature through radiant elements such as floors, ceilings, and interior partitions. WFG can control the surface temperatures and provide an excellent thermal environment with indoor air temperatures outside the comfort range. Several difficulties and system misfunction must be discussed. WFG technology is hindered by a high initial expenditure that includes other required equipment. A storage tank is needed to avoid overheating in summer, and dimensioning this tank is of paramount importance to the correct performance of the system. WFG technology has to be implemented at the first stage of the design process to bring down the initial expense and payback period. The lack of reliable simulation tools for this technology might slow down the market adoption because of the many parameters that influence thermal performance in different orientations.

Author Contributions: Conceptualization, B.M.S., F.d.A.G., J.A.H.R.; methodology, B.M.S., F.d.A.G.; software, J.A.H.R.; formal analysis, M.G., F.d.A.G.; data curation, F.d.A.G., J.A.H.R., J.A.F.G.; writing—original draft preparation, B.M.S., F.d.A.G., J.A.H.R.; writing—review and editing, M.G., B.M.S., J.A.F.G.; visualization, M.G., B.M.S., J.A.F.G.; supervision, J.A.H.R.; project administration, F.d.A.G.; funding acquisition, F.d.A.G., J.A.F.G. All authors have read and agreed to the published version of the manuscript.

Funding: This research received no external funding.

Institutional Review Board Statement: Not applicable.

Informed Consent Statement: Not applicable.

Acknowledgments: Special thanks to the Central Laboratory of Solar Energy and New Energy Sources, of the Bulgarian Academy of Science (CL SENES-BAS) for providing measured data of the solar radiation in the different facades of the test facility in Sofia, Bulgaria. This work was supported by Keene State College Faculty Development Grant program.

Conflicts of Interest: The authors declare that they have no conflict of interest.

References

1. Kolokotsa, D.; Rovas, D.; Kosmatopoulos, E.; Kalaitzakis, K. A roadmap towards intelligent net zero- and positive-energy buildings. *Sol. Energy* **2011**, *85*, 3067–3084. [CrossRef]
2. AbuGrain, M.Y.; Alibaba, H.Z. Optimizing Existing Multistory Building Designs towards Net-Zero Energy. *Sustainability* **2017**, *9*, 399. [CrossRef]
3. European Union. Directive (EU) 2018/844 of the European Parliament and of the Council of 30 May 2018. *Amending Directive 2010/31/EU on the Energy Performance of Buildings and Directive 2012/27/EU on Energy Efficiency*. 2018. Available online: https://eur-lex.europa.eu/legal-content/EN/TXT/PDF/?uri=CELEX:32018L0844&from=EN (accessed on 21 March 2020).
4. Kotarela, F.; Kyritsis, A.; Papanikolaou, N. On the Implementation of the Nearly Zero Energy Building Concept for Jointly Acting Renewables Self-Consumers in Mediterranean Climate Conditions. *Energies* **2020**, *13*, 1032. [CrossRef]
5. Kerdan, I.G.; Raslan, R.; Ruyssevelt, P.; Gálvez, D.M. An exergoeconomic-based parametric study to examine the effects of active and passive energy retrofit strategies for buildings. *Energy Build.* **2016**, *133*, 155–171. [CrossRef]

6. Hu, M.; Qiu, Y. A comparison of building energy codes and policies in the USA, Germany, and China: Progress toward the net-zero building goal in three countries. *Clean Technol. Environ. Policy* **2019**, *21*, 291–305. [CrossRef]
7. Marino, C.; Nucara, A.; Panzera, M.F.; Pietrafesa, M.; Pudano, A. Economic Comparison between a Stand-Alone and a Grid Connected PV System vs. Grid Distance. *Energies* **2020**, *13*, 3846. [CrossRef]
8. Luo, Y.; Zhang, L.; Bozlar, M.; Liu, Z.; Guo, H.; Meggers, F. Active building envelope systems toward renewable and sustainable energy. *Renew. Sustain. Energy Rev.* **2019**, *104*, 470–491. [CrossRef]
9. Arpino, F.; Cortellessa, G.; Frattolillo, A. Experimental and numerical assessment of photovoltaic collectors performance dependence on frame size and installation technique. *Sol. Energy* **2015**, *118*, 7–19. [CrossRef]
10. Sartori, I.; Napolitano, A.; Voss, K. Net zero energy buildings: A consistent definition framework. *Energy Build.* **2012**, *48*, 220–232. [CrossRef]
11. Romano, R.; Gallo, P. The SELFIE Project Smart and efficient envelope' system for nearly zero energy buildings in the Mediterranean Area. *GSTF J. Eng. Technol.* **2018**, *4*, 562–569.
12. Maturi, L.; Belluardo, G.; Moser, D.; Del Buono, M. BiPV System Performance and Efficiency Drops: Overview on PV Module Temperature Conditions of Different Module Types. *Energy Procedia* **2014**, *48*, 1311–1319. [CrossRef]
13. Lu, L.; Yang, H. Environmental payback time analysis of a roof-mounted building-integrated photovoltaic (BIPV) system in Hong Kong. *Appl. Energy* **2010**, *87*, 3625–3631. [CrossRef]
14. Ordóñez Mendieta, Á.J.; Hernández, E.S. Analysis of PV Self-Consumption in Educational and Office Buildings in Spain. *Sustainability* **2021**, *13*, 1662. [CrossRef]
15. Freitas, S.; Brito, M. Non-cumulative only solar photovoltaics for electricity load-matching. *Renew. Sustain. Energy Rev.* **2019**, *109*, 271–283. [CrossRef]
16. Talavera, D.; Muñoz-Rodríguez, F.J.; Jimenez-Castillo, G.; Rus-Casas, C. A new approach to sizing the photovoltaic generator in self-consumption systems based on cost–competitiveness, maximizing direct self-consumption. *Renew. Energy* **2019**, *130*, 1021–1035. [CrossRef]
17. European Commission; Joint Research Centre; Institute for Environment and Sustainability. *International Reference Life Cycle Data System (ILCD) Handbook—General Guide for Life Cycle Assessment—Detailed Guidance*; First edition March 2010; EUR 24708 EN; Publications Office of the European Union: Luxembourg, 2010.
18. Ballif, C.; Perret-Aebi, L.-E.; Lufkin, S.; Rey, E. Integrated thinking for photovoltaics in buildings. *Nat. Energy* **2018**, *3*, 438–442. [CrossRef]
19. IEC. *Standard IEC 60904-3: Photovoltaic Devices. Part 3: Measurement Principles for Terrestrial Photovoltaic (PV) Solar Devices with Reference Spectral Irradiance Data*; IEC: Geneva, Switzerland, 2019.
20. Honrubia-Escribano, A.; Ramirez, F.J.; Gómez-Lázaro, E.; Garcia-Villaverde, P.M.; Ruiz-Ortega, M.J.; Parra-Requena, G. Influence of solar technology in the economic performance of PV power plants in Europe. A comprehensive analysis. *Renew. Sustain. Energy Rev.* **2018**, *82*, 488–501. [CrossRef]
21. Buonomano, A.; Calise, F.; Palombo, A.; Vicidomini, M. BIPVT systems for residential applications: An energy and economic analysis for European climates. *Appl. Energy* **2016**, *184*, 1411–1431. [CrossRef]
22. Lugo-Laguna, D.; Arcos-Vargas, A.; Nuñez-Hernandez, F. A European Assessment of the Solar Energy Cost: Key Factors and Optimal Technology. *Sustainability* **2021**, *13*, 3238. [CrossRef]
23. Kim, A.A.; Reed, D.A.; Choe, Y.; Wang, S.; Recart, C. New Building Cladding System Using Independent Tilted BIPV Panels with Battery Storage Capability. *Sustainability* **2019**, *11*, 5546. [CrossRef]
24. Aaditya, G.; Pillai, R.; Mani, M. An insight into real-time performance assessment of a building integrated photovoltaic (BIPV) installation in Bangalore (India). *Energy Sustain. Dev.* **2013**, *17*, 431–437. [CrossRef]
25. Skoplaki, E.; Palyvos, J.A. On the temperature dependence of photovoltaic module electrical performance: A review of efficiency/power correlations. *Sol. Energy* **2009**, *83*, 614–624. [CrossRef]
26. Al-Aboosi, F.Y.; Al-Aboosi, A.F. Preliminary Evaluation of a Rooftop Grid-Connected Photovoltaic System Installation under the Climatic Conditions of Texas (USA). *Energies* **2021**, *14*, 586. [CrossRef]
27. Yuan, Z.; Zhou, J.; Qiao, Y.; Zhang, Y.; Liu, D.; Zhu, H. BIM-VE-Based Optimization of Green Building Envelope from the Perspective of both Energy Saving and Life Cycle Cost. *Sustainability* **2020**, *12*, 7862. [CrossRef]
28. Al-Hammad, A.-M.; Hassanain, M.A.; Juaim, M.N. Evaluation and selection of curtain wall systems for medium-high rise building construction. *Struct. Surv.* **2014**, *32*, 299–314. [CrossRef]
29. Nastasi, B. Renewable Hydrogen Potential for Low-carbon Retrofit of the Building Stocks. *Energy Procedia* **2015**, *82*, 944–949. [CrossRef]
30. Cui, Z.; Mizutani, A. Research on the Reduction Effect of Transparent Glass on Cooling Power Energy Consumption Research on the Reduction of Cooling and Heating Loads by Transparent Solar Heat Absorbing Glass Panels (Part 2). *J. Asian Archit. Build. Eng.* **2016**, *15*, 651–658. [CrossRef]
31. Tavares, P.; Bernardo, H.; Gaspar, A.; Martin, A. Control criteria of electrochromic glasses for energy savings in Mediterranean buildings refurbishment. *Sol. Energy* **2016**, *134*, 236–250. [CrossRef]
32. Ghosh, A.; Norton, B.; Duffy, A. Daylighting performance and glare calculation of a suspended particle device switchable glazing. *Sol. Energy* **2016**, *132*, 114–128. [CrossRef]

33. Hemaida, A.; Ghosh, A.; Sundaram, S.; Mallick, T.K. Evaluation of thermal performance for a smart switchable adaptive polymer dispersed liquid crystal (PDLC) glazing. *Sol. Energy* **2020**, *195*, 185–193. [CrossRef]
34. Arıcı, M.; Karabay, H.; Kan, M. Flow and heat transfer in double, triple and quadruple pane windows. *Energy Build.* **2015**, *86*, 394–402. [CrossRef]
35. Tao, Q.; Jiang, F.; Li, Z.; Zheng, J. A model of heat gain calculation for buildings with shuttle louvers: Verification and a case study. *J. Build. Eng.* **2020**, *29*, 101101. [CrossRef]
36. Ismail, K.A.R.; Salinas, C.T.; Henriquez, J.R. A comparative study of naturally ventilated and gas filled windows for hot climates. *Energy Convers. Manag.* **2009**, *50*, 1691–1703. [CrossRef]
37. Chow, T.T.; Li, C. Liquid-filled solar glazing design for buoyant water-flow. *Build. Environ.* **2013**, *60*, 45–55. [CrossRef]
38. Moreno Santamaria, B.; del Ama Gonzalo, F.; Lauret Aguirregabiria, B.; Hernandez Ramos, J.A. Evaluation of Thermal Comfort and Energy Consumption of Water Flow Glazing as a Radiant Heating and Cooling System: A Case Study of an Office Space. *Sustainability* **2020**, *12*, 7596. [CrossRef]
39. Gutai, M.; Kheybari, A.G. Energy consumption of hybrid smart water-filled glass (SWFG) building envelope. *Energy Build.* **2021**, *230*, 110508. [CrossRef]
40. Moreno Santamaria, B.; del Ama Gonzalo, F.; Pinette, D.; Gonzalez-Lezcano, R.-A.; Lauret Aguirregabiria, B.; Hernandez Ramos, J.A. Application and Validation of a Dynamic Energy Simulation Tool: A Case Study with Water Flow Glazing Envelope. *Energies* **2020**, *13*, 3203. [CrossRef]
41. Gutai, M.; Kheybari, A.G. Energy consumption of water-filled glass (WFG) hybrid building envelope. *Energy Build.* **2020**, *218*, 110050. [CrossRef]
42. Lyu, Y.; Chow, T.-T. Economic, energy and environmental life cycle assessment of a liquid flow window in different climates. *Build. Simul.* **2020**, *13*, 837–848. [CrossRef]
43. Gutai, M.; Lee, S.T.; Magori, B.; Morishita, Y.; Kheybari, A.; Spencer, J. Construction Aspects of Hybrid Water-Filled Building Envelopes. *J. Facade Des. Eng.* **2020**, *8*, 127–152.
44. Sierra, P.; Hernandez, J.A. Solar heat gain coefficient of water flow glazing. *Energy Build.* **2017**, *139*, 133–145. [CrossRef]
45. Lyu, Y.-L.; Chow, T.-T.; Wang, J.-L. Numerical prediction of thermal performance of liquid-flow window in different climates with anti-freeze. *Energy* **2018**, *157*, 412–423. [CrossRef]
46. Vasisht, M.S.; Srinivasan, J.; Ramasesha, S.K. Performance of solar photovoltaic installations: Effect of seasonal variations. *Sol. Energy* **2016**, *131*, 39–46. [CrossRef]
47. Priarone, A.; Silenzi, F.; Fossa, M. Modelling Heat Pumps with Variable EER and COP in EnergyPlus: A Case Study Applied to Ground Source and Heat Recovery Heat Pump Systems. *Energies* **2020**, *13*, 794. [CrossRef]
48. National Renewable Energy Laboratory. PVWatts Calculator. Available online: https://pvwatts.nrel.gov/ (accessed on 20 April 2021).
49. Koffi, B.; Cerutti, A.; Duerr, M.; Iancu, A.; Kona, A.; Janssens-Maenhout, G. CoM Default Emission Factors for the Member States of the European Union—Version 2017. European Commission, Joint Research Centre (JRC). Available online: http://data.europa.eu/89h/jrc-com-ef-comw-ef-2017 (accessed on 3 March 2021).
50. Huld, T.; Jäger-Waldau, A.; Ossenbrink, H.A.; Szabó, S.; Dunlop, E.; Taylor, N.G. *Cost Maps for Unsubsidised Photovoltaic Electricity*; JRC 91937; Joint Research Centre of the European Commission: Brussels, Belgium, 2014.
51. Fthenakis, V.; Frischknecht, R.; Raugei, M.; Kim, H.C.; Alsema, E.; Held, M.; de Wild-Scholten, M. Methodology Guidelines on Life Cycle Assessment of Photovoltaic Electricity. *IEA PVPS Task* **2011**, *12*, 6–21.
52. Ahmad Ludin, N.; Ahmad Affandi, N.A.; Purvis-Roberts, K.; Ahmad, A.; Ibrahim, M.A.; Sopian, K.; Jusoh, S. Environmental Impact and Levelised Cost of Energy Analysis of Solar Photovoltaic Systems in Selected Asia Pacific Region: A Cradle-to-Grave Approach. *Sustainability* **2021**, *13*, 396. [CrossRef]
53. Lai, C.S.; McCulloch, M.D. Levelized cost of electricity for solar photovoltaic and electrical energy storage. *Appl. Energy* **2017**, *190*, 191–203. [CrossRef]
54. Solar Energy Industries Association (SEIA). Solar Soft Costs. Available online: https://www.seia.org/resources (accessed on 7 April 2021).
55. IRENA. *Renewable Power Generation Costs in 2019*; International Renewable Energy Agency: Abu Dhabi, United Arab Emirates, 2020.
56. Jordan, D.C.; Kurtz, S.R.; VanSant, K.; Newmiller, J. Compendium of photovoltaic degradation rates. *Prog. Photovolt. Res. Appl.* **2016**, *24*, 978–989. [CrossRef]

energies

MDPI

Article

Pandemic of Childhood Myopia. Could New Indoor LED Lighting Be Part of the Solution?

David Baeza Moyano [1,*] and Roberto Alonso González-Lezcano [2]

[1] Department of Chemistry and Biochemistry, Campus Montepríncipe, Universidad San Pablo CEU, 28668 Alcorcón, Madrid, Spain
[2] Arquitecture and Design Depertment, Escuela Politécnica Superior, Campus Montpríncipe, Universidad San Pablo CEU, 28668 Alcorcón, Madrid, Spain; rgonzalezcano@ceu.es
* Correspondence: baezams@ceu.es

Abstract: The existence of a growing myopia pandemic is an unquestionable fact for health authorities around the world. Different possible causes have been put forward over the years, such as a possible genetic origin, the current excess of children's close-up work compared to previous stages in history, insufficient natural light, or a multifactorial cause. Scientists are looking for different possible solutions to alleviate it, such as a reduction of time or a greater distance for children's work, the use of drugs, optometric correction methods, surgical procedures, and spending more time outdoors. There is a growing number of articles suggesting insufficient natural light as a possible cause of the increasing levels of childhood myopia around the globe. Technological progress in the world of lighting is making it possible to have more monochromatic LED emission peaks, and because of this, it is possible to create spectral distributions of visible light that increasingly resemble natural light in the visible range. The possibility of creating indoor luminaires that emit throughout the visible spectrum from purple to infrared can now be a reality that could offer a new avenue of research to fight this pandemic.

Keywords: daylighting; circadian lighting; indoor lighting; dopamine; myopia

Citation: Baeza Moyano, D.; González-Lezcano, R.A. Pandemic of Childhood Myopia. Could New Indoor LED Lighting Be Part of the Solution? *Energies* **2021**, *14*, 3827. https://doi.org/10.3390/en14133827

Academic Editor: Alessandro Cannavale

Received: 23 May 2021
Accepted: 21 June 2021
Published: 25 June 2021

1. Introduction

The development of the components of the visual system during infancy and early childhood appears to be biphasic, with emmetropization occurring within the first 2 years of infancy during a rapid exponential phase [1].

Myopia in humans consists of decompensation of the refractive power of the cornea and lens compared to the axial length, such that images lie in front of the retina [2,3]. The components of the ocular system (axial length and anterior and posterior chamber depths) during emmetropization of children aged 3 to 9 months are currently longer than those measured in previous studies [4]. Some authors define myopia as a multifactorial disorder controlled by genetic interactions and environmental risk factors [5].

Myopia is a growing major public health problem and is the world's largest refractive problem [6–9]. The richest countries in the Asia–Pacific region have the highest prevalence in the world. The are multiple studies on the incidence of myopia worldwide according to ethnic and geographic parameters, with myopia prevalence at 80–90% in young adults and high myopia rates of 10–20% [10] in the urban populations in East Asia, especially in China and South Korea [11,12], Taiwan [13], and Singapore [14], while more than one-third of the people from the United States suffer from it [15–17]. Rates differ among people living in urban or rural communities, with similarities in other respects [6].

How myopia and high myopia are defined depends on the prevalence studies selected. Some projections state that 50% of the population will have myopia and 10% will have high myopia within thirty years [2,6,17].

It has been associated with an increased risk of retinal detachment, macular degeneration, early-onset glaucoma, and cataracts [6,8,12,18] and is the leading cause of blindness worldwide [6,19,20].

Data vary depending on the source; however, this social health problem is of gigantic proportions. In Europe, the above-mentioned levels have not yet been reached; however, the percentage of children and adolescents with myopia continues to increase [21,22].

The scientific community, to date, has not been able to obtain conclusive results regarding its cause, making it difficult to find a solution.

Because of this, there has been an abundance of studies from around the world. A representative example is the Cochrane Database of Systematic Reviews, which includes 41 studies (6772 participants) evaluating the effects acting on myopia progression in children with optometric corrections, and pharmaceutical procedures, such as muscarinic receptor antagonists, cycloplegic eye drops, and intraocular pressure-lowering medications [2].

The aim of this paper is to analyze the different possible causes of this pandemic, which scientists have highlighted in recent years, and to raise the possibility that something can be done to attenuate its progression based on new discoveries about the influence of light absorbed through our eyes on multiple biochemical mechanisms in our organisms, including the retina.

2. Materials and Methods

A bibliographic search was conducted in the biomedical databases of PubMed and ScienceDirect. Variations of the words "myopia", "nearsightedness", "short-sight", "progression", "refractive error", "Dopamine", "prevalence" were used as search criteria. These were used in combination with the inclusion criteria "risk factors", "outside", "outdoor", "atropine", "schoolchildren", "eye growth, "circadian rhythms", "indoor LED lighting".

Figure 1 below shows the flow diagram of the process carried out for the development of this article.

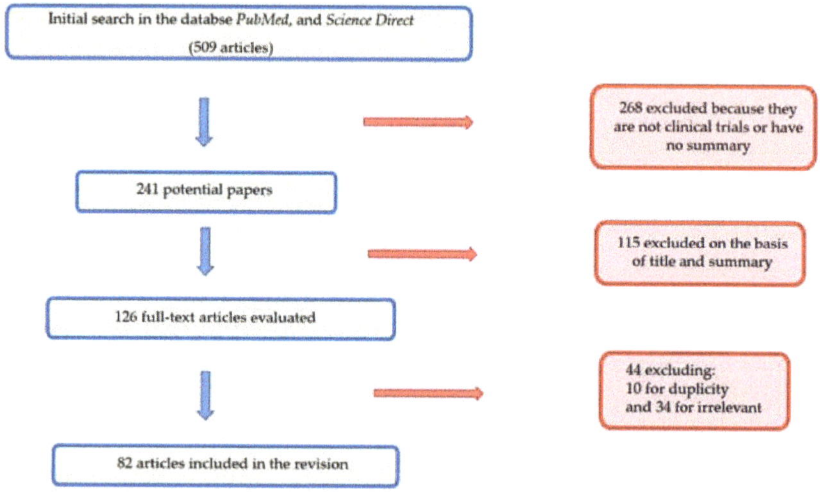

Figure 1. PRISMA flow chart for literature search and selection of articles included in the literature review Pandemic of Childhood Myopia. Could new indoor LED lighting be part of the solution?

The following inclusion criteria were used for the selection of selected articles: Scientific articles from the last two years, publications dealing with indoor lighting and new luminaires, indoor lighting standards, and magazines and articles written in English.

With regard to the exclusion criteria taken into account for the preparation of the work, the following were excluded publications that do not have the keyword "myopia" or "LED".

3. Results

Heritability is one of the factors most linked with young myopia together with increased near-distance work, increased school performance, and decreased sporting activity; however, there is no evidence that children inherit a myopathic environment or a susceptibility to the effects of near-distance work performed by their parents [23]. Children with myopic parents who do less sport and spend less time receiving natural light have a greater risk of becoming myopic than children who are in the same situation but for whom one or neither of their parents who are myopic [24]. The existence of this pandemic may be due to the pressure of increased school performance and limited outdoor time rather than a heightened sensitivity to these factors [10]. Increasing prevalence in many parts of the world makes it unlikely that its cause is related to a genetic factor [25].

Multiple factors have been raised that could influence the development of myopia in children, such as access to education, access to outdoor activities, and lack of exposure to natural light [6,26,27].

3.1. Myopia and Overwork at Close Distances

Increased working hours at close distances could increase the prevalence of myopia. Near-distance work has been defined as the group of activities that are carried out in proximal and to intermediate distances such as watching TV [14,27].

A higher incidence has been found in Chinese children living in Singapore compared to others living in Xiamen, because of positive relationships between the number of hours of reading, hours measured in tasks at close distances, use of electronic devices, and having high myopia; however, it is not certain that other factors do not play a role [14].

The dominant theory supported by numerous epidemiological studies about the origin of myopia is associated with near-distance work during eye development; however, there are doubts about this, as there are experimental animal studies in which no involvement of accommodation in the development of myopia is found. The site of action of atropine, which blocks myopia progression in humans, does not act on accommodation [3,10].

There are studies suggesting that the relationship between mass schooling and the onset and progression of myopia lies in the short focal distance required for reading and writing [22,28], whereby increased time spent in education may inadvertently increase the chances of suffering from myopia [29].

Others are of the opinion that, generally, no association has been found between working at close distances and myopia, except for the subgroup of people with high levels of nearby work and moderate levels of outdoor activity. A weak protective effect of outdoor activity on myopia was observed in children in rural China [30]. A high percentage of refractive problems was found in Jewish children attending orthodox schools. It has not been seen in other population subgroups, and it is very similar to that seen in East and Southeast Asia. Similar refractive distributions have also been reported for recent cohorts of young adults in Singapore [31] and South Korea [10,32].

3.2. Medication to Treat the Progression of Myopia

The use of atropine [16] and anticholinergic blockers has been shown to be suitable in the control of the progression myopia; however, there is considerable debate about their effectiveness and the possible long-term side effects [3].

Many researchers thought that excess accommodation was responsible for myopia and that atropine causes temporary paralysis of the smooth ciliary muscle [8].

Anticholinergics are blockers of the action of acetylcholine at muscarinic receptors (MRs). Acetylcholine is a neurotansmitter that plays an important role in retinal development and regulates eye growth. The most effective treatment to slow the progression of myopia is topical antimuscarinic medication, namely the use of low-dose atropine (0.01–0.05%) eye drops. However, the side effects of this medication include light sensitivity and blurring [2], so it is rarely prescribed [8].

The topical application of Pirenzepine does not block accommodation while decreasing the progression of childhood myopia [8,26].

3.3. *Light in the Lives of Children and Adolescents*

Outdoor illumination levels depend on weather, altitude, and latitude. The illuminance of a sunny day can be 150–130,000 lx, 50,000 lx on a hazy sunny day, and 15,000 lx on an overcast day. However, these levels are in stark contrast to typical indoor illumination levels, with values from around 1000 lx to values of 500–100 lx [27,33]. The effects of illumination on human refractive development occur within the context of the changing refractive state in the years after birth [33]. Circadian responses to light are thought to be mediated primarily by melanopsin-containing retinal ganglion cells, not rods or cones. Melanopsin cells are intrinsically blue-light-sensitive but also receive input from visual photoreceptors [34,35]. The scientific literature contains a large number of studies that relate circadian, neuroendocrine, and neurobehavioral responses to calibrated light exposures [35].

The light that enters through the human eye not only has the function of image formation but also influences the health and well-being of human beings by producing non-imaging effects in the long and short term (acute) [36]. From a light spectrum point of view, melanopsin in ipRGCs has a maximum of between 460 and 500 nm, while the visual system is more sensitive to mid-wavelengths of the visible spectrum at around 555 nm [37,38]. The duration, timing, spatial distribution, intensity, and power of the spectrally distributed light reaching the eyes can influence circadian rhythms and thus health [38,39].

Light does not affect children and adolescents in the same way as it does adults. For many days throughout the year, when children and adolescents come to school in the morning, they do so in total or partial darkness. The intensity of natural light entering classrooms through windows varies throughout the day and during different seasons. Students work for many hours in classrooms with lighting that does not resemble natural light in either composition or intensity. When they leave school, they usually spend very little time receiving natural light. This results in students living virtually every day with insufficient light, not receiving the proportion and amount of visible light for which we humans are genetically designed [40].

Teenagers usually go to sleep and get up several hours later than normal, and they have difficulty waking up in the morning for school. The reason could be the hormonal changes that occur at puberty. Schools do not have enough global lighting (artificial light and daylight) to stimulate their circadian system. This is more important in the months of less natural light. As teenagers spend more time indoors, they may lose the morning light necessary for circadian resetting. It would be advisable for the protection of the health of adolescents that they receive higher levels of morning light (or daylight) in schools and lower levels of night light (or daylight) at home [39].

School building spaces are the most important non-residential indoor environments, which often have, among other things, inadequate lighting [41]. The structural characteristics of building installations have a profound influence on learning. Inadequate conditions in classrooms are mentioned as factors relevant to poor student progress [4]. Ensuring good quality lighting in educational environments is complicated [42]. Student performance is better in classrooms with higher light intensity. Children can differentiate their light needs according to the task at hand [43]. In rooms where lighting is not uniform, with more luminosity over the immediate task area than the surrounding area, discomfort effects may be important [44]. There is extensive evidence of damage being caused to children's vision due to poorly lit classrooms [45].

A study was carried out with students who were fitted with a filter that did not allow the most important part of the light for the elimination of melatonin to pass through. Eleven students at a North Carolina school with unusually high levels of daylight were fitted with orange filter glasses to remove short-wavelength circadian light for five consecutive days. It was observed that the orange filter glasses allowed them to perform their activities; however, their dim light melatonin onset (DLMO) was delayed by half an hour compared to the previous week. Another study compared the behavior of students between two seasons, with significant differences in the amount of natural light received during the day. In this study of 16 students in New York, their DLMO was found to be delayed by 20 min and sleep output by 16 min in the spring compared to the winter [39].

3.4. Myopia and Light Insufficiency

Both the time spent outdoors and in sports are related to incident myopia. Researchers believe that outdoor time has the largest effect, independent of distance work activity [46] and physical activity levels [21].

In a comparison of myopia among children, it was concluded that the effects are related to educational pressure and time spent outdoors. Sports activities for hours in the summer months by both myopic and nonmyopic children can contribute to delay the excessive growth of the eye [47]. It has been observed that the progression of myopia is much faster in months with less daylight (winter) than in months with more daylight (summer) [48]. Seasonal effects (quantity of daylighting) are more powerful than more potent medicaments or optical treatments to try to reduce the progression of myopia [10].

Myopes have lower vitamin D levels than non-myopes (Donal Mutti et al., 2011). Studies to date have suggested that the relationships between factors that may lead to the onset of myopia, and vitamin D levels may be minimal [33].

A comparison of progression rates between Taiwanese school children in urban and rural areas showed that the average progression rate in urban areas was higher than in rural areas. Environmental factors such as urban development and academic level may be important factors contributing to myopic progression [13]. Se ha observado una disminución de la incidencia de la miopía en niños cuyas eddes estaban entre uno y tres años [9] después de aumentar el tiempo que estuvieron entre 1 y 2 horas diarias [49,50].

A decrease in the incidence of myopia has been observed in children whose ages were between one and three years [9] after increasing the time they were outdoors between 1 and 2 h per day [49,50].

Children living in countries like Singapore with a high prevalence of short-sightedness spend less time outdoors than children in countries like Australia with a lower prevalence of short-sightedness [27]. It has been observed that the prevalence of myopia in Chinese children aged 6–7 living in Sydney (3.3%) is significantly lower than in Singapore (29.1%). The reason for this difference in the prevalence of myopia is that children living in Sydney are much longer outdoors each week. Children in Taiwanese schools are required to perform at least 11 h of outdoor activities per week, and as a result, Taiwanese health authorities claim that shortsightedness has been reduced in all children. There is no need to receive high solar intensity for the prevention of myopia; however, longer outdoor activities with less intense sunlight intensity in corridors or under trees are preferable [12].

The approach of increasing time outdoors has been validated in school-based interventional research, with increases from 25 to 50%. The Myopia Control Programme in China specifies a period of one to two hours outdoors every day [9,50].

The spectral power distribution of the Sun is continuous, while that of the luminaires that we use are discontinuous. It has been found in experimental tests with animals that there are wavelengths of light that affect the refractive development of the eye and the growth of the axial length of the eye. Researchers say that when chicks receive intense light, whether it is daylighting or artificial light, there is a delay in the development of experimental myopia [17]. Exposure to violet light (VL, 360–400 nm wavelength) in particular, which is the shortest wavelength range of visible light, has a protective effect against myopia by suppressing the elongation of the axial length (AL) [17,51]. As VL acts on the myopia-suppressive gene EGR1, it suppresses myopia progression and is therefore an important external environmental factor in controlling myopia. It induces a significantly higher up-regulation of EGR1 in chick chorioretinal tissues than blue light under the same conditions [51]. Violet light is abundantly present in sunlight, while it is rarely detected in indoor ambient light. If the VL content of natural light at different times of the day is compared with that entering through glass into indoor environments, the absence of VL behind the glass can be seen [17]. These results suggest that violet light is important for not only the prevention of myopia progression but also the onset of myopia [51]. Excessive ultraviolet (UV) protection and the absence of VL in artificial light may be possible reasons for the global myopia pandemic [17].

A systematic review of articles on this subject revealed great heterogeneity in the results, and it can be stated that increased outdoor time is effective both in preventing the onset of myopia and in slowing down the progression of myopic refractive error; however, paradoxically, outdoor time does not slow down the progression of myopia in eyes that are already myopic [52]. Most but not all prospective studies and cross-sectional surveys find an antimyopic effect with increasing outdoor time [26].

There have been research studies, although not statistically significant, in which red lighting illumination is provided to people during the afternoon (i.e., 15:00). People said they perceive a significant reduction in the feeling of drowsiness and a greater subjective sense of vitality. The use of luminaires that combine white LED with red light could give a sufficient stimulus to be able to work during the afternoon [53].

According to Mardaljevic's research, 24 h a day can be divided into three phases in relation to the dark–light cycle. The phase 6:00–10:00 a.m. is defined as circadian resetting, 10:00–18:00 is the period of alerting effects of daylight, and 18:00–6:00 the time of bright light avoidance and dim light only. The most important thing to get a circadian reset is to receive intense light during the morning. It is desirable that people are exposed to bright light during 10:00–18:00 for its potential to increase alertness [54,55]. It should be noted that there are important differences in the intensity in each part of the light spectrum of the day that we receive depending on altitude, latitude, and the season of the year. The color of the furniture is also a factor to be taken into account. The combination of interior lighting and daylight at each workplace will depend on the position of the worker in relation to the windows [41,56].

Although light reaching the retina may be the key, it is difficult to measure the W/cm^2 of each wavelength band entering through the pupil and reaching the retina [33]. Objective studies on the association between light exposure and myopia, not time spent outdoors, are needed to assess how much bright light is necessary for myopia prevention [57].

3.5. The Non-Visual Effects of Light on Humans

Light influences the hormonal secretion in our behavior and the regulation of circadian cycles [37]. The suprachiasmatic nucleus (SCN) is responsible for generating circadian rhythms that receive inputs from the intrinsically photosensitive retinal ganglion cells (ipRGCs) through the activation of the photopigment melanopsin. Dopamine (DA) is a neurotransmitter that has the ability to modulate the refractive development of the human eye depending on its concentration, which varies depending on the amount of light received, especially in blue range. DA has a circadian internal retinal clock regulating function [26].

Diurnal rhythms in ocular dimensions, rhythms in retinal signaling, and molecular biology are linked to refractive development. DA acts as an inhibitor of axial elongation [9]. Retinal dopamine (DA) messages affect many intraretinal processes of the light–dark cycle, including the entire light–dark adaptive state of the retina. The possible link of circadian rhythms in the control of ocular refraction could explain the effects of outdoor and light exposures on refractive development. Researchers believe that the retinal endogenous clock could have a connection with the light entering our visual system and the retinal Zeitgebers, and therefore it could influence the development of the eyeball and the possibility of producing ametrophies if the appropriate amount of ambient light is not received at every moment of the day. This connection could give a biological explanation of the apparent positive action to curb the nearsightedness of daylight exposure [25].

Dopamine may play an important role in the refractive development of the human eye [58]. D2 (D2R and D4R) receptors are considered the most important myopia-related receptors within the two groups of dopamine receptors in the retina (D1 and D2) [17].

A significant number of publications state that exposure to sunlight and being outdoors stimulate the release of dopamine in the retina [17,27,58]. The release of dopamine is increased by receiving daylight or indoor lighting and decline without light [25,58]. Dopamine levels in the retina increase if we are exposed to intense or bright light. This could be an option to control the progress of myopia [11].

The dopamine metabolite dihydroxyphenylacetic acid (DOPAC) is generally accepted as an indicator of dopamine release. Midday retinal levels of this metabolite are 30% higher in our "elevated" light-level condition (15,000 lx·7.75 h per day) [33].

Chronodisruption is a common problem in modern societies because of our habits of life. The most developed societies have the highest prevalence of child myopia. The study of chronodisrupion could improve the knowledge of the mechanisms that produce myopia and open new fields of research and treatment to reduce the growth of the childhood myopia pandemic [25].

Since the discovery about 20 years ago of the non-visual methods of light absorption, it has been known that apart from the image-forming effects (IF) of light from which the criteria for correct lighting were developed, non-image-forming effects (NIF) of light exist. The discovery of NIF has enhanced researchers' belief in the importance of daylighting [59] and has raised new criteria to be taken into account for proper interior lighting [60]. Due to all of the factors mentioned above, the parameters to be met by a luminaire and its environment for the proper lighting of the workstation have to be modified and expanded [46].

3.6. Regulatory Lighting Framework

The technical characteristics of the luminaires and the lighting conditions at the workplaces where they are used are clearly defined in international standards [61,62]. The standards ISO 8995:2002 "Lighting in Workplaces" (2201/220) [61] and prEN 12464-1:2019 "Lighting in Workplaces" [62] are in the process of being reviewed. International regulations are beginning to take account of the NIF effects of light [62].

The recommended illuminances for classrooms are between 300, 500, and 750 lx, with recommendations for blackboards to be maintained at an illuminance of 500 lx in order to avoid specular reflections. Classrooms and tutorial rooms should be maintained at an illuminance of 300 lx.

Lighting requirements are determined by the satisfaction of three basic human needs: visual comfort, where workers have a sense of well-being, and indirectly also contributing to a high level of productivity; visual performance, where workers are able to perform their visual tasks, even in difficult circumstances and for longer periods; and safety [62].

The Illuminating Engineering Society of North America (IESNA) [63] has written indoor lighting proposals, in which the intensities vary depending on age, which is not found in the ISO and European standards.

Table 1 shows the recommendations for target illuminances (lx) based on the visual ages of the observers (years) for certain indoor education situations.

Table 1. Indoor lighting at workplaces. Lx per group of ages recommended.

<25 Years Old	25 to 65 Years Old	>65 Years Old
200 lx	400 lx	800 lx
250 lx	500 lx	1000 lx
375 lx	750 lx	1500 lx

The illumination must be controllable. Classrooms used for evening classes and adult education should be maintained at an illuminance of 500 lx [63].

The most important parameters to be taken into account for the manufacture and sale of luminaires can be taken from the CIE Technical Reports and the Commission's Delegate Regulation (EU) N° 874/2012 from 12 July 2012 [64].

The photometric and chromatic tests outlined in EN 13032-4 [65] are some of the most important parameters to be taken into account, including the light emission or luminous flx emission, luminaire opening angle, correlated color temperature (CCT), color rendering index (CRI), and risk of blue light (according to EN 62471) [66]. Other factors are the energy consumption indicators and light source service life parameters. In a study carried out on LED luminaires purchased by consumer inspectors in commercial establishments in the Community of Castilla la Mancha in Spain, it was found that they meet the requirements set out by international standards. All of the luminaires analyzed were found to be in risk group 0 (no risk) [67,68].

3.7. New Indoor LED Lighting Luminaires

The differences between indoor and outdoor ambient lighting, such as the intensity and wavelength of modern electronic lighting equipment, may be a way to control myopia as an environmental factor [17].

Scientists around the world are proposing new luminaires, the spectral power distribution (SPD) of which is intended to resemble that of the sun in order to match our circadian cycles [69].

Several prototypes developed by researchers who claim that these different combinations of LED light improve performance are shown below and that they could serve to not alter our circadian rhythm.

Remote-controllable light source grouping, together with a quadruple-chip light-emitting diode driven by pulse width modulation currents to provide healthy lighting: RGB tricolor LEDs can emit pure white light. The color rendering index is of worse quality when the CCT is changed. The combination of white LED with the red–green–blue–white (RGBW). The authors of this research believe that the best solution for good interior lighting are RGBW LED (Figure 2) [70].

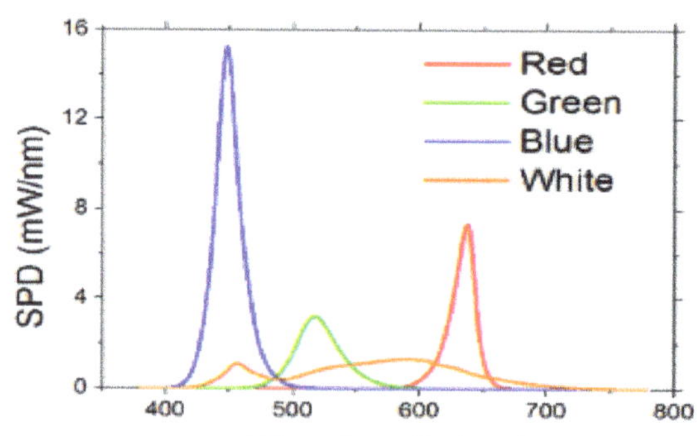

Figure 2. Parameters of the RGBW LED.

Other researchers think that the best option is the tunable LED lighting with five channels of RGCWW: red, green, cyan, warm white, and cool white (RGCWW). LEDs are individually controlled in this prototype. The researchers claim that, with this new source of indoor lighting, they managed to obtain dynamic daylight spectrum and complete mixing monochromatic thanks to the width and shape of the LED spectra, the green gap, and the computation capability. They can be very useful to illuminate interiors where not enough daylight comes in (Figure 3) [71].

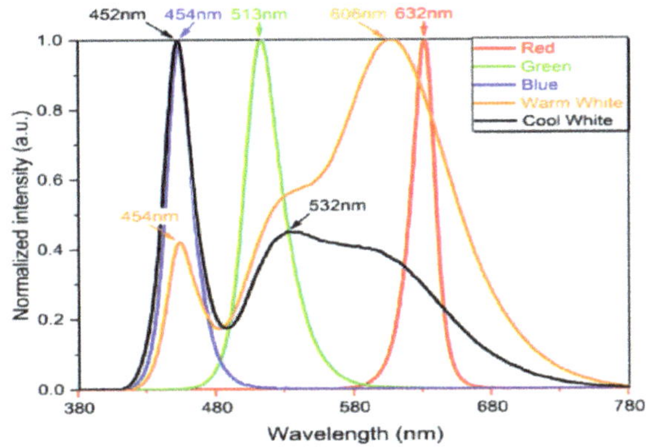

Figure 3. Tunable LED lighting with five channels of RGCWW.

Another approach involves spectral energy distribution with qualities close to daylight with a high CRI, with a continuous and balanced spectrum in the visible emission range, ideally harmonized with human sensitivity. The high blue energy part of the total blue range can be reduced to minimize the ups and downs of the spectral energy distribution. The spectral intensities in the visible range of three developed LED combinations are shown below (Figure 4) [72].

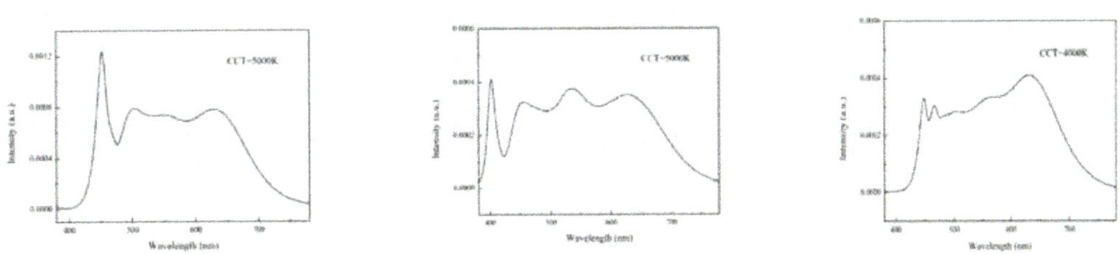

(**a**) The emission spectrum of 5000 K LED package under blue LED excitation.

(**b**) The emission spectrum of 5000 K LED package under violet LED excitation.

(**c**) The emission spectrum of 4000 K LED package under dual-wavelength blue LED excitation.

Figure 4. Intensity and spectral distribution of different combinations of LED visible light emitters [72].

The CIE, in its Technical Report CIE 015:2018, showed different spectral distributions. The next figure shows the relative SPD of each of the nine LED illuminants proposed in CIE Publication 015:2018, indicating their corresponding CCTs and distances to the Planckian locus in the UV space [73].

Figure 5 shows the SPD Standard D65 and D50, together with their corresponding indoor illuminants, ID65 and ID50. Also shown in the figure is the spectral transmittance of the average glass used to define the two indoor CIE illuminants [74].

Researchers have tried to reproduce daylight by creating light sources with five different types of light emitting diodes: red, green, blue LEDs; warm white light emitting LEDs and cold white light emitting LEDs (RGBWW) [71,75–77]. The circadian action factor (CAF), circadian stimulus (CS), or equivalent melanopic lx (EML) are some of the NIF parameters that researchers take into account for the design of new interior light sources [40,75,76,78].

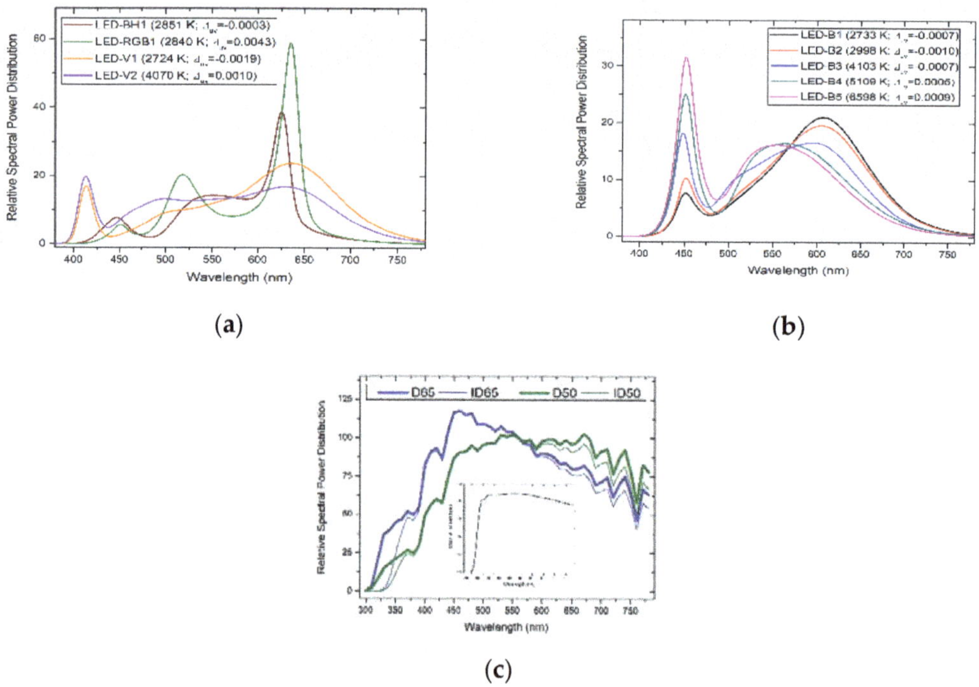

(a) (b) (c)

Figure 5. LED illuminants spectral power distribution of (**a**) the SPD, (**b**) the Standards D65 and (**c**) D50 [79].

If we compare daylight with new LED light sources, new indoor luminaires cannot produce an exact correlation with NIF effects of the light and the circadian cycle because of the differences between people and multiple environmental factors that cannot be controlled [68,77,79].

4. Discussion and Conclusions

This paper has shown the possible causes that the scientific world considers to be responsible for the current myopia pandemic. The authors of this article have found an extensive bibliography for each section presented in this review. It is not possible to mention them all, and the authors have tried to refer to the articles of the first quartiles.

There is increasing evidence implicating daylight and circadian cycles in the process of the emmetropization of the eye. The possible influence of light on myopia progression has been taken into consideration by the developed countries in Asia most affected by this pandemic [9].

These measures seem to us to be appropriate and novel in the face of this important child health problem; however, even so, we believe that they may not be sufficient due to the periods of the day when they are inevitably in enclosed spaces, both because they are in class or studying at home due to educational pressure and because of the use at close distances of electronic devices of all kinds in their leisure time. These activities are generally carried out in poorly lit environments, and therefore we believe that it is in this area that solutions should be sought.

Some authors claim that in addition to increasing outdoor time, the use of small doses of atropine, the use of rigid contact lenses to flatten corneal surfaces at night, and the use of prescription glasses and special contact lenses can reduce the progression of the myopia by 50.0% or more [2,9]. Our aim is to propose possible solutions that make their use and that of optometric correctors unnecessary.

For more than a decade, studies have proposed the importance of the blue part of light as a regulator of DA synthesis in the retina as a key factor in the appearance or otherwise of myopia [11,26,35,64], while others have claimed that it is found in the VL part of the spectrum [17,51]. The blue part centered at 490 nm is key in the regulation of circadian cycles; however, as outlined in recent CIE publications, the five photoreceptors present in the retina influence this regulation in a way that is still largely unknown [37].

Numerous studies by scientists studying IF and NIF effects of the light absorbed through our eyes have found an important role of each part of the visible light spectrum [33]. Scientific experts have recommended the amount and composition of light we are to receive in each part of the day [80]. Photoreceptors of the eye absorb light between the wavelengths 380 to 780 nm [81].

We propose that the education authorities consider that children should spend at least one hour a day in the open-air mornings performing sports or educational activities, especially during autumn and winter.

By simple evolutionary logic, the entire range of the electromagnetic spectrum to which our retinal photoreceptors are sensitive must have a positive function, whether known to researchers or not. In order to try to help slow down the current myopia pandemic, we believe that we should neither increase the intensity of indoor light sources nor increase the blue part of the spectrum in a massive and uncontrolled way, but rather try to replicate the spectral distribution and intensity of each part of the livable solar spectrum in the least imperfect way possible, and to analyze, in each interior space, the lack of intensity and spectral distribution, ranging from purple to the limit of red to infrared, from the natural light that can enter through windows (daylighting) and complement it with new combined LED light sources for interiors. It would be interesting to study how this can be replicated from 380 to 780 nm continuously with existing LED light sources, as well as to consider replacing window panes with materials that allow light to pass through from 380 nm. According to the progress of LED technology and studies of the emission of existing LED luminaires, we believe that it is possible to develop luminaires with a regulated irradiance so that their emission is higher in the blue range during the morning, in compliance with current international regulations.

The use of luminaires that combine white LED with red light could justify reducing the amount of blue light in the luminaires

On the basis of the research carried out for this study together with that of other publications currently under review, the research group of the degree in Architecture of the University San Pablo CEU is developing a utility model that will have an emission in the entire range of the visible in which the irradiance of blue and VL can be regulated for at least three different levels.

There are an enormous amount of variables that influence the light that enters through our eyes at every moment of the day, depending on such a large number of factors that it is not possible to manufacture a luminaire that can replace daylight [82], but if new LED indoor lighting is not developed with this range of visible light to see the biological responses, we will not be able to see if they can be a part of the solution for the pandemic of myopia.

Author Contributions: Conceptualization, D.B.M.; methodology, D.B.M.; formal analysis, D.B.M.; investigation, D.B.M.; resources, D.B.M.; data curation, D.B.M. and R.A.G.-L.; writing—original draft preparation, D.B.M. and R.A.G.-L.; writing—review and editing, D.B.M. and R.A.G.-L.; visualization, D.B.M. and R.A.G.-L.; supervision, R.A.G.-L.; project administration, D.B.M.; funding acquisition, D.B.M. and R.A.G.-L. All authors have read and agreed to the published version of the manuscript.

Funding: The authors wish to thank CEU San Pablo University Foundation for the funds dedicated to the Project Ref. USP CEU-CP20V12 provided by CEU San Pablo University.

Conflicts of Interest: The authors declare no conflict of interest.

References

1. Mutti, D.O.; Sinnott, L.T.; Mitchell, G.L.; Jordan, L.A.; Friedman, N.E.; Frane, S.L.; Lin, W.K. Ocular Component Development during Infancy and Early Childhood. *Optom. Vis. Sci.* **2018**, *95*, 976–985. [CrossRef]
2. Walline, J.J.; Lindsley, K.B.; Vedula, S.S.; Cotter, S.A.; Mutti, D.; Ng, S.M.; Twelker, J.D. Interventions to slow progression of myopia in children. *Cochrane Database Syst. Rev.* **2020**, *1*, CD004916. [CrossRef]
3. Morgan, I.; Megaw, P. Using natural stop growth signals to prevent excessive axial elongation and ther development of Myopia. *Ann. Acad. Singap.* **2004**, *33*, 16–20.
4. Mutti, D.O.; Mitchell, G.L.; Jones, L.A.; Friedman, N.E.; Frane, S.L.; Lin, W.K.; Moeschberger, M.L.; Zadnik, K. Axial Growth and Changes in Lenticular and Corneal Power during Emmetropization in Infants. Investig. *Opthalmol. Vis. Sci.* **2005**, *46*, 3074–3080. [CrossRef] [PubMed]
5. Saw, S.-M. How blinding is pathological myopia? *Br. J. Ophthalmol.* **2006**, *90*, 525–526. [CrossRef]
6. Holden, B.A.; Fricke, T.R.; Wilson, D.A.; Jong, M.; Naidoo, K.S.; Sankaridurg, P.; Wong, T.Y.; Naduvilath, T.; Resnikoff, S. Global Prevalence of Myopia and High Myopia and Temporal Trends from 2000 through 2050. *Ophthalmology* **2016**, *123*, 1036–1042. [CrossRef] [PubMed]
7. Bourne, R.R.; Stevens, G.A.; White, R.A.; Smith, J.L.; Flaxman, S.R.; Price, H.; Jonas, J.B.; Keeffe, J.; Leasher, J.; Naidoo, K.; et al. Causes of vision loss worldwide, 1990–2010: A systematic analysis. *Lancet Glob. Health* **2013**, *1*, e339–e349. [CrossRef]
8. Upadhyay, A.; Beuerman, R.W. Biological Mechanisms of Atropine Control of Myopia. *Eye Contact Lens Sci. Clin. Pract.* **2020**, *46*, 129–135. [CrossRef] [PubMed]
9. Jan, C.; Li, L.; Keay, L.; Stafford, R.S.; Congdon, N.; Morgan, I. Prevention of myopia, China. *Bull. World Health Organ.* **2020**, *98*, 435–437. [CrossRef]
10. Morgan, I.G.; French, A.N.; Ashby, R.S.; Guo, X.; Ding, X.; He, M.; Rose, K.A. The epidemics of myopia: Aetiology and prevention. *Prog. Retin. Eye Res.* **2018**, *62*, 134–149. [CrossRef]
11. Zhou, X.; Pardue, M.T.; Iuvone, P.M.; Qu, J. Dopamine signaling and myopia development: What are the key challenges. *Prog. Retin. Eye Res.* **2017**, *61*, 60–71. [CrossRef]
12. Wu, P.-C.; Chang, L.-C.; Niu, Y.-Z.; Chen, M.-L.; Liao, L.-L.; Chen, C.-T. Myopia prevention in Taiwan. *Ann. Eye Sci.* **2018**, *3*, 12. [CrossRef]
13. Shih, Y.-F.; Chiang, T.-H.; Hsiao, C.K.; Chen, C.-J.; Hung, P.-T.; Lin, L.L.-K. Comparing myopic progression of urban and rural Taiwanese schoolchildren. *Jpn. J. Ophthalmol.* **2010**, *54*, 446–451. [CrossRef] [PubMed]
14. Saw, S.-M.; Zhang, M.-Z.; Hong, R.-Z.; Fu, Z.-F.; Pang, M.-H.; Tan, D.T.H. Near-Work Activity, Night-lights, and Myopia in the Singapore-China Study. *Arch. Ophthalmol.* **2002**, *120*, 620–627. [CrossRef] [PubMed]
15. Vitale, S.; Sperduto, R.D.; Ferris, F.L. Increased Prevalence of Myopia in the United States Between 1971–1972 and 1999–2004. *Arch. Ophthalmol.* **2009**, *127*, 1632–1639. [CrossRef]
16. Walline, J.; Smith, M. Controlling myopia progression in children and adolescents. *Adolesc. Health Med. Ther.* **2015**, *6*, 133–140. [CrossRef]
17. Jiang, X.; Kurihara, T.; Torii, H.; Tsubota, K. Progress and Control of Myopia by Light Environments. *Eye Contact Lens Sci. Clin. Pract.* **2018**, *44*, 273–278. [CrossRef]
18. Wu, P.-C.; Chen, C.-T.; Lin, K.-K.; Sun, C.-C.; Kuo, C.-N.; Huang, H.-M.; Poon, L.Y.-C.; Yang, M.-L.; Chen, C.-Y.; Huang, J.-C.; et al. Myopia Prevention and Outdoor Light Intensity in a School-Based Cluster Randomized Trial. *Ophthalmology* **2018**, *125*, 1239–1250. [CrossRef]
19. Flanagan, J.; Fricke, T.; Morjaria, P.; Yasmin, S. Myopia: A growing epidemic. *Community Eye Health* **2019**, *32*, 9. [PubMed]
20. Cao, K.; Wan, Y.; Yusufu, M.; Wang, N. Significance of Outdoor Time for Myopia Prevention: A Systematic Review and Meta-Analysis Based on Randomized Controlled Trials. *Ophthalmic Res.* **2020**, *63*, 97–105. [CrossRef]
21. Guggenheim, J.A.; Northstone, K.; McMahon, G.; Ness, A.; Deere, K.; Mattocks, C.; Pourcain, B.S.; Williams, C. Time Outdoors and Physical Activity as Predictors of Incident Myopia in Childhood: A Prospective Cohort Study. *Investig. Opthalmol. Vis. Sci.* **2012**, *53*, 2856–2865. [CrossRef]
22. Pärssinen, O. The increased prevalence of myopia in Finland. *Acta Ophthalmol.* **2011**, *90*, 497–502. [CrossRef]
23. Mutti, D.; Mitchell, G.; Moeschberger, M.L.; Jones, L.A.; Zadnik, K. Parental Myopia, Near Work, School Achievement, andChildren's Refractive Error. *Investig. Ophthalmol. Vis. Sci.* **2002**, *43*, 3633–3640.
24. Jones, L.A.; Sinnott, L.T.; Mutti, D.O.; Mitchell, G.L.; Moeschberger, M.L.; Zadnik, K. Parental History of Myopia, Sports and Outdoor Activities, and Future Myopia. *Investig. Opthalmol. Vis. Sci.* **2007**, *48*, 3524–3532. [CrossRef]
25. Stone, R.A.; McGlinn, A.M.; Chakraborty, R.; Lee, D.C.; Yang, V.; Elmasri, A.; Landis, E.; Shaffer, J.; Iuvone, P.M.; Zheng, X.; et al. Altered ocular parameters from circadian clock gene disruptions. *PLoS ONE* **2019**, *14*, e0217111. [CrossRef] [PubMed]
26. Chakraborty, R.; Ostrin, L.A.; Nickla, D.L.; Iuvone, P.M.; Pardue, M.T.; Stone, R.A. Circadian rhythms, refractive development, and myopia. *Ophthalmic Physiol. Opt.* **2018**, *38*, 217–245. [CrossRef] [PubMed]
27. Lanca, C.; Teo, A.; Vivagandan, A.; Htoon, H.M.; Najjar, R.; Spiegel, D.P.; Pu, S.-H.; Saw, S.-M. The Effects of Different Outdoor Environments, Sunglasses and Hats on Light Levels: Implications for Myopia Prevention. *Transl. Vis. Sci. Technol.* **2019**, *8*, 7. [CrossRef] [PubMed]
28. Huang, H.-M.; Chang, D.S.-T.; Wu, P.-C. The Association between Near Work Activities and Myopia in Children—A Systematic Review and Meta-Analysis. *PLoS ONE* **2015**, *10*, e0140419. [CrossRef]

29. Mountjoy, E.; Davies, N.; Plotnikov, D.; Smith, G.D.; Rodriguez, S.; Williams, C.; Guggenheim, J.; Atan, D. Education and myopia: Assessing the direction of causality by mendelian randomisation. *BMJ* **2018**, *361*, k2022. [CrossRef]
30. Lin, Z.; Gao, T.Y.; Vasudevan, B.; Ciuffreda, K.J.; Liang, Y.B.; Jhanji, V.; Fan, S.J.; Han, W.; Wang, N.L. Near work, outdoor activity, and myopia in children in rural China: The Handan offspring myopia study. *BMC Ophthalmol.* **2017**, *17*, 1–8. [CrossRef]
31. Koh, V.; Yang, A.; Saw, S.M.; Chan, Y.H.; Lin, S.T.; Tan, M.M.H.; Tey, F.; Nah, G.; Ikram, M.K. Differences in Prevalence of Refractive Errors in Young Asian Males in Singapore between 1996–1997 and 2009–2010. *Ophthalmic Epidemiol.* **2014**, *21*, 247–255. [CrossRef] [PubMed]
32. Kim, E.C.; Morgan, I.G.; Kakizaki, H.; Kang, S.; Jee, D. Prevalence and Risk Factors for Refractive Errors: Korean National Health and Nutrition Examination Survey 2008–2011. *PLoS ONE* **2013**, *8*, e80361. [CrossRef] [PubMed]
33. Norton, T.T.; Siegwart, J.T. Light levels, refractive development, and myopia—A speculative review. *Exp. Eye Res.* **2013**, *114*, 48–57. [CrossRef]
34. Gooley, J.J.; Rajaratnam, S.; Brainard, G.C.; Kronauer, R.E.; Czeisler, C.A.; Lockley, S.W. Spectral Responses of the Human Circadian System Depend on the Irradiance and Duration of Exposure to Light. *Sci. Transl. Med.* **2010**, *2*, 31ra33. [CrossRef]
35. Lucas, R.J.; Peirson, S.N.; Berson, D.M.; Brown, T.M.; Cooper, H.M.; Czeisler, C.A.; Figueiro, M.G.; Gamlin, P.D.; Lockley, S.W.; O'Hagan, J.B.; et al. Measuring and using light in the melanopsin age. *Trends Neurosci.* **2014**, *37*, 1–9. [CrossRef]
36. Khademagha, P.; Aries, M.; Rosemann, A.; van Loenen, E. Implementing non-image-forming effects of light in the built environment: A review on what we need. *Build. Environ.* **2016**, *108*, 263–272. [CrossRef]
37. Stevens, R.G. Light-at-night, circadian disruption and breast cancer: Assessment of existing evidence. *Int. J. Epidemiol.* **2009**, *38*, 963–970. [CrossRef] [PubMed]
38. Bellia, L.; Pedace, A.; Barbato, G. Lighting in educational environments: An example of a complete analysis of the effects of daylight and electric light on occupants. *Build. Environ.* **2013**, *68*, 50–65. [CrossRef]
39. Figueiro, M.G. An Overview of the Effects of Light on Human Circadian Rhythms: Implications for New Light Sources and Lighting Systems Design. *J. Light Vis. Environ.* **2013**, *37*, 51–61. [CrossRef]
40. Moyano, D.B.; Lezcano, R.A.G. The importance of light in our lives. towards new lighting in schools. In *Advancements in Sustainable Architecture and Energy Efficiency*; IGI Global: Hershey, PA, USA, 2021; ISBN 9781799870234. [CrossRef]
41. López-Chao, V.; Lorenzo, A.A.; Saorín, J.L.; De La Torre-Cantero, J.; Melián-Díaz, D. Classroom Indoor Environment Assessment through Architectural Analysis for the Design of Efficient Schools. *Sustainability* **2020**, *12*, 2020. [CrossRef]
42. Nocera, F.; Faro, A.L.; Costanzo, V.; Raciti, C. Daylight Performance of Classrooms in a Mediterranean School Heritage Building. *Sustainability* **2018**, *10*, 3705. [CrossRef]
43. Moreno, M.B.P.; Labarca, C.Y. Methodology for Assessing Daylighting Design Strategies in Classroom with a Climate-Based Method. *Sustainability* **2015**, *7*, 880–897. [CrossRef]
44. Winterbottom, M.; Wilkins, A. Lighting and discomfort in the classroom. *J. Environ. Psychol.* **2009**, *29*, 63–75. [CrossRef]
45. Ho, M.-C.; Chiang, C.-M.; Chou, P.-C.; Chang, K.-F.; Lee, C.-Y. Optimal sun-shading design for enhanced daylight illumination of subtropical classrooms. *Energy Build.* **2008**, *40*, 1844–1855. [CrossRef]
46. Rose, K.A.; Morgan, I.G.; Ip, J.; Kifley, A.; Huynh, S.; Smith, W.; Mitchell, P. Outdoor Activity Reduces the Prevalence of Myopia in Children. *Ophthalmology* **2008**, *115*, 1279–1285. [CrossRef]
47. Deng, L.; Gwiazda, J.; Thorn, F. Children's Refractions and Visual Activities in the School Year and Summer. *Optom. Vis. Sci.* **2010**, *87*, 406–413. [CrossRef]
48. Donovan, L.; Sankaridurg, P.; Ho, A.; Chen, X.; Lin, Z.; Thomas, V.; Smith, E.L.; Ge, J.; Holden, B. Myopia Progression in Chinese Children is Slower in Summer Than in Winter. *Optom. Vis. Sci.* **2012**, *89*, 1196–1202. [CrossRef]
49. He, M.; Xiang, F.; Zeng, Y.; Mai, J.; Chen, Q.; Zhang, J.; Smith, W.W.; Rose, K.; Morgan, I.G. Effect of Time Spent Outdoors at School on the Development of Myopia Among Children in China. *JAMA* **2015**, *314*, 1142–1148. [CrossRef]
50. Wu, P.-C.; Tsai, C.-L.; Wu, H.-L.; Yang, Y.-H.; Kuo, H.-K. Outdoor Activity during Class Recess Reduces Myopia Onset and Progression in School Children. *Ophthalmology* **2013**, *120*, 1080–1085. [CrossRef]
51. Torii, H.; Kurihara, T.; Seko, Y.; Negishi, K.; Ohnuma, K.; Inaba, T.; Kawashima, M.; Jiang, X.; Kondo, S.; Miyauchi, M.; et al. Violet Light Exposure Can Be a Preventive Strategy Against Myopia Progression. *EBioMedicine* **2017**, *15*, 210–219. [CrossRef]
52. Xiong, S.; Sankaridurg, P.; Naduvilath, T.; Zang, J.; Zou, H.; Zhu, J.; Lv, M.; He, X.; Xu, X. Time spent in outdoor activities in relation to myopia prevention and control: A meta-analysis and systematic review. *Acta Ophthalmol.* **2017**, *95*, 551–566. [CrossRef]
53. Figueiro, M.; Steverson, B.; Heerwagen, J.; Yucel, R.; Roohan, C.; Sahin, L.; Kampschroer, K.; Rea, M. Light, entrainment and alertness: A case study in offices. *Light. Res. Technol.* **2019**, *52*, 736–750. [CrossRef]
54. Konis, K. A novel circadian daylight metric for building design and evaluation. *Build. Environ.* **2017**, *113*, 22–38. [CrossRef]
55. Andersen, M.; Mardaljevic, J.; Lockley, S.W. A framework for predicting the non-visual effects of daylight—Part I: Photobiology—Based model. *Light. Res. Technol.* **2012**, *44*, 37–53. [CrossRef]
56. Moyano, D.B.; Lezcano, R.A.G. Indoor lighting worplaces. Towards new indoor lighting. *Health Well Being.* **2021**. [CrossRef]
57. Landis, E.G.; Yang, V.; Brown, D.; Pardue, M.T.; Read, S.A. Dim Light Exposure and Myopia in Children. *Investig. Opthalmol. Vis. Sci.* **2018**, *59*, 4804–4811. [CrossRef] [PubMed]
58. Dong, F.; Zhi, Z.; Pan, M.; Xie, R.; Qin, X.; Lu, R.; Mao, X.; Chen, J.-F.; Willcox, M.D.P.; Zhou, X. Inhibition of experimental myopia by a dopamine agonist: Different effectiveness between form deprivation and hyperopic defocus in guinea pigs. *Mol. Vis.* **2011**, *17*, 2824–2834.

59. Zielinska-Dabkowska, K.M.; Xavia, K. Protect our right to light. *Nat. Cell Biol.* **2019**, *568*, 451–453. [CrossRef]
60. Zielinska-Dabkowska, K.M. Make lighting healthier. *Nature* **2018**, *553*, 274–276. [CrossRef]
61. ISO 8995-1. *Lighting of Indoor Workplaces (CIE S008/2001)*; ISO: Geneva, Switzerland, 2002.
62. European Committee for Standardization. *Light and Lighting—Lighting of Work Places—Part 1: Indoor Work Places*; CEN: Brussels, Belgium, 2019; prEN 12464-1.
63. DiLaura, D.L.; Harrold, R.M.; Houser, K.; Mistrick, R.G.; Steffy, G.R. A Procedure for Determining Target Illuminances. *J. Illum. Eng. Soc.* **2011**, *7*, 145–158. [CrossRef]
64. Commission Delegated Regulation (EU) No 874/2012 of 12 July 2012 Supplementing Directive 2010/30/EU of the European Parliament and of the Council with Regard to Energy Labelling of Electrical Lamps and Luminaires Text with EEA Relevance. Available online: https://eur-lex.europa.eu/legal-content/EN/TXT/?uri=celex%3A32012R0874 (accessed on 15 April 2021).
65. European Committee for Standardization. *Light and Lighting—Measurement and Presentation of Photometric Data of Lamps and Luminaires—Part 4: LED Lamps, Modules and Luminaires*; CEN: Brussels, Belgium, 2016; EN 13032-4.
66. European Committee for Standardization. *Photobiological Safety of Lamps and Lamp Systems*; CEN: Brussels, Belgium, 2009; EN 62471.
67. Moyano, D.B.; Moyano, S.B.; López, M.G.; Aznal, A.S.; Lezcano, R.A.G. Nominal risk analysis of the blue light from LED luminaires in indoor lighting design. *Optik* **2020**, *223*, 165599. [CrossRef]
68. Moyano, D.B.; Moyano, S.B.; López, M.G.; Aznal, A.S. Photometric and colorimetric analysis of light emitting diode luminaires for interior lighting design. *Color. Res. Appl.* **2021**, *46*, 791–807. [CrossRef]
69. Nie, J.; Zhou, T.; Chen, Z.; Dang, W.; Jiao, F.; Zhan, J.; Chen, Y.; Chen, Y.; Pan, Z.; Kang, X.; et al. Investigation on entraining and enhancing human circadian rhythm in closed environments using daylight-like LED mixed lighting. *Sci. Total. Environ.* **2020**, *732*, 139334. [CrossRef]
70. Wu, T.; Lin, Y.; Zhu, H.; Guo, Z.; Zheng, L.; Lu, Y.; Shih, T.-M.; Chen, Z. Multi-function indoor light sources based on light-emitting diodes–a solution for healthy lighting. *Opt. Express* **2016**, *24*, 24401–24412. [CrossRef] [PubMed]
71. Nie, J.; Wang, Q.; Dang, W.; Dong, W.; Zhou, S.; Yu, X.; Zhang, G.; Shen, B.; Chen, Z.; Jiao, F.; et al. Tunable LED Lighting with Five Channels of RGCWW for High Circadian and Visual Performances. *IEEE Photon. J.* **2019**, *11*, 1–12. [CrossRef]
72. Liu, J.G.; Tang, W.; Shen, C.; Ke, Y.; Sun, G. Advances in higher color quality and healthier white LEDs. In Proceedings of the 14th China International Forum on Solid State Lighting: International Forum on Wide Bandgap Semiconductors, Beijing, China, 1–3 November 2017; pp. 13–16.
73. Carter, E.C.; Schanda, J.D.; Hirschler, R.; Jost, S.; Luo, M.R.; Melgosa, M.; Ohno, Y.; Pointer, M.R.; Rich, D.C.; Viénot, F.; et al. *Colorimetry*, 4th ed.; CIE Central Bureau: Vienna, Austria, 2018; ISBN 9783902842138.
74. Manuel Melgosa Latorre. Revisión de algunos conceptos de ciencias del color relacionados con la luminación. *Luces CEI* **2020**, *70*, 16–21.
75. Dai, Q.; Huang, Y.; Hao, L.; Lin, Y.; Chen, K. Spatial and spectral illumination design for energy-efficient circadian lighting. *Build. Environ.* **2018**, *146*, 216–225. [CrossRef]
76. Oh, J.H.; Yang, S.J.; Do, Y.R. Healthy, natural, efficient and tunable lighting: Four-package white LEDs for optimizing the circadian effect, color quality and vision performance. *Light. Sci. Appl.* **2014**, *3*, e141. [CrossRef]
77. Cajochen, C.; Freyburger, M.; Basishvili, T.; Garbazza, C.; Rudzik, F.; Renz, C.; Kobayashi, K.; Shirakawa, Y.; Stefani, O.; Weibel, J. Effect of daylight LED on visual comfort, melatonin, mood, waking performance and sleep. *Light. Res. Technol.* **2018**, *51*, 1044–1062. [CrossRef]
78. Wende, B.; Fischer, J. 2.1 Definition and measurement of radiometric quantities. In *Subvolume A*; Springer Science and Business Media LLC: Berlin/Heidelberg, Germany, 2005; pp. 45–51.
79. Rea, M.; Figueiro, M. Light as a circadian stimulus for architectural lighting. *Light. Res. Technol.* **2018**, *50*, 497–510. [CrossRef]
80. Brown, T.; Brainard, G.; Cajochen, C.; Czeisler, C.; Hanifin, J.; Lockley, S.; Lucas, R.; Munch, M.; O'Hagan, J.; Peirson, S.; et al. Recommendations for Healthy Daytime, Evening, and Night-Time Indoor Light Exposure. *Preprints* **2020**. [CrossRef]
81. CIE. *Commission Internationale de l'Eclairage Proceedings*; Cambridge University Press: Cambridge, UK, 1931.
82. Münch, M.; Wirz-Justice, A.; Brown, S.A.; Kantermann, T.; Martiny, K.; Stefani, O.; Vetter, C.; Wright, J.K.P.; Wulff, K.; Skene, D.J. The Role of Daylight for Humans: Gaps in Current Knowledge. *Clocks Sleep* **2020**, *2*, 61–85. [CrossRef] [PubMed]

energies

Article

The Spirit of Time—The Art of Self-Renovation to Improve Indoor Environment in Cultural Heritage Buildings

Coline Senior [1,*], Alenka Temeljotov Salaj [1], Milena Vukmirovic [2], Mina Jowkar [1] and Živa Kristl [3]

[1] Department of Civil and Environmental Engineering, Norwegian University of Science and Technology (NTNU), 7491 Trondheim, Norway; alenka.temeljotov-salaj@ntnu.no (A.T.S.); mina.jowkar@ntnu.no (M.J.)
[2] Department of Landscape Architecture and Horticulture, University of Belgrade, 11000 Belgrade, Serbia; milena.vukmirovic@sfb.bg.ac.rs
[3] European Faculty of Law, New University, 5000 Nova Gorica, Slovenia; ziva.kristl@epf.nova-uni.si
* Correspondence: coline.senior@ntnu.no

Abstract: The purpose of this paper is to explore the challenges of an old low-standard urban district with a strong historical and cultural heritage and propose more sustainable renovation solutions, acceptable for the residents and municipality. The challenges of physical renovation or refurbishment are complex due to poor condition of the buildings, municipal ownership and governance, mixed management with community and low rents, which are insufficient to cover the costs. The paper discusses the proposed solutions of living standards, supported by the research in two directions: (i) available resources and reuse of materials, (ii) developing a renovation guidance for inhabitants from the building physics perspective, including indoor environment quality. Challenges related to energy efficiency are addressed from the decision-making perspective to overcome the barrier of lack of motivation to invest in energy-efficient measures at the individual and community level. The interdisciplinary approach complements engineering-focused studies with a focus on the comfort conditions and the influence of occupant habits in sustainable buildings. The methods used were literature review, case studies with observations and survey, looking to cover all technical, social, and historical aspects of sustainable renovation of cultural heritage buildings with the same level of importance. Results show that to keep a sustainable, low-cost urban living model, instructions for self-renovation are a valuable guidance for non-professional actors to make more sustainable choices. In conclusion, we can emphasize that inhabitants are accustomed to lower living standards, so the project is aimed to present the proper solutions for improvement as a balance between new sustainable technical solutions, personal self-renovation skills, habits, and health.

Keywords: self-renovation; habits and comfort; sustainable building material; cultural heritage buildings

Citation: Senior, C.; Salaj, A.T.; Vukmirovic, M.; Jowkar, M.; Kristl, Ž. The Spirit of Time—The Art of Self-Renovation to Improve Indoor Environment in Cultural Heritage Buildings. *Energies* **2021**, *14*, 4056. https://doi.org/10.3390/en14134056

Academic Editors: Boris Igor Palella and Roberto Alonso González Lezcano

Received: 28 May 2021
Accepted: 28 June 2021
Published: 5 July 2021

1. Introduction

An increasing body of scientific literature has put the focus on energy efficiency measures for cultural heritage buildings, but this approach tends to undermine cultural values which are often described only as constrains [1]. In the extensive systematic literature review on sustainable refurbishment of historical buildings, Loli and Bertolin (2018) [2] pointed out the "Scandinavian paradox", which often shows the Scandinavian countries as frontrunners when it comes to implementing and operationalizing sustainability goals [3], but on the other hand, the scientific production on methods for sustainable maintenance and renovation of cultural heritage buildings in Scandinavia is relatively scarce [2]. Norway is quite advanced in terms of energy-efficiency renovation with a yearly rate of 2.5% of the existing building stock compared to the 13 other EU countries where data are available, which have rates between 0.5% and 2.0% [2]. However when it comes to "deep renovation" defined as interventions that fundamentally affect the buildings performance, the European rate is stagnating at 0.2% [4,5].

The renovation of cultural heritage buildings can produce positive effects for the socio-economic regeneration of the cities [6] and boost the application of contemporary living models, sustainable management strategies, and maintenance procedures, which can balance the up-to-date requirements of energy efficiency, human comfort, and operating cost reduction [7] ensuring also energy and economic benefits [6,8].

In this regard, different studies indicated that the most critical factors for deep renovation of existing housing concern the non-technical barrier related to social, logistic, legislative, and financial constraints [9], while behavior and local cultural factors, in particular, can reduce the efficiency of the renovation initiatives, impacting on energy use, conservation state, management, and maintenance operations [6,9,10]. The importance of cultural heritage for the community and the specific character it has built over time are two sensitive elements that need to be combined with the contemporary expectations and responsible and sustainable lifestyle.

For policy makers, supporting self-renovation projects of residential buildings can contribute to lessen social exclusion and spatial segregation and encourage tenants to feel a sense of responsibility for their dwellings [11]. However, the lack of technical and legal knowledge among residents combined with the high costs are still major barriers in the implementation of energy efficient renovation measures [12,13]. In addition, some of the researchers emphasize the lack of motivation as a barrier for more sustainable renovation solutions [14–16], and proposing the focus on social sustainability aspects [17], more concretely: 'factors affecting participation' [18,19], 'relationship between participants' [20–22], 'engagement strategy' [23–25], and 'influence of participation' [26–28]. As stated by Esmaeilpoorarabi et al. (2020) [29], it is important to mobilize resources, improve relationships, promote cooperation, and ultimately achieve community engagement and trust [30,31]

The art of self-renovation in low-income urban communities is quite unique in the sense that it transcends a purely technical approach and often constitutes an opportunity for them to feel more ownership of their homes, promote social inclusion, and strengthen community bonds [32]. Financing and education of workforce are also central challenges that need to be addressed [12]. The social aspects are widely underrepresented in energy research, and although there is no question that technological engineering studies are of utmost importance to improve technical performances of buildings, due to the fact that energy-efficiency relies on human actions and informed choices, more interdisciplinary research is needed to include a softer approach and address social barriers and drivers to a successful energy transition [33–35]

1.1. National Regulations for Conservation of Cultural Heritage Buildings

The purpose of building protection is to preserve our sources of cultural heritage and history. Building protection became a formal matter in Norway in 1913 when the first national antiquarian was hired. Since then, it has been the Directorate for cultural heritage national antiquarian that has been responsible for cultural heritage policy [36]. In Norway, there is a broad political will to preserve cultural heritage buildings [37]. It points out further that the preservation of cultural heritage can contribute with knowledge and input to sustainable resource management by providing insight into how environmental problems have affected our patrimony and give a better understanding of how they can be solved. About listed buildings, it says: *"From a long-term socio-economic perspective, preserving valuable parts of the building stock rather than demolishing and rebuilding can be significantly more profitable"* [37].

The Norwegian Planning- and Building Act § 31-1 states that when doing renovation or rehabilitation, the municipality shall ensure that historical, architectural, or other cultural value associated with a building is preserved as far as possible. Nevertheless, there is a challenge associated with the maintenance and preservation of the protected building stock due to the ever-increasing demands for lower greenhouse gas emissions [38].

The existing building stock in Norway accounts for around 40% of the total energy consumption in the country, 22% of which is for residential and 18% for non-residential buildings [39]. In addition, most buildings built before 1950 were without insulation, which makes both protection and improvement of this building stock challenging. Fulfilling the requirements related to energy savings could mean major structural interventions and may lead to changes that affect cultural historical values of the building [36].

1.2. Improving Energy Efficiency in Existing Buildings

According to Ugarte et al. (2016) two of the main problems that low-standard households need to face as a consequence of low energy efficiency are extreme indoor temperatures and high humidity levels. Rehabilitation could address these problems by taking measures with three main outcomes: improvements in indoor air quality, humidity levels, and indoor temperature [40]. For this matter, retrofit programs may focus on a weatherization based on changes in the shell of the building, such as insulation of facade wall, ceiling and basement, facade painting, or replacement of windows and doors [41,42].

Given the environmentally friendly nature of wood, its low maintenance cost, and broad accessibility, wood is one of the most renewable materials in construction [43]. However, when it is used in building facades, it needs careful treatment and maintenance to minimize thermal and humidity discomfort sources in buildings. Energy wastage in the facade tends to be derived from humidity, since it provokes expansions and contractions of the material and mold growth, causing degradation of timber. From an environmental point of view, one of the best ways to protect wood is with linseed oil [44]. It saturates the external layer of the wood, so that it is impossible for water to penetrate the structure. Moreover, it allows water kept inside to evaporate [45]. It is important to plan the functioning of the building envelope for the future degraded climate conditions; in northern Europe, climate adaptation has to take into account moisture resistance (due to increased precipitation) and slight increase of temperatures [46,47].

When refurbishing the wooden structure, two main aspects need to be considered. First, if replacement of a timber member needs to be made, the new item must belong to the same species of wood and have the same quality. Secondly, it is also recommended to use the same (or similar) techniques that the original craftsmen used to convert the timber and assemble it [45].

This way, the identity of the building can be preserved. Moreover, some materials should be avoided or, at least, reduced to a minimum. Such as epoxy resins, steel reinforcements and plastic paint for coating [45].

Existing retrofit programs for low-standard buildings are based on weatherization of households; this way comfort improves and costs are reduced [48]. Weatherization is usually focused on changes in the structure, as insulation of ceilings and walls, air sealing, and duct sealing, and the electricity savings can reach up to almost half the original price [41]. Some of the actions that can be taken in order to achieve a good weatherization are [42,49]:

- Insulation of roof or upper ceiling.
- Insulation of façade wall.
- Insulation of basement ceiling.
- Painting and mending the façade.
- Replacement of doors and windows.
- Use of hybrid HVAC systems.

Regarding the refurbishment of doors and windows, they play an undoubtable role in the ventilation and weatherization of a household, but sometimes their high prices might keep the owners from fixing them. Anyway, it is a key element to address, as old doors can be leaky and have extremely poor insulation; similarly, single glaze windows are inefficient and should be switched to double-glazing or interior storm windows which provide moderate insulation and avoid air infiltration [41].

Insulation is a key element in order to achieve a good energy efficiency level. Some natural and renewable materials have been proved to have a great performance as insulating materials, presenting important advantages regarding costs, ecology, and energy savings, if compared to traditional materials.

Natural fibers, such as technical hemp, jute, and flax have very good mechanical, acoustic, and thermal insulation properties. Moreover, they can be combined in different proportions and be useful in façades, roofs, floors, partition walls, and external walls [50]. Natural fibers, such as technical hemp, jute, and flax, have very good mechanical, acoustic, and thermal insulation properties. Moreover, they can be combined in different proportions and be useful in façades, roofs, floors, partition walls, and external walls [50].

Some additional costs can derive form the insulation of the façade, since activities as scaffolding and painting cannot be overlooked and homeowners rehabilitating a building exterior need to pay for them no matter what [42].

Other non-structure related improvements that could be applied are [42]:

- Installation of a low temperature boiler.
- Installation of wood pellet or wood chip heating systems (biomass heating systems).
- Insulation of heating pipes.
- Installation of solar heating collectors.
- Installation of heat pumps.

This said, it is clear that some good weatherization methods can be decisive for inhabitants of the building, not only in economic terms but also healthwise, as inadequate heating, cooling, or ventilation of the spaces can be a life-or-death issue [51,52].

1.3. The Scope of This Study

The scope of the study, considering the transformation of historical and cultural background of Svartlamon, poor housing condition, reuse of materials, insufficient funding, and existing self-renovation model, aligns with the New Bauhaus European initiative which vouches for a softer approach to the challenges in reaching sustainability goals. The European Commission's initiative aims to bridge the world of science and technology with the world of art and culture by developing affordable and accessible living spaces, striving to improve quality of life and highlight simplicity, functionality, and circularity of materials while accounting for the need for comfort and attractiveness in citizens' daily life [53]. Building upon this initiative, this study focuses on three principles:

(i) To keep the alternative community acting as a pioneer in terms of sustainable, low-cost urban living models and retaining the strong identity of the district;
(ii) To advance the relationship between the community and the municipality by improving the co-management model facing the safety building level as a foundation for self-renovation outset;
(iii) To self-renovate the houses more in accordance with sustainable refurbishment principles.

The paper aims to address some of the challenges linked to unsupervised self-renovation work that can result in alteration of the cultural heritage buildings, negative consequences of the maintenance backlog, and unsustainable technical solutions. At the same time, this study looks into approaches that will increase the sustainable renovation rate that is currently lagging behind due to socio-cultural barriers [54]. The case study proposed in this paper focuses on "soft approach" to energy renovation. This paper deals with energy efficiency from the decision-making perspective addressing the challenge of lacking motivation to invest in energy-efficient measures. The interdisciplinary approach complements engineering-focused studies with a focus on the comfort conditions and the influence of occupant habits in sustainable buildings. It proposes a practical self-renovation guidance for residents with a focus on sustainable technical and technological solutions, focusing on indoor environment, health, and wellbeing. This is motivated by a gap identified in previous studies that revealed communication challenges between public authorities and residents in the field of energy retrofitting of cultural heritage buildings that resulted in decaying buildings rather than improving their sustainability [13]. This can be also seen as an urban experimentation (Newton and Frantzeskaki 2021) case which can be replicable for similar circumstances.

The main research question is *how to raise the self-renovation standards of cultural protected though poorly maintained area*. The aim is to explore how it is possible to refurbish the houses, keeping all the existing aspects of the community, the history, the culture for re-use of materials, and the low budget and adding new ones like thermal insulation and general improvements of the indoor environment. The analysis of the physical properties of buildings such as ventilation, lighting, thermal weaknesses, and thermal insulation will be explained, in order to propose appropriate measures to meet the growing demand for a pleasant indoor climate while safeguarding the cultural heritage values. In addition, a survey conducted among residents of a low-standard urban community is presented to understand the opportunities and challenges of sustainable self-renovation of cultural heritage buildings.

This paper is structured as such: Section 2 presents the overview of the material and methods used for the study; Section 3 shows the study results; Section 4 presents a guidance for sustainable self-renovation in accordance with the theoretical and empirical background, and Section 5 concludes with the main findings.

2. Materials and Methods

To address the research question, we investigate in three directions: (i) improving the living standards and culture of a low-income intentional community, (ii) widening the reuse of materials as a part of circular economy solution, and (iii) finding high quality LCC solutions for specific problems. A number of buildings with a focal point on observation in the intentional community of Svartlamon in Trondheim, Norway, were chosen as the case study. This was combined with multiple on-site visits and a questionnaire survey. The case of Svartlamon was selected for its uniqueness and self-organized maintenance and operation system. The economic, socio-cultural, and environmental challenges posed by its ownership structure and eventful history as well as the subversive nature of Svartlamon will be presented in the following section. The present study is a result of several projects running at the Norwegian University of Science and Technology with master's students since 2018.

The data collection and observation have been conducted in collaboration between master's students and the authors of this paper during the spring and fall semesters, from 2018 to 2020 [55,56]. Initial data about Svartlamon's historical and cultural background were collected through informal meetings with Svartlamon housing foundation (Svartlamon boligstiftetse) and desk research at Trondheim Municipal Archive Center (Trondheim Byarkiv). Observation of indoor has been organized by Svartlamon housing foundation several times per year for the students and the research team. The observations of the outdoor environment were taken and documented whenever was needed.

A questionnaire survey was designed to investigate the living conditions in Svartlamon, the self-renovation work organized and done by residents, and their perception of sustainable technologies. The questionnaires included 4 sections; the first section collected demographic information such as age, employment status, composition of the household, living period in Svartlamon, the affiliation to a sub-community, and types of facilities respondents shared with others (bathrooms, toilets, kitchen, living room, laundry room, etc.) The second section on the living conditions consisted of questions about social aspects of the community life, motivation factors for living in Svartlamon, opinions on possible changes or improvements that could be made, and their perception of the core-values of the community. The third section collected data on self-renovation habits and culture, the respondents' involvement in such work, and their will to participate more in the future. The last section of the questionnaire addressed the available technologies within the households as well as their perception of innovative sustainable technologies.

Further on, the zoning plan states regulations in regard to refurbishment at the area. According to § 3–5, in the zoning plan, it is specified that: "*Existing buildings shall be preserved except for the buildings marked as demolition objects in the planning map of the area. For existing buildings, no major reconstruction, extensions or facade changes can be made without this being presented to the Cultural Heritage Management Office in advance.*" [68]

In this study, we highlight both the technical, social, and cultural aspects with equal importance. This combined approach of sustainable renovation of cultural heritage buildings is seen as an added value to tackle the challenges in this domain.

3. Results

3.1. Case Study Description—Observation

At Svartlamon, there are a total of 25 residential buildings with 130 dwellings/tenancies. The total gross area of the residential buildings is approximately 7000 m^2. About 200 people live in the dwellings rented by svartlamoen boligstiftelse by the municipality. There are also four commercial buildings, and the gross area that the culture and industry foundation rents from the municipality constitutes ca 2500 m^2 (Table 1).

Table 1. Characteristics of Svartlamon.

Aerial picture of Svartlamon (©Blom)	area	Ca. 27 km^2
	population	Ca. 200 inhabitants
	type and number of buildings (year of construction)	25 residential buildings and 4 commercial buildings
	type and number of flats	130 dwellings
	main infrastructure	Kindergarten, culture house, shops, railways, and roads
	maintenance funding (per year)	Housing Foundation ca. 1.3 M. NOK + Municipality ca. 1.2 M NOK

About 80 people work in varying positions in Svartlamon. In addition, the kindergarten is managed by the municipality. The area is regulated as special conservation area and has about 35 antiquarian classified buildings (residential buildings and outbuildings) [69].

Svartlamon is a district located to the east of the Trondheim city center, which hosts in addition to the residential buildings, a reusable shop, free shop, cultural festival, concert venue, kindergarten and book café. Most residents share some facilities like bathrooms (toilets and showers) and kitchens. The old houses are wooden and brick buildings. Inhabitants tried to renovate some parts of the buildings since their construction in the 19th century, but the buildings are in a general bad condition now. Lack of money and complex organization between the municipality and inhabitants lead to insufficient investments in maintenance of the buildings and regeneration of the area. The buildings are owned by the municipality, and since 2001, the housing foundation has been responsible for collecting rents and ensuring the maintenance. This was formalized in the contract that binds the foundation and the municipality, by which the municipal landlord transferred all its responsibilities for maintenance and preservation to the foundation. Today, the foundation manages 151 leases divided into 35 houses with about 200 people living there. The foundation has three people working full-time, a manager, a carpenter, and an electrician. The money collected by the foundation finances the salaries of the three employees, renovations, and maintenance operations. In order to keep the rent as low as possible, residents are expected to do as much work as possible by themselves. The Housing Foundation has a permanent office and meeting place on site. In addition, in 1990, Svartlamon residents' association was founded with the purpose to preserve the houses and to defend low-income people's right to live downtown. Everyone who has a lease with the Housing Foundation is automatically a member of the Residents' Association. They have monthly meetings and govern with a flat structure and consensus. The residents of Svartlamon have a leading role in the renovation. They manage to do almost everything by themselves with low budget, reusing materials from the old houses from other parts of Trondheim and trying to keep the buildings in their original state. The focus of this study is mostly on the technical and technological aspects, but reflecting also the economic, organizational, social, cultural, and historical ones.

Svartlamon is located close to the city center of Trondheim, with a climate type dominated by the winter season, cold period with short daylight, relatively little precipitation which is mostly in the form of snow, and low humidity. It is located north of the humid continental climate. There is an average of 272.0 days of precipitation. The average temperature for the year in Trondheim is 4.8 °C. The warmest month is July with an average temperature of 13 °C. The coolest month is January, with an average temperature of −3 °C.

Wooden buildings (Figure 4) are houses with bedrooms and kitchen. Each house does not necessarily have a bathroom. The community shares few bathrooms. Most of the houses were built at the end of 19 century; they are now deteriorated even though people tried to keep them in a satisfactory condition. Most of them are two-story houses. In the area, we can notice a sloping terrain, some gardens, pebbles on the ground, and brick floor in some paths. There are no paved roads or sidewalks between houses. The site is located near the Strandveien road with cars and bus. There is some lighting next to the main road but not in the smaller paths between houses. There is also the railway right next to houses. Most of the houses do not have any basements except the storage room for the ancient milk shop.

Figure 4. Svartlamon wooden buildings (NTNU).

Foundations are in bad condition. Houses have many windows, but they are damaged, showing problems with condensation and humidity. All the facades are painted; however, the painting is very old and is deteriorating (Figure 5). Roofs are not insulated, and all houses have a steep roof. Many houses have tile roof. Clay tiles are a traditional kind of roofing material, it is a material that is both widely available and easy to shape into forming a channel to direct the flow of water. Some of the houses have a corrugated metal roof which is inexpensive and quick to install. Mainly, the rainwater gutters and downpipes are in good condition close to the roof, but the lower parts are worse (leakages), causing constant humidity on the walls, allowing moisture to seep into the houses, so molds, fungus, and mosses are spotted. They affect the health and well-being of people. The façades are complicated to insulate due cultural protection, so the design of the facades and windows should be original. Walls, doors, windows, partitions, and finishes are wooden and old.

Figure 5. Deteriorating façade (NTNU).

Many walls are not straight, and they lean (Figure 6), likely because the houses were built on poorly compacted embankments or on a heterogeneous soil. No drainage systems could be seen around them. Some of the load bearing elements are rotten. Even some entry doors are not straight. Most of the houses do not have any bathroom, and the only sink in their house is the kitchen one. The houses with bathrooms share them. The plumbing system is old and partly maintained but not to the standards that we can find in newer houses. The electrical system is at its full capacity, old but safe.

Figure 6. Leaning house (NTNU).

A typical example of housing block is presented: two buildings Strandveien 19 and 21 (Figure 7), from 1893, with four flats each, shared facilities in the common basement, and a shop in the ground floor. The foundation material is stone; basement walls are made of brick, and upper floor levels have a wooden structure. During the time only some small interventions were done, and foundations were stabilized in 2001; a wind barrier on the north façade was installed in 2016, and the damaged wooden panels were replaced (Figure 8). Both houses need major refurbishments as some foundations and walls are falling apart, and the structure is in a very bad condition, showing many cracks.

A comparison of the floor plans shows some smaller changes in the usability of the space in the floors and basement. An apartment in the basement is changed to a bathroom and storage place for inhabitants, and some internal walls were removed to create a larger common space. In one of the flats in the second floor, an extra bedroom and own bathroom was built.

An analysis of technical condition is presented, based on the observation (Figure 9). The foundations of the two buildings are connected and have the same structure, consisting of stones. They support brick walls with a thickness around 0.40 m and a height equal to the basement height, which is up to 3.00 m. There is no drainage system on the outer side of the walls. The inner walls structure from the basement on is carried by wooden columns. The wooden façade is made from stacked timber with a thickness of 7 cm.

Figure 7. Strandveien 19 and 21 (Trondheim Municipality, 2016).

Figure 8. Renovation of the façade (Trondheim Municipality, 2016).

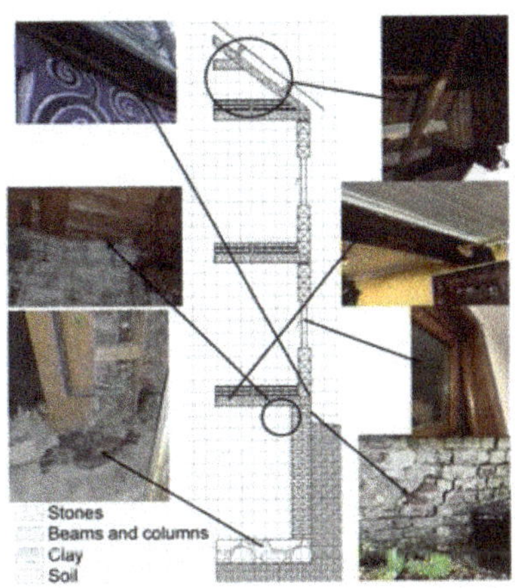

Figure 9. Illustrated section of the building in Strandveien 19 (Sekkal, 2019).

The structure of the slabs is made of primary beams resting on the columns and the brick wall. The wooden slab is made in classic form: secondary layer of beams with space of 50 cm between each, another layer for wooden panel floors, and clay or gravel in between as insulation. The frame of the buildings is classic with a wooden truss, which is supported by the columns in the walls. Above the trusses, there are purlins that support perpendicular wooden boards. The distance between the trusses is more than 6 m.

Windows are not the original ones, and they are in simple or double glazing. However, we saw that some frames are damaged mainly due to humidity.

Almost no insulation is found, and it is only in the basement, in front of the brick wall (15 cm of glass wool). Some of the residents put an internal insulation.

The walls have many cracks both inside and outside the buildings. A large crack in the cladding of the north façade between the two houses extends along the entire height of the ground floor, and the thickness is from 1 to 2 cm. From documented material, it is there from 2012 and is still the same size, without any changes (Figure 10). There are many other smaller cracks in the brick walls, missing cement binders and in the wooden walls as well. On the north façade several humidity traces could be seen (Figure 11).

Figure 10. Crack between the buildings Strandveien 19 and 21 (NTNU).

Figure 11. Humidity damage (NTNU).

The situation extended for a longer period could cause the structural and health problems due to humidity indoor (Figure 12), which could freeze during the winter, or formation of molds and fungi. In addition, the cladding contact between the brick and wooden wall is not tightened enough (Figure 11, upper left corner). There are some damages on the south wall: some pieces of bricks are falling; there are cracks between the bricks with missing binders and traces of humidity coming from the outside, i.e., from the soil (the bedrock in Svartlamon is very shallow). The analysis shows various causes of humidity in the wall: cracks outdoor, poor cladding condition, no drainage, capillary rise, bad condition of the gutters causing too much water streaming on the façade, and the wooden board above the wall of the ground floor does not play its part to 'drop' off water. The humidity traces are located at the top of the wall or on the ground, behind the cracks, in the contact area of two different cladding materials, and behind the wooden façade, panels are not tightened enough and have no vapor barrier.

The ventilation is not efficient in the buildings. The air exchange is only through the doors and windows, somewhere through leaks in the walls, and small extractors in bathrooms (Figure 13). There is no ventilation system in the kitchen.

Figure 12. Indoor wall in one of the rooms (NTNU).

Figure 13. Integrated humidity control, occupancy sensor, and downtime (NTNU).

The acoustic is problematic, and there is a lot of sound transmission. During the visit to one of the apartments, the discussion from another one was heard as well as a washing machine from the common stairs. Inhabitants made an agreement between the residents to respect a certain schedule of "making noise". The noise could be heard through the windows of the north façade, which is facing to the railway.

The electrical system is old, safe, but insufficient, so they should consider new equipment. They use mostly electrical heaters. The lighting conditions are quite poor.

The municipality conducted an evaluation of the maintenance backlog of the building stock of Svartlamon in 2016. Their findings, which align with our on-site observation (Table 2), support the evidence that the available funds for maintenance were very low compared to the required renovation work (examples given in Table 3 for Strandveien 19 and 21).

Table 2. Problems of the building stock based on site-observation.

Foundation, Basement	Facade	Walls	Floors	Windows	Basic Infrastructure
humidity, stability, cracks, missing drainage	cracks on the wall, rotten wooden panels, deteriorating paint	cracks on the walls, molds, fungus, rotten elements, humidity	noise disturbance between floors	simple glazing, humidity damages on wood frames	missing toilets, baths

Table 3. Estimated renovation costs compared to available funds (Trondheim Kommune, 2016).

	Further Improvement Costs	Available Funds
Strandveien 19	709,090 NOK	126,270 NOK
Strandveien 21	591,936 NOK	105,408 NOK

In the case study of Svartlamon, different local resources were found. The observations revealed several workshops available at no cost to the residents; photos of the workshop are shown in Figures 14 and 15. Here, the renters have both materials and tools that can make it easier to maintain their home and houses. There is also a dedicated group that manages the workshops.

Figure 14. Workshop session on window restauration (Svartlamon residents association, 2016).

Figure 15. Workshop at Svartlamon (NTNU).

The workshop voluntary group manages the workspaces (Figure 15) at Svartlamon with associated tools, materials, and a recycling stock that is available both to the residents and to the residents association.

They receive a lot of recycled materials from both outside and inside Svartlamon. From time to time, Svartlamon also receives old material and equipment from houses that the municipality is tearing down. The idea is that access to tools and materials should stimulate reuse and local refurbishment. It is stated as a cautious estimate that the voluntary group organizing the workshop produces about 250–300 h of work per year [69].

Based on their own observations and this information, the students decided to investigate solutions that could improve the living standards in the buildings in a more sustainable way with respect for the low-budget and self-renovation culture deeply anchored in the Svartlamon community.

3.2. Results of the Survey

Based on the results of the questionnaire, relevant information regarding the living conditions in Svartlamon are presented in this section. 54.2% of the respondents share bathrooms and toilets with other households. The respondents rated their relationship to their neighbors above average (on a scale from 0–6, 100% rated it above 3; with 58.3% rating it as a 6/6). The main factors that influenced their choice to move Svartlamon were Social aspect/sense of belonging (83.3%), Freedom (79.2%), Budget/economic reasons (66.7%), Having friends/family that were already living in Svartlamon (66.7%). To the open question on what they liked the most about living in Svartlamon, the social aspects came out the most, including community feeling, local democracy, shared facilities and green areas, economic freedom (non-binding leases, possibility to settle in a mobile home, no mortgage, cheap rent that allow to focus personal investments elsewhere, e.g., "pursuing their dreams"), general feeling of freedom (to set up art installations, start new activities and events, DIY projects). To the open question on whether they would like to change anything in Svartlamon, most respondents would like to have an even bigger community engagement, i.e., that more people would actively participate in the collective activities and decisions.

The interesting thing about the self-renovation culture in Svartlamon is that a majority of the respondents have experience in renovation work (Figure 16).

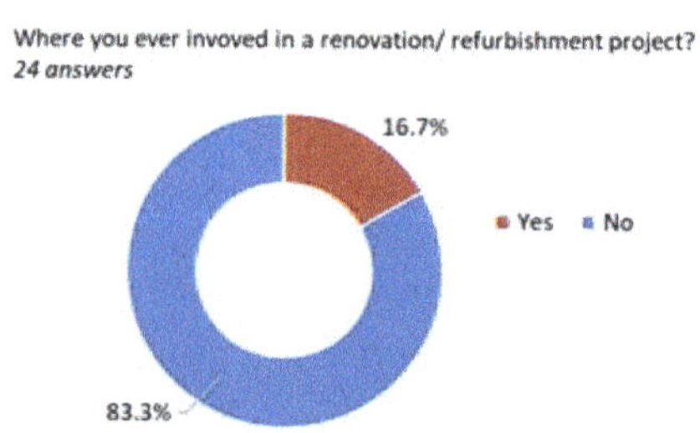

Figure 16. Participants' involvement in renovation/refurbishment projects.

Furthermore, results point towards a strong community aspect of these projects with respondents being involved not only in their own house renovation but also those of other residents (Figure 17), and a further engagement of the community at large beyond the household (Figure 18).

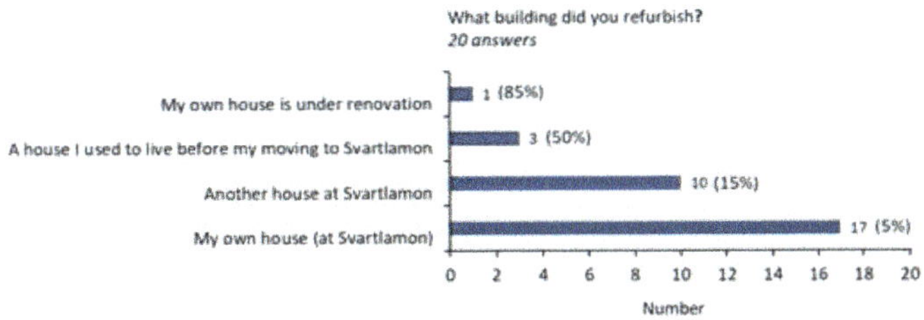

Figure 17. The type of buildings and the participants involvement in renovation plans.

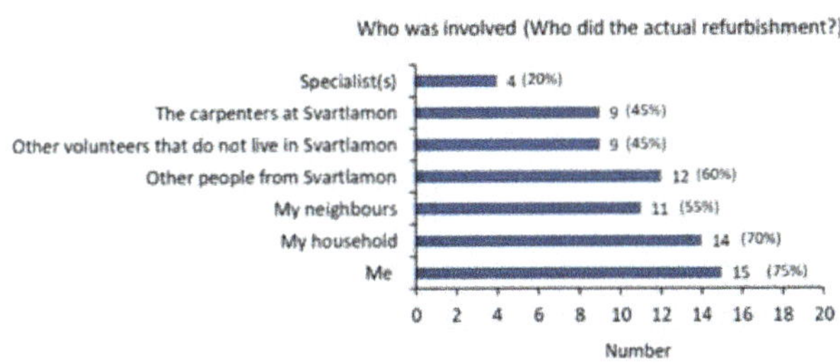

Figure 18. Involved groups in renovation plans.

However, most of them did not discuss the project with the community beforehand (Figure 19), and a vast majority expressed their will to be involved in future projects (Figure 20).

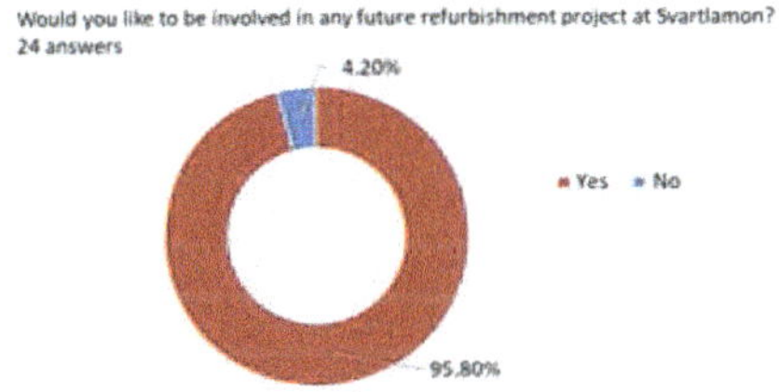

Figure 19. Participants interest in involvement in future renovation plans.

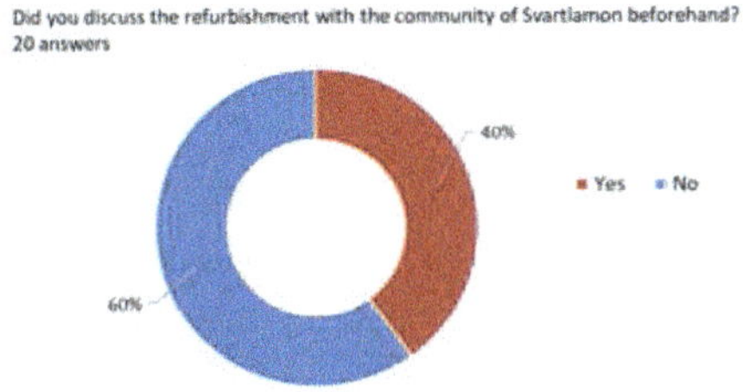

Figure 20. Respondents' engagement with community before renovations.

Regarding the technologies available to their household, results show that the most basic needs are met, while more sustainable technologies are not (Figure 21). However, a majority of respondents expressed no need for new technologies (Figure 22).

From the open question about the use for new technologies in their daily life, some respondents expressed a will to have more sustainable energy sources such as solar panels combined with thermoelectric generator, heat pumps, and bio toilets.

> *"We are saving up to buy a thermoelectric generator which will help us utilize our solar energy system better during the dark months of winter, via the wood stove. We are also saving up for a bio/compost toilet"*. (Survey respondent)

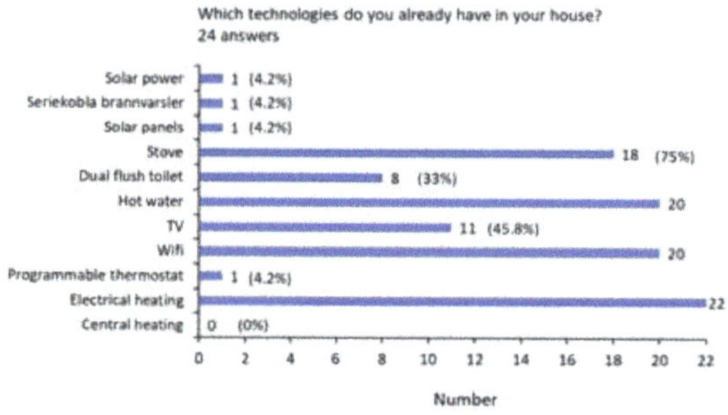

Figure 21. The technologies available in the respondents' houses.

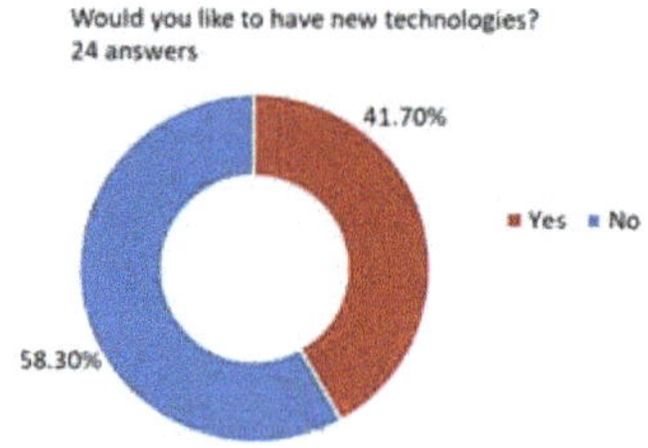

Figure 22. Participants' interest in applying new technologies.

Others expressed a need for keeping up the basic comfort with examples of kitchen fans and floor heating systems. For the majority who expressed no need for new technologies, respondents expressed a will to keep the simple living standards as they are for cultural reasons or lack of space in their dwellings.

> *"I have what I need"*, *"My house doesn't need anything else, nor is our house big enough"*, *"The rooms are original 1880s, and I like the quiet atmosphere in the rooms"*. (Survey respondents)

An analysis is designed to address challenges and find solutions from different perspectives that could be useful for inhabitants with low budget and a 'do it yourself' mindset.

The students covered different areas, such as indoor temperature, air quality, noise disturbance, energy consumptions, cracked walls, fire safety, and lack of private bathrooms.

4. Guidance for Sustainable Refurbishment for Improving Indoor Environment

As a result of the observations and the survey, the students developed under supervision of academic and professional experts, a practical guide in form of a small magazine for residents in which they addressed each part of the sustainable renovation work (Figure 23, Trouillon et al. (2019)).

Figure 23. Practical guide developed by NTNU Students for Svartlamon residents.

The first part of this magazine recalled information about the cultural heritage background and communicated the results of the survey. Further on, each element of the sustainable renovation guide was addressed in the most practical way yet easily understandable for a non-expert audience. This practical guide provided background information on the buildings' symptoms, the importance of addressing them, the potential for reusing materials, and the possible renovation measures to be taken.

The following section goes into the details of each element covered more succinctly in the magazine.

4.1. Reuse of Materials

While global concerns are saving energy, protecting the environment, as well as the scarcity of resources, re-use and recycling take place in the very first steps towards a promising future. The inhabitants and the entire community of Svartlamon have a good understanding of these issues, and many of them explain that they are living in this neighborhood not only for the low rent price, but rather to live a simple life and have a low environmental impact. Furthermore, this practice is the preferred one in the community, both for economic and environmental reasons. Svartlamon could also benefit from a collaboration with Loopfront, a digital platform created in Trondheim that enables the reuse of materials in the construction industry, promoting a circular economy [70].

More specific information about the reuse of construction materials is presented below in order to provide a basis for technical guidance for non-professional residents who engage in self-renovation work.

Bricks

Brick is a robust material with a long service life. Thus, recycled bricks have great potential for use in new constructions. Recycled bricks can be used in different contexts depending on the technical characteristics of bricks. Reclaimed bricks that are frost-proof can be used for cladding in facades while those that are not frost-proof are used in a plastered façade. Bricks used over several floors must have sufficient compressive strength. The reuse potential of bricks varies for different periods of time. Bricks covered with paint or treated with surface treatment that may contain PCBs, chlorine paraffins, heavy metals and other hazardous substances should be avoided [71,72].

A project at Lilleborg in Norway shows the following figures on the cost of unit price for reuse of bricks:

- Demolition and cleaning: approx. 5.5 NOK/brick;
- Re-burning of brick with low frost resistance: approx. 3NOK/brick;
- Price for new brick: 3kr/stone.

Results of the Lilleborg project indicate that due to the costs associated with sample extraction and quality control, the reuse volume per demolition object should be at least 50,000 bricks [72].

Wood

Results from the project "Reuse House Trondheim" showed that wood is one of the most popular materials for reuse [73]. Re-used material groups include all types of treated and untreated wood, glued-laminated timber, and wood fiber products, where spruce and pine are the most commonly used types of wood in Norway. Wood is most often used in constructive elements. It makes up about 30–40% of the total waste during a demolition. In the "Reuse House Trondheim", approximately 85% of the timber-frame and cladding were made of reclaimed wood, and smaller wood elements like doors, window frames, and kitchen interiors were also reused [74]. It is important to keep a record of quality assurance process when re-using or repurposing wood so that the material keeps its quality and usability. Wood that should be avoided is CCA- and creosote impregnated wood which belongs to the group impregnated wood and is considered hazardous waste.

Metal

Metals such as steel, zinc, copper, and aluminum components are sorted for reuse. One should be aware that materials that belong to the hazardous waste category should not be reused. Surface treatment must be assessed against limits for hazardous substances and whether there may be a risk of leakage [74]. One must be careful about the metal components with surface treatments that contain sensing substances (asbestos, heavy metals, Pcb, chlorine paraffins).

4.2. Indoor Environment

In the case of low-standard buildings, retrofit recommendations are based on weatherization of households, to improve indoor comfort and reduce energy-consumption costs. Weatherization is usually focused on changes in the structure, such as insulation of ceilings and walls, air sealing, and duct sealing.

Ventilation

To have a healthy indoor environment, fresh air is required in buildings to lighten and minimize odors, to improve the oxygen level for respiration, and to increase thermal comfort [75]. Natural ventilation could be a good environmentally friendly solution in Svartlamon buildings to improve air quality as the budget is very tight and the cultural heritage must be preserved [76]. This ventilation system can not only bring in fresh air using the natural force of wind and provide a high ventilation rate, which can be cost and energy efficient compared to mechanical systems, but also (if applied properly) can provide the opportunity of accessing daylight in buildings [77].

The proposed solution implies having straight pipes (re-used when applicable) through the roof and the ceilings of the apartments in order to exhaust the polluted air from the secondary rooms (kitchens in this case). Each apartment would have its own pipe as shown in Figure 24 which refers to a possible solution for the house of Strandveien 21. Putting vents through the walls is also a good solution for air change. However, the location and orientation of these pipes are crucial in terms of controlling the airflow, pollution release rate, and the amount of fresh air coming in [78], not having vents close to the exhaust pipe, but at the opposite side to improve airflow and preferably placed facing the wind, but also not close to the windows which is a cold area.

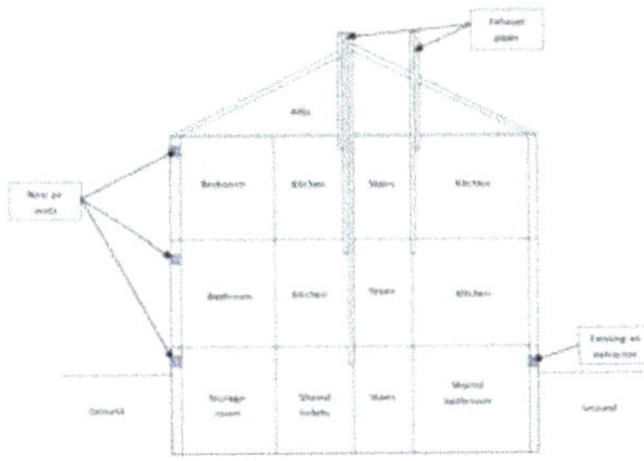

Figure 24. Simplified drawing of the proposed ventilation system for Strandveien 21 (Practical guide by Gendarme, 2019).

Ventilation outlets would be a vertical duct (at the opposite side of the air inlets) from the ceiling of the apartment to the roof of the house but also air extractors. The vertical straight pipe will evacuate hot air through the roof due to the low density of heated air.

Sound insulation

Materials which will be used for sound insulation need to be selected carefully. The goal is to prevent, or at least reduce, the sound coming from other apartments since it is the most annoying sound source. For keeping a low budget, only the walls (separating the common stairs and the apartments) and the floors (separating the first floor and the second floor) are addressed. Sound absorbent materials act as springs in a mass–spring–mass system. The sound absorbent is a soft, porous material that dampens sound and which is placed between two rigid walls considered as the masses [79]. These rigid walls stop a part of the sound according to the mass law: the heavier and denser a material, the higher the acoustic insulation is. The growing environmental awareness triggered a shift towards more environmentally friendly materials from renewable resources, such as waste wool and recycled polyester fibers [80]. Wooden fiber is good for acoustic since it is quite heavy but also mineral wool that has a good absorption factor at all frequencies [81,82].

Frames can be created in front of the existing walls separating the apartments from the common stairs/area, in which the insulation will be placed. For optimal insulation, the frame must be detached from the wall. That way, if one wall vibrates, it does not transmit its vibration to the second wall.

Lighting conditions

Ensuring good natural lighting conditions in a building is important to improve the health and wellbeing of residents [83]. A solution to enhance natural lighting in buildings would have been putting larger windows, but the cultural heritage regulations make this impossible. The alternative solution is to choose colors that reflect natural light. The reflection value of light can vary from very high percentage for white glossy finish to almost null for mat black. This can affect daylight illuminance in the room and impact the visual comfort of residents [84]. Finally, improving the artificial light equipment in the rooms can contribute to better conditions.

Windows

One solution will be to use windows with the lower U-value than windows already installed in order to reduce heat loss through window, especially in the north façade where there is the least solar gain [85–87]. However, windows with low U-Value are expensive and have a lower transmittance which deteriorates the properties of natural lighting [88]. A better performing frame like an insulated frame allows to reduce heat loss through the frame, and if they are well installed, they will highly reduce air leakage [89]. However, some of them could be expensive, and the design of the building does not allow to put any windows, as the external visual aspect must be kept. In the cheapest way, it is recommended to stop air leakage around the windows just with a seal and the use of reclaimed material to replace damaged parts.

Wind barrier

Airtight building envelope in lightweight constructions in cold and moderate climates is normally realized by a continuous interior air- and vapor-tight barrier. However, a proper interior air barrier that fulfills the stricter environmental requirements is usually labor intensive due to many internal joints in buildings (e.g., interior walls, perforations necessary for electrical and plumbing devices) [90,91].

Given the high possibility of unwanted infiltration of cold air in buildings through forced convection, using "wind barrier" can help to protect the outside insulation layer from such cold air infiltration into the constructions [92]. The wind barriers can also serve as a drainage plane to prevent water infiltration into the structure [92].

Air leakage

Uncontrolled infiltration of air through the building results from holes in the envelope (chimneys and ducts), gaps between building components especially at the roof, joints around movable elements such as doors and windows, and penetration of air through building components under pressure from wind. These infiltration means a high heat loss (up to 30%). Reducing infiltration will therefore significantly reduce heat loss and reduce overconsumption [93]. The following actions can be taken to reduce infiltration:

- Replace air windows with low leakage units;
- Seal around door and window frames and other components with flexible mastic seal;
- Seal cracks and holes in the envelope, mainly around duct, pipe passage.

An old building breaths due to its air infiltration weaknesses, ventilation is mainly due to low airtightness. Therefore, in reducing air leaks, the building will be more airtight and this type of natural ventilation will be less efficient. It is important to consider ventilation in the renovation planning, in order to avoid reduction of inside air quality causing moisture, material degradation and health problems for the residents thereafter [94].

Insulation of exterior components

Given the considerable influence of building envelope properties on the construction energy performance, one of the most common methods to improve energy efficiency in constructions is to insulate the internal and external walls [95]. To minimize energy use in buildings, providing thermal and moisture insulation in the layers tend to be very efficient. Thermal insulation can be installed on the external or internal side of the building envelope. However, in case of historical buildings, applying insulation in the external walls should be done in a more careful way due to the necessity of preserving the ancient and distinctive appearance in such buildings [96].

However, the influence of moisture transfer and condensation inside the walls (which would affect the thermal performance of the materials as well) is usually overlooked. For instance, it is shown by Barbosa and Mendes (2008) [97] that ignoring the influence of moisture in thermal insulation layers may lead to underestimation of yearly heat flux in buildings, which will result in huge waste of energy. Hence, it is crucial to view both thermal and humidity aspects in installing insulation in both internal and external wall layers.

The given circumstances in Svartlamon offer both advantages and disadvantages in terms of wall insulation of external walls. The usual measures used in the performance of insulation techniques are usually characterized by a high-cost factor. However, Svartlamon has to deal with limited budgets. Due to this fact as well as the historical and heritage characteristics of buildings in this area, it is not advisable to install insulation in the outer shell of the building, as a composite thermal insulation system. On the contrary, techniques can be used which may differ slightly from the state of the art or which do not fully comply with the regulations. The loosened requirements for building standards in Svartlamon offer a great advantage in this respect. Thus, apart from fire protection, the energy standards for new buildings do not necessarily have to be fully met. In the following section, different possibilities will be explained which can lead to an enhancement of the living situation and comfort.

Internal insulation is characterized by numerous disadvantages, as poor execution can result in extensive consequential damage [85]. Internal insulation in historical buildings can also lead to reduction of floor area, change of spatial room pro-portions, loss of historic fabric, and influence on the hygrothermal behavior of the insulated walls [98]. However, with careful execution it also offers several advantages with regard to Svartlamon. For instance, single rooms such as the living room can be insulated as desired to reduce heating times and energy loss. Although this means that valuable living space is lost, it also saves costs as no scaffolding is necessary, and the wooden cladding of the façade does not have to be removed and replaced.

Due, among other things, to the widespread use of timber-/timber-framed constructions in Norway, blow-in insulation should also be taken into consideration. With the help of this technique, all cavities are filled with loose bulk material and the insulating effect of the wall is significantly increased without a change in the external appearance of the building being visible after the measure. However, the installation should be carried out or at least supervised by a specialized company, since a deficient execution leads to large consequential damages. Nevertheless, it is a very cost-effective and above all very fast method with few working hours. There are also numerous sustainable and natural materials available for the choice of insulation material. For example, old newspapers can be collected and recycled to provide a much cheaper alternative to manufactured raw materials.

The structural and physical properties of the exterior walls are influenced by the right choice of materials and therefore also by the choice of insulating materials. Nowadays, many different insulating materials are used in the building industry. The Tables 4 and 5 were adapted from Gabriel and Ladener (2018) [99] and Kolb (2014) [100] and developed by one of the student [101] to be incorporated in the practical guidance and each give an overview of the individual properties of the selected insulating materials and techniques.

Table 4. Insulation materials and their properties (Beck, 2019).

Material	**Thickness [cm]**	**Aprox.Cost* [NOK/m²] W/m²k**	**Primary Energy Consumption [kwh/m³]**	**Heat Insulation Effect**	**Summer Heat Protection**	**Moisture Control**
	For U-Value= 0.2 W/m²K					
Flax	20	250–300	200–400	●●●●○	●●●○○	●●●●○
Hemp	22.5	200–300	200–400	●●●●○	●●●○○	●●●●○
Mineral bonded wood wool board	45	750–950	450	●●●◐○	●●●●○	●●●●○
Sheep wool	20	350–600	40–80	●●●●○	●●◐○○	●●●●●
Cellulose (a) loose (b) panels	20–22.5	150–200 250–300	25–30 100–600	●●●●○	●●●◐○	●●●●●
Calcium silicate boards	50–100	250 (50 mm)	–	●●●○○	-	●●●●●
Mineralfiber (a) glass wool (b) rock wool	17.5–25	100–300	95	●●●●●	●●○○○	○○○○○
Expanded perlite (a) thermal insulation (b) impact sound insulation	22.5–30	200–400	300–500	●●●○○	●●●○○	○○○○○
Polystyrene panels (a) EPS (b) XPS	12.5–20	100–150 400–500	360–600 600–900	●●●●●	●●○○○	○○○○○

Table 5. Insulation techniques and their potential for self-renovation (Beck, 2019).

	Sustainable Raw Materials Usable	Change in Facade Appearance	Thermal Bridge Reduction to a Minimum	Fire Protection Improvement	Personal Contribution Possible	Area of Application	Costs [NOK/m²] at Thickness t [cm]
Curtain wall	●●●●●	●●●●●	●●●●○	●○○○○	●●●●◐	All building types	1300–1700; t = 16
Thermal insulation composite system	●●●○○	●●●○○	●●●●●	●●◐○○	●○○○○	All building types	1100–1500; t = 10
Internal insulation	●●○○○	○○○○○	●●○○○	●◐○○○	●●○○○	Listed facades	900–1200; t = 10
Blow-in insulation	●●●●●	○○○○○	●◐○○○	○○○○○	○○○○○	Double-shell wall (cavities needed)	150–300;
Insulation plaster	○○○○○	●●●●○	●●●●○	●●●◐○	●○○○○	All building types	600–800;
Above rafter insulation	●●●●●	●●●○○	●●●●●	●○○○○	●●●●○	Rafter roof construction	2000–2500;
Between-/under- rafter insulation	●●●●●	●○○○○	●●◐○○	●○○○○	●●●●○	Rafter roof construction	400–600;
Attic insulation	●●●●●	○○○○○	●●●●○	●○○○○	●●●●●	All building types	400–600;

Provided information consists of approximate costs per square meter for the materials and their technical parameters as well as indications regarding insulation techniques that could be more or less easily applicable for self-renovation work. A point system is used to describe the insulating efficiency, summer heat protection, and moisture control (Table 4), whereby five filled out points stand for very high (very good) and five empty points for low (very bad). The same point system is used to assess the insulation techniques (Table 5).

5. Discussion and Conclusions

The purpose of this study was to explore the challenges related to sustainable refurbishment of an experimental urban ecological area with a strong historical and cultural heritage. We proposed possible renovation solutions and instructions with respect to the self-renovation culture that dominates in the community while also accounting for the complex ownership structure, governance, and insufficient funds for the maintenance backlog. Findings from this paper provided insight in two aspects of self-renovation:

(i) Available resources and reuse of materials;
(ii) Renovation guidance for inhabitants from the building physics perspective and improvements to the indoor comfort.

The research prepared together with master's students from NTNU is based on on-site observation and a close collaboration with the housing foundation of Svartlamon, thus providing solid basis for recommending real-life practical solutions that reflect the needs and the will of the community while trying to improve the overall sustainability of the building stock.

Based on the research, this study presents answers to the research question *How to raise the self-renovation standards of a culturally protected although poorly maintained area?*

Results show that in order to keep a sustainable, low-cost urban living model, instructions for self-renovation including information on their impact on sustainability are valuable as a guidance for non-professionals to make informed choices. We can emphasize that inhabitants are used to lower living standards, so the project is aimed to present the proper solutions for improvement as a balance between new technical solutions, personal self-renovation skills, habits, and health. This study addresses the challenges related to lack of motivation, knowledge, and communication that constitute major barriers to the implementation of sustainable renovation measures by proposing concrete actions based on state-of-the-art energy-retrofit interventions that account for the pre-existing cultural and economic factors influencing community decisions and capabilities.

The community has had a long history of fighting for the preservation of the area as a low-standard independent neighborhood. Despite of the maintenance backlog, the will of the community to work together to renovate the buildings is an asset. Having only a low budget and little knowledge, they constantly found creative ways to address their problems, for example, the infrastructure and network systems for re-using materials are already well implemented. However, the local storage of the materials is a limit to their ambitions, and the community as well as the municipality could benefit from being part of a wider network as proposed by Loopfront. This pre-existing culture should be combined with more sustainable solutions such as improving the thermal insulation, upgrading the windows with better glazing and frames, etc. With an improved maintenance strategy and access to the proper tools and resources, the community could develop their self-renovation culture towards more efficient and environmentally friendly practices.

Recommendations for future self-renovation work should provide some key-knowledge to the inhabitants about insulation, structure, indoor climate, building physics, and energy efficiency. This would allow them to make better-informed choices when engaging in self-renovation projects. All aspects of a renovation plan are closely linked and should be considered simultaneously. For instance, repairment, sealing of cracks and sealing of air infiltration weaknesses, adding new layers like wind barrier and insulation lead to buildings more airtight which means that ventilation system must be considered and improved. Regarding the indoor comfort in the building, it is relevant to point out that they were not initially designed for the modern lifestyle and hygiene habits; inhabitants' activities and use of the kitchens and bathrooms have caused humidity damages that should be fixed in order to insure an acceptable indoor air quality.

Furthermore, when developing a practical maintenance guide, the effects and the actual cost-benefit factor for the residents of Svartlamon must also be considered, and a balance must be found between technical solutions and user behavior.

To improve the living standards in Svartlamon, thermal comfort and building physics, the proposed solutions and recommendations aforementioned might be a promising start to sustainably improve the houses in the long-term.

Author Contributions: Conceptualization, C.S., A.T.S. and M.V.; formal analysis, C.S., A.T.S. and M.V.; investigation, C.S. and Ž.K.; methodology, C.S. and A.T.S.; resources, Ž.K.; supervision, A.T.S.; visualization, M.J.; writing—original draft, C.S., A.T.S. and M.V.; writing—review and editing, C.S., A.T.S., M.V., M.J. and Ž.K. All authors have read and agreed to the published version of the manuscript.

Funding: This research received no external funding.

Institutional Review Board Statement: Not applicable.

Informed Consent Statement: Not applicable.

Acknowledgments: The authors would like to thank all the master students who took part in the course "Refurbishment technology Specialized course" at NTNU Faculty of Civil ad Environmental Engineering since 2018 for their work and their innovative thinking which are the foundations of this paper.

Conflicts of Interest: The authors declare no conflict of interest.

References

1. Lidelöw, S.; Örn, T.; Luciani, A.; Rizzo, A. Energy-efficiency measures for heritage buildings: A literature review. *Sustain. Cities Soc.* **2019**, *45*, 231–242. [CrossRef]
2. Loli, A.; Bertolin, C. Towards zero-emission refurbishment of historic buildings: A literature review. *Buildings* **2018**, *8*, 22. [CrossRef]
3. Terama, E.; Peltomaa, J.; Mattinen-Yuryev, M.; Nissinen, A. Urban Sustainability and the SDGs: A Nordic Perspective and Opportunity for Integration. *Urban Sci.* **2019**, *3*, 69. [CrossRef]

4. Esser, A.; Dunne, A.; Meeusen, T.; Quaschning, S.; Denis, W. *Comprehensive Study of Building Energy Renovation Activities and the Uptake of Nearly Zero-Energy Buildings in the EU Final Report*; European Commission: Brussels, Belgium, 2019; 87p.
5. Burman, E.; Mumovic, D. *Measurement and Verification Models for Cost-Effective Energy-Efficient Retrofitting*; Elsevier: Amsterdam, The Netherlands, 2017; ISBN 9780081011287.
6. Lucchi, E.; Delera, A.C. Enhancing the historic public social housing through a user-centered design-driven approach. *Buildings* **2020**, *10*, 159. [CrossRef]
7. Abdellatif, M.; Al-Shamma'a, A. Review of sustainability in buildings. *Sustain. Cities Soc.* **2015**, *14*, 171–177. [CrossRef]
8. Semprini, G.; Gulli, R.; Ferrante, A. Deep regeneration vs shallow renovation to achieve nearly Zero Energy in existing buildings: Energy saving and economic impact of design solutions in the housing stock of Bologna. *Energy Build.* **2017**, *156*. [CrossRef]
9. Gianfrate, V.; Piccardo, C.; Longo, D.; Giachetta, A. Rethinking social housing: Behavioural patterns and technological innovations. *Sustain. Cities Soc.* **2017**, *33*. [CrossRef]
10. McCabe, A.; Pojani, D.; van Groenou, A.B. The application of renewable energy to social housing: A systematic review. *Energy Policy* **2018**, *114*. [CrossRef]
11. Guarini, M.R. Self-renovation in Rome: Ex ante, in itinere and ex post evaluation. In *Computational Science and Its Applications, Proceedings of the ICCSA 2016 Conference, Beijing, China, 4–7 July 2016*; Lecture Notes in Computer Science (including subseries Lecture Notes in Artificial Intelligence and Lecture Notes in Bioinformatics); Springer: Berlin/Heidelberg, Germany, 2016; Volume 9789, pp. 204–218. ISBN 9783319420882.
12. Risholt, B.; Berker, T. Success for energy efficient renovation of dwellings—Learning from private homeowners. *Energy Policy* **2013**, *61*, 1022–1030. [CrossRef]
13. Godbolt, Å.L.; Flyen, C.; Hauge, Å.L.; Flyen, A.C.; Moen, L.L. Future resilience of cultural heritage buildings—How do residents make sense of public authorities' sustainability measures? *Int. J. Disaster Resil. Built Environ.* **2018**, *9*, 18–30. [CrossRef]
14. Xue, Y.; Temeljotov-Salaj, A.; Engebø, A.; Lohne, J. Multi-sector partnerships in the urban development context: A scoping review. *J. Clean. Prod.* **2020**, *268*, 122291. [CrossRef]
15. Hauge, Å.L.; Hanssen, G.S.; Flyen, C. Multilevel networks for climate change adaptation—What works? *Int. J. Clim. Chang. Strateg. Manag.* **2019**, *11*. [CrossRef]
16. Kobal Grum, D. Interactions between human behaviour and the built environment in terms of facility management. *Facilities* **2018**, *36*. [CrossRef]
17. Grum, B.; Kobal Grum, D. Concepts of social sustainability based on social infrastructure and quality of life. *Facilities* **2020**, *38*. [CrossRef]
18. Bissonnette, J.F.; Dupras, J.; Messier, C.; Lechowicz, M.; Dagenais, D.; Paquette, A.; Jaeger, J.A.G.; Gonzalez, A. Moving forward in implementing green infrastructures: Stakeholder perceptions of opportunities and obstacles in a major North American metropolitan area. *Cities* **2018**, *81*. [CrossRef]
19. Gan, X.; Chang, R.; Wen, T. Overcoming barriers to off-site construction through engaging stakeholders: A two-mode social network analysis. *J. Clean. Prod.* **2018**, *201*. [CrossRef]
20. Le Feuvre, M.; Medway, D.; Warnaby, G.; Ward, K.; Goatman, A. Understanding stakeholder interactions in urban partnerships. *Cities* **2016**, *52*. [CrossRef]
21. Hein, A.M.; Jankovic, M.; Feng, W.; Farel, R.; Yune, J.H.; Yannou, B. Stakeholder power in industrial symbioses: A stakeholder value network approach. *J. Clean. Prod.* **2017**, *148*. [CrossRef]
22. Yang, A.; Bentley, J. A balance theory approach to stakeholder network and apology strategy. *Public Relat. Rev.* **2017**, *43*. [CrossRef]
23. Tyl, B.; Vallet, F.; Bocken, N.M.P.; Real, M. The integration of a stakeholder perspective into the front end of eco-innovation: A practical approach. *J. Clean. Prod.* **2015**, *108*. [CrossRef]
24. Aoki, N. Sequencing and combining participation in urban planning: The case of tsunami-ravaged Onagawa Town, Japan. *Cities* **2018**, *72*. [CrossRef]
25. Ferguson, L.; Chan, S.; Santelmann, M.; Tilt, B. Exploring participant motivations and expectations in a researcher-stakeholder engagement process: Willamette Water 2100. *Landsc. Urban Plan.* **2017**, *157*. [CrossRef]
26. Li, H.; de Jong, M. Citizen participation in China's eco-city development. Will 'new-type urbanization' generate a breakthrough in realizing it? *J. Clean. Prod.* **2017**, *162*. [CrossRef]
27. Temeljotov Salaj, A.; Gohari, S.; Senior, C.; Xue, Y.; Lindkvist, C. An interactive tool for citizens' involvement in the sustainable regeneration. *Facilities* **2020**, *38*, 859–870. [CrossRef]
28. Temeljotov Salaj, A.; Lindkvist, C.M. Urban facility management. *Facilities* **2021**, *39*, 525–537. [CrossRef]
29. Esmaeilpoorarabi, N.; Yigitcanlar, T.; Kamruzzaman, M.; Guaralda, M. How can an enhanced community engagement with innovation districts be established? Evidence from Sydney, Melbourne and Brisbane. *Cities* **2020**, *96*. [CrossRef]
30. Fleiß, E.; Hatzl, S.; Seebauer, S.; Posch, A. Money, not morale: The impact of desires and beliefs on private investment in photovoltaic citizen participation initiatives. *J. Clean. Prod.* **2017**, *141*. [CrossRef]
31. Swapan, M.S.H. Who participates and who doesn't? Adapting community participation model for developing countries. *Cities* **2016**, *53*. [CrossRef]
32. Foultier, C. Empowering the Unprivileged: The Case of Self-renovation in Disadvantaged Areas. In *Current Issues in European Cultural Studies, Linköping Electronic Conference Proceedings, Norrköping, Sweden, 15–17 June 2011*; Linköping University Electronic Press: Linköping, Sweeden, 2011; pp. 545–551.

33. Royston, S.; Foulds, C. The making of energy evidence: How exclusions of Social Sciences and Humanities are reproduced (and what researchers can do about it). *Energy Res. Soc. Sci.* **2021**, *77*, 102084. [CrossRef]
34. Overland, I.; Sovacool, B.K. The misallocation of climate research funding. *Energy Res. Soc. Sci.* **2020**, *62*, 101349. [CrossRef]
35. Baum, C.M.; Bartkowski, B. It's not all about funding: Fostering interdisciplinary collaborations in sustainability research from a European perspective. *Energy Res. Soc. Sci.* **2020**, *70*, 101723. [CrossRef]
36. Norwegian Government Bygningsvern i Norge. Available online: https://www.regjeringen.no/no/tema/klima-og-miljo/kulturminner-og-kulturmiljo/innsiktsartikler-kulturminner/bygningsvern-i-norge/id2343540/ (accessed on 20 May 2021).
37. Det Kongelige Miljøverndepartementet. *Living with our Cultural Heritage St.meld. nr. 16 (2004–2005)*; Norwegian Government: Oslo, Norway, 2005; Volume 16.
38. *Bystrategi 2017–2020*; Riksantikvaren: Oslo, Norway, 2017.
39. Igor, S.; Wachenfeldt, B.J.; Anne Grete, H. Energy demand in the Norwegian building stock: Scenarios on potential reduction. *Energy Policy* **2009**, *37*, 1614–1627. [CrossRef]
40. Ugarte, S.; Van Der Ree, B.; Voogt, M.; Eichhammer, W.; Ordoñez, J.A.; Reute, M.; Schlomann, B.; Llore, P.; Villafáfila, R. *Policy Department: Economic and Scientific Policy-Energy Efficiency for Low-Income Households*; European Parliament: Brussels, Belgium, 2016.
41. Talbot, J. *Mobilizing Energy Efficiency in the Manufactured Housing Sector*; American Council for an Energy-Efficient Economy: Washington, DC, USA, 2012; Volume 20045.
42. Stieß, I.; Dunkelberg, E. Objectives, barriers and occasions for energy efficient refurbishment by private homeowners. *J. Clean. Prod.* **2013**, *48*, 250–259. [CrossRef]
43. Humar, M.; Lesar, B.; Kržišnik, D. Moisture Performance of Façade Elements Made of Thermally Modified Norway Spruce Wood. *Forests* **2020**, *11*, 348. [CrossRef]
44. Zafar, F.; Sharmin, E.; Ashraf, S.M.; Ahmad, S. Ambient-cured polyesteramide-based anticorrosive coatings from linseed oil—A sustainable resource. *J. Appl. Polym. Sci.* **2005**, *97*, 1818–1824. [CrossRef]
45. Larsen, K.E.; Nils, M. *Conservation of Historic Timber Structures an Ecological Approach*; Butterworth-Heinemann Series in Conservation and Museology; Reed Educational and Professional Publishing Ltd.: Oxford, UK, 2000.
46. Lisø, K.R.; Kvande, T.; Time, B. Climate Adaptation Framework for Moisture-resilient Buildings in Norway. *Energy Procedia* **2017**, *132*, 628–633. [CrossRef]
47. Kristl, Ž.; Senior, C.; Temeljotov Salaj, A. Key challenges of climate change adaptation in the building sector. *Urbani Izziv* **2020**, *31*, 101–111. [CrossRef]
48. Bradshaw, J.L.; Bou-Zeid, E.; Harris, R.H. Comparing the effectiveness of weatherization treatments for low-income, American, urban housing stocks in different climates. *Energy Build.* **2014**, *69*, 535–543. [CrossRef]
49. Zavadskas, E.K.; Kaklauskas, A.; Tupenaite, L.; Mickaityte, A. 2008. Decision-Making Model for Sustainable Buildings Refurbishment. Energy Efficiency Aspect. In Proceedings of the 7th International Conference on Environmental Engineering, ICEE, Vilnius, Lithuania, 22–23 May 2008; pp. 894–901.
50. Korjenic, A.; Zach, J.; Hroudová, J. The use of insulating materials based on natural fibers in combination with plant facades in building constructions. *Energy Build.* **2016**, *116*, 45–58. [CrossRef]
51. Pye, M. *Energy Efficiency Programs for Low-Income Households: Successful Approaches for a Competitive Environment*; American Council for an Energy-Efficient Economy: Washington, DC, USA, 1996; Volume 20036.
52. Underhill, L.J.; Milando, C.W.; Levy, J.I.; Dols, W.S.; Lee, S.K.; Fabian, M.P. Simulation of indoor and outdoor air quality and health impacts following installation of energy-efficient retrofits in a multifamily housing unit. *Build. Environ.* **2020**, *170*, 106507. [CrossRef]
53. New European Bahaus-Initiative European Commission. *Why the New European Bauhaus (NEB)? What Are the Next Concrete Steps?* European Commission: Brusels, Belgium, 2021.
54. Hermelink, A.; Schimschar, S.; Offermann, M.; John, A.; Reiser, M.; Pohl, A.; Grözinger, J. *Comprehensive Study of Building Energy Renovation Activities and the Uptake of Nearly Zero-Energy Buildings in the EU*; European Commission: Cologne, Germany, 2019.
55. Trouillon, D.; Gendarme, E.; Mouton, L.; Dehon, O.; Pauchon, S.; Beck Heremans, V.; Sekkal, Y. *How to Improve the Living Standard of Inhabitants of Svartlamon in a Sustainable Way?* Report for TBA4178—Refurbishment Technology Specialised Course; NTNU: Trondheim, Norway, 2019.
56. Berge, I.C.M.; Horten, H.; Hua, C.R.R.; Knudsen, A.H.V.; Solli, C.J.; Pérez Celaya, N. *Sustainable Refurbishment at Biskop Darres Gate 10, Svartlamon*; Report for TBA4178 Course "Refurbishment Technology Specialised"; NTNU: Trondheim, Norway, 2020.
57. EUROPAN 10 Comittee. EUROPAN10 Trondheim, Norway 2010. Available online: http://europan.no/ukategorisert/e10-the-sites/ (accessed on 20 May 2021).
58. Øverland Gater, P. *Historieglimt 7*; Lademoen Kirkes Småskrifter: Trondheim, Norway, 2011.
59. Leuraers, C.; van Dooren, A.; Bjørberg, S.; Temeljotov Salaj, A. Alternative Community—Promotor or Inhibitor of Sustainable Development. In *Keeping up with Technologies to Turn Built Heritage into the Places of Future Generations, Conference Proceedings, 6th International Academic Conference on Places and Technologies, Pecs, Hungary, 9–10 May 2019*; University of Pecs Faculty of Engineering and Information Technology: Pecs, Hungary, 2019; pp. 582–590.
60. Sager, T. Planning by intentional communities: An understudied form of activist planning. *Plan. Theory* **2018**, *17*, 449–471. [CrossRef]

61. Kozeny, G. Intentional Communities: Lifestyles Based on Ideals. In *Communities Directory: A Guide to Cooperative Living*; Fellowship for Intentional Community: Langley, DC, USA, 1995; pp. 18–24.
62. Lowenstein, O. Svartlamoen's Ragged Glory—Where Alt-Culture Meets Avant Timber Architecture. Available online: http://www.fourthdoor.co.uk/unstructured/unstructured_08/svartlamon.php (accessed on 10 May 2021).
63. Svartlamon Webgruppe Hva er Svartlamon-Organisering. Available online: https://svartlamon.org/organisering/ (accessed on 10 May 2021).
64. Vukmirović, M.; Nikolić, M. Industrial heritage preservation and the urban revitalisation process in Belgrade. *J. Urban Aff.* **2021**. [CrossRef]
65. Salaj, A.T.; Lindkvist, C.; Jowkar, M. Social needs for sustainable refurbishment in Trondheim. In Proceedings of the 19th EuroFM Research Symposium 2020, Online. 3–4 June 2020; pp. 51–61.
66. Jowkar, M.; Temeljotiv-Salaj, A.; Lindkvist, C.M.; Støre-Valen, M. Sustainable Building Renovations: Barriers and Potential Motivations. *Constr. Manag. Econ.* **2021**, in press.
67. Kulenović, R. *MOGUĆNOSTI (Re)Aktivacije Napuštenih Industrijskih Objekata: Raskršća i Putokazi (Possibilities for (Re)Activation of Vacant Industrial Buildings: Crossroads and Road Signs)*; Kulturklammer-Centar za Kulturne Interakcije: Belgrade, Serbia, 2008. (In Serbian)
68. Trondheim Municipality. *R 219b-Bestemmelser til Endret Reguleringsplan for Svartlamoen (Reinaområdet)*; Trondheim Municipality: Trondheim, Norway, 2006; pp. 1–10.
69. Municipal Council of Trondheim. *Saksdokumenter—Sak PS 0193/16-Evaluaering Av Svartlamon*; Municipal Council of Trondheim: Trondheim, Norway, 2016.
70. Wahlström, M.; Castell-Rüdenhausen, M.; Astrup, T.F.; Oberender, A.; Jensen, C.; Johansson, P.; Wærner, E. *Strategies and Methods for Implementing CE in Construction Activities in the Nordic Countries—Supporting Cases*, 508th ed.; Nordic Council of Ministers, Ed.; Nordic Energy Research, Nordisk Ministerråd: Copenhagen, Denmark, 2021; ISBN 9789289369299.
71. Madsø, F. *Lilleborg Gjenbruk av Tegl Raporter*; NCC Statsbygg: Oslo, Norway, 2001.
72. Kilvær, L.; Sunde, O.W.; Eid, M.S.; Rydningen, O.; Fjeldheim, H. Forsvarlig Ombruk av Byggevarer. 2019. Available online: https://dibk.no/globalassets/02.-om-oss/rapporter-og-publikasjoner/forsvarlig-ombruk-av-byggevarer_resirqel-2019.pdf (accessed on 20 May 2021).
73. Pettersen, N. *Gjenbrukshus i Trondheim*; Miljøenheten in Trondheim Kommune: Trondheim, Norway, 2005; ISBN 8277270917. (In Norwegian)
74. Sørnes, K.; Nordby, A.S.; Fjeldheim, H.; Hashem, S.M.B.; Mysen, M.; Dahl Schlanbusch, R. *Anbefalinger ved Ombruk av Byggematerialer*; Sintef: Trondheim, Norway, 2014.
75. Walker, A. Natural Ventilation. 2016. Available online: https://www.wbdg.org/resources/natural-ventilation (accessed on 20 May 2021).
76. Nunes de Freitas, P.; Guedes, M.C. The use of windows as environmental control in "Baixa Pombalina's" heritage buildings. *Renew. Energy* **2015**, *73*, 92–98. [CrossRef]
77. Atkinson, J.; Yves, C.; Pessoa-Silva, C.-L.; Jensen, P.; Li, Y.; Seto, W.-H. *Natural Ventilation for Infection Control in Health-Care Settings*; Atkinson, J., Chartier, Y., Pessoa-Silva, C.L., Jensen, P., Li, Y., Seto, W.-H., Eds.; WHO Press: Geneva Switzerland, 2009.
78. Lin, Y.-P.; Iba, C.; Lai, C.-M. Natural Ventilation Effectiveness of Round Wall-Mounted Vent Caps in Residential Kitchens. *Energies* **2018**, *11*, 1230. [CrossRef]
79. Rasmussen, B. Sound insulation between dwellings—Requirements in building regulations in Europe. *Appl. Acoust.* **2010**, *71*, 373–385. [CrossRef]
80. Patnaik, A.; Mvubu, M.; Muniyasamy, S.; Botha, A.; Anandjiwala, R.D. Thermal and sound insulation materials from waste wool and recycled polyester fibers and their biodegradation studies. *Energy Build.* **2015**, *92*, 161–169. [CrossRef]
81. Zhao, J.; Wang, X.M.; Chang, J.M.; Yao, Y.; Cui, Q. Sound insulation property of wood-waste tire rubber composite. *Compos. Sci. Technol.* **2010**, *70*, 2033–2038. [CrossRef]
82. Zhu, X.; Kim, B.J.; Wang, Q.W.; Wu, Q. Recent advances in the sound insulation properties of bio-based materials. *BioResources* **2014**, *9*, 1764–1786. [CrossRef]
83. Dovjak, M.; Košir, M.; Kristl, Ž.; Krainer, A. Assessment of Sustainability Aspects of Daylighting in Buildings. In Proceedings of the Euro Sun Conference, Aix-les-Bains, France, 16–17 September 2015; ISBN 9783981465938.
84. Pajek, L.; Košir, M.; Kristl, Ž.; Kacjan Žgajnar, K.; Dovjak, M. Indoor evnironmental quality (IEQ) in Slovenian children daycare centres. Part I: Results of in-situ measurements. *Int. J. Sanit. Eng. Res.* **2017**, *11*, 27.
85. Blecich, P.; Franković, M.; Kristl, Ž. Energy retrofit of the Krsan Castle: From sustainable to responsible design—A case study. *Energy Build.* **2016**, *122*, 23–33. [CrossRef]
86. Košir, M.; Gostiša, T.; Kristl, Ž. Influence of architectural building envelope characteristics on energy performance in Central [European climatic conditions. *J. Build. Eng.* **2018**, *15*, 278–288. [CrossRef]
87. Kristl, Z.; Zbasnik-senegacnik, M. Energy Renovation of Large Neighbourhoods in Slovenia. In Proceedings of the Fourth ISES-Europe Solar Congress, Bologna, Italy, 23–26 June 2002; Renewable Energy for Local Communities of Europe, Toward Rio+10. International Solar Energy Society: Trento, Italy, 2000; pp. 1–8.
88. Garnier, C.; Muneer, T.; McCauley, L. Super insulated aerogel windows: Impact on daylighting and thermal performance. *Build. Environ.* **2015**, *94*, 231–238. [CrossRef]

89. Burattini, C.; Nardecchia, F.; Bisegna, F.; Cellucci, L.; Gugliermetti, F.; De Lieto Vollaro, A.; Salata, F.; Golasi, I. Methodological approach to the energy analysis of unconstrained historical buildings. *Sustainability* **2015**, *7*, 10428–10444. [CrossRef]
90. Aho, H.; Vinha, J.; Korpi, M. Implementation of airtight constructions and joints in residential buildings. In Proceedings of the 8th Nordic Building Physics Symposium, Copenhagen, Denmark, 16–18 June 2008; Technical University of Denmark: Twente, Danemark, 2008.
91. Kalamees, T. Air tightness and air leakages of new lightweight single-family detached houses in Estonia. *Build. Environ.* **2007**, *42*, 2369–2377. [CrossRef]
92. Langmans, J.; Klein, R.; Eykens, P.; De Paepe, M.; Roels, S. Feasibility of using wind barriers as air barriers in wood frame construction. In Proceedings of the International Conference on Thermal Performance of the Exterior Envelopes of Whole Buildings, Buildings XI, Clearwater, FL, USA, 5–10 December 2010.
93. Martín-Garín, A.; Millán-García, J.A.; Hidalgo-Betanzos, J.M.; Hernández-Minguillón, R.J.; Baïri, A. Airtightness analysis of the built heritage–field measurements of nineteenth century buildings through blower door tests. *Energies* **2020**, *13*, 6727. [CrossRef]
94. Boro, M. *Veileder-Råd om Energisparing i Gamle Hus*; Riksantikvaren: Oslo, Norway, 2013; ISBN 9788275740807.
95. Yan, L.; Liu, C. Techno-economic analysis for constructing solar photovoltaic projects on building envelopes. *Build. Environ.* **2018**, *127*, 37–46.
96. Xu, C.; Shuhong, L.; Zou, K. Study of heat and moisture transfer in internal and external wall insulation configurations. *Build. Eng.* **2019**, *24*, 100724. [CrossRef]
97. Barbosa, R.M.; Mendes, N. Combined simulation of central HVAC systems with a whole-building hygrothermal model. *Energy Build.* **2008**, *40*, 276–288. [CrossRef]
98. Harrestrup, M.; Svendsen, S. Internal insulation applied in heritage multi-storey buildings with wooden beams embedded in solid masonry brick façades. *Build. Environ.* **2016**, *99*, 59–72. [CrossRef]
99. Gabriel, I.; Ladener, H. *Vom Altbau zum Effizienzhaus*; Ökobuch: Staufen im Breisgau, Germany, 2018; ISBN 9783936896756.
100. Kolb, B. *Altbausanierung mit Nachwachsenden Rohstoffen (Renovation of Old Buildings with Renewable Raw Materials)*; Fachagentur Nachwachsende Rohstoffe: Gülzow, Germany, 2014.
101. Beck, S. *Insulation Improvement Considering Sustainable and Ecological Aspects*; Report for TBA4178 Course "Refurbishment Technology Specialised"; NTNU: Trondheim, Norway, 2019.

energies

Article

Airflow Analysis of the Haida Plank House, a Breathing Envelope

Roberto Alonso González Lezcano [1] and María Jesús Montero Burgos [2,*]

1 Escuela Politécnica Superior, Montepríncipe Campus de Boadilla del Monte, Universidad San Pablo-CEU, CEU Universities, 28040 Madrid, Spain; rgonzalezcano@ceu.es
2 Facultad de Humanidades y Ciencias de la Comunicación, Campus de Moncloa, Universidad San Pablo-CEU, CEU Universities, 28040 Madrid, Spain
* Correspondence: mar.montero@ceindo.ceu.es

Abstract: The Haida plank house is one of the most important models built by the native American Indians. Built on the southwest coast of Canada, it adapts the tradition of the ancient pit houses to the requirements of the humid and cold climate characteristic of the Haida Gwaii Islands. This construction is composed by two main pieces: the central pit covered by a wooden envelope. Both protect its dwellers and their hearths. The ventilation system is based on two solutions: the gaps between the wall planks and a smoke hole that can be opened or closed in the roof at will. The aim of the present research is to analyze the way these two elements arrange the indoor airflow in order to ensure the comfortability of the house. Four cases have been proposed, according to four different dimensions for the gaps: 1, 2, 3 and 4 cm. Each case has been doubled in order to determine how the state of the smoke hole affected the corresponding results. This way, it has been concluded that if the gaps' width becomes higher than 4 cm, the airflow velocity comfort level would be exceeded. It is been possible to observe how the state of the smoke hole influences the way the air moves around the dwelling.

Keywords: ventilation; CFD analysis; archaeology; architecture; native American Indians; traditional architecture; vernacular architecture

Citation: Lezcano, R.A.G.; Burgos, M.J.M. Airflow Analysis of the Haida Plank House, a Breathing Envelope. *Energies* **2021**, *14*, 4871. https://doi.org//10.3390/en14164871

Academic Editor: Alessandro Cannavale

Received: 28 May 2021
Accepted: 3 August 2021
Published: 10 August 2021

1. Introduction

The dwellings built by the Native American Indians are one of the most interesting examples of primitive architecture that can be found. When analyzing them, it can be seen that they were designed in the search for one main goal: to achieve a construction that takes advantage of its environment and of the available resources in the most efficient way [1]. In order to succeed, it was necessary to have a deep knowledge about the way that the environment worked and the advantages and disadvantages offered by those resources. Such was the case of these communities.

When the European explorers arrived in North America by the end of the 15th century, they found a bunch of cultures that represented a period of history that was already lost in their own continent [2–4]. Among the members of those expeditions, there were not only sailors but also artists, historians, engineers, botanists, or anthropologists, whose work is extremely valuable nowadays [5]. Specifically in the architecture field, they gathered a great amount of information about the dwellings they had the opportunity to visit. Dimensions, building materials, building processes, or indoor distribution are some of the features described in the documentation they created.

The present research is focused on one of these dwellings, the plank house built by the communities settled in the coast of southwest of Canada and northwest of the United States. The goal of this work is to analyze the passive ventilation strategies that are contained in its design.

The study case was built by the Haida, one of the most important groups of the zone. Two types of houses were built by this community, according to two different structural solutions. The use of six or four beams was the main decision to make (Figure 1).

Figure 1. (**a**) Four beams model based on [1] (p. 270); (**b**) Six beams model based on [1] (p. 270).

The first one is probably the eldest [1] (p. 271). The house that has been chosen for the present research, dwelling number 3 from the village of Ninstints, corresponds to this structural solution.

The Haida name for the village was "Red Cod Island Town", which clearly indicates the way of life of its inhabitants. Its first European visitors changed its name from the one of the chief of the tribe, Nañ stîns, "He who is two". It contained twenty dwellings, although at present, just seventeen of them can be located, as can be seen in Figure 2 [6] (p. 103).

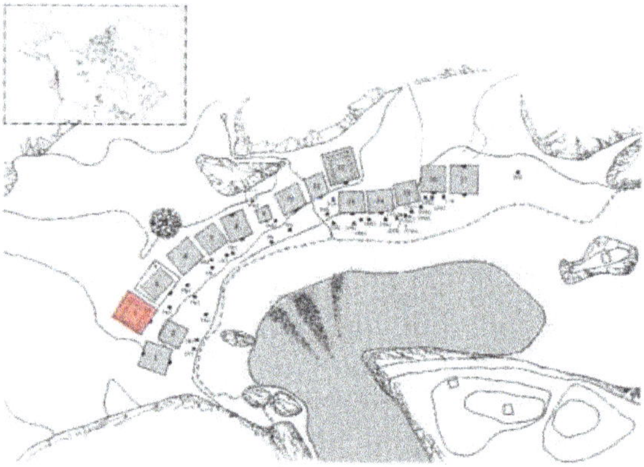

Figure 2. Plan of Ninstints (dwelling number 3 in red) based on [6] (p. 10).

As usually happens in this type of villages, the line of dwellings was arranged along the coast line in order to ease the entry to the ocean and to take advantage of the airflow that arose between the mountains and the ocean. This mechanism was a very useful feature of these locations, since this airflow kept the wooden structures of the dwellings from being rotten. The red cedar wood (*Thuja plicata* Donn ex D. Don) was chosen by the Haida, since it is naturally covered by a special oil that also protects it from moisture [7] (p. 63).

Walls were composed by a line of wooden planks (Figure 3). It was necessary to leave a little gap between them, so the airflow could enter into the house. As told by Underhill [8] (p. 84), one of these dwellings was rebuilt some time ago by modern Indians. When it was finished, they lit a hearth indoors, and smoke filled the house completely. They could not understand how their ancestors could have lived in such a smoky place. An Indian elder man who had seen all the process explained them how important those gaps between the planks were: "Why, it was the cracks. With enough cracks, you have fine circulation of air".

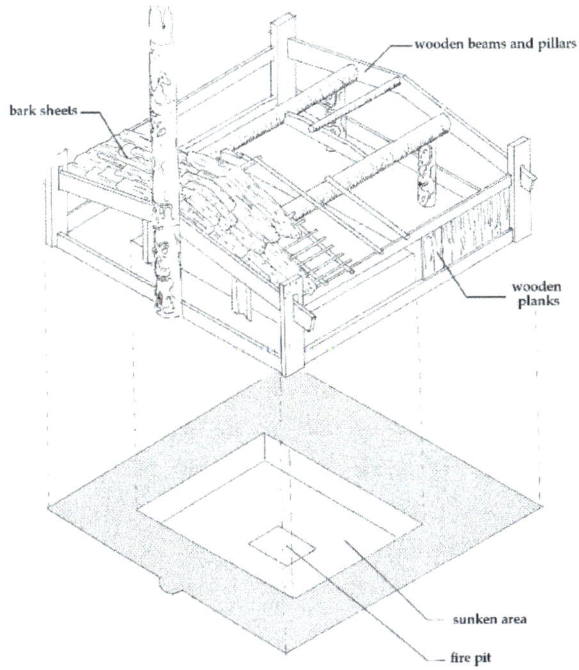

Figure 3. Exploded axonometric of a Haida dwelling construction system based on [9] (p. 171).

The door consisted of a hole carved in a totem pole, and it usually was covered with another plank, hung from a cedar bark rope, and blocked with some poles fit in the structure [10] (p. 21).

This way, the structure of the dwelling was solved by means of wooden beams and poles, and its envelope was composed of wooden planks. Roofs were built in two different ways: wooden planks or bark sheets. The first option was used by the richest families, while the second possibility was chosen by the rest of the community [10] (p. 21). Whatever solution was chosen, these pieces were overlapped, such as the tiles used in modern roofs, in order to keep rainwater away from the house; finally, they were weighted down with stones [11] (p. 72).

The indoor space was arranged around a central terraced pit [12] (p. 146B). It was the place where the hearth was lit. There could be several hearths, if there was more than one family living in the house. The roof planks could be removed in order to let the smoke out.

As explained before, the target of the present research work is to analyze the airflow system that triggers indoor ventilation. There are three elements to focus on: the dimensions of the gaps between the planks, the influence of the smoke hole in the indoor airflow, and the way the airflow behaves in the pit.

2. Materials and Methods

2.1. Materials

The dwelling analyzed was built by the Haida in the village of Ninstints, Haida Gwaii Islands, British Columbia. Specifically, it is dwelling number three. It was built according to the oldest type of structure, as explained before, and it had a central pit, which was the usual piece of this type of dwelling [10] (p. 107). Its floor plan was 14.1 per 14.85 m. The pit (9.45 per 9.9 m) was 1.8 m deep; it was delimited by two steps that were each 75 cm high. The height of the house has been obtained by taking the Haida dwelling drawn by Niblack in 1890 (Figure 4); [13] (p. 42) as reference. This way, its height has been estimated as 3.5 m for the side walls and as 5.03 m for the ridge of the two main façades.

Figure 4. Cutaway section of Haida dwelling based on [13] (detail of plate XXXV).

2.2. Methods

First of all, the ventilation rate of the dwelling has been calculated in order to determine if it was high enough. This way, the 2nd Florida Solar Energy Center method has been used [14] (pp. 101–103). This method is used to dimension the inlet and outlet areas in cross-ventilated spaces. It is based on the pressure increase between the inlet and the outlet openings. The results must be read taking into account that they are not influenced by the ambient temperature and that this method assumes that every inlet opening has the same positive pressure value, and every outlet has the same negative pressure value.

According to this method, the ventilation rate is calculated according to the following formula:

$$Ach = [A \cdot W \cdot \sqrt{(f_3 \cdot f_4 \cdot PD)}]/[4.3 \times 10^{-4} \cdot V]. \quad (1)$$

"A" is equal to the total effective area of the openings (m^2) and is determined following the next expression:

$$A = (A_o \cdot A_i)/(A_o^2 + A_i^2)^{0.5}. \quad (2)$$

The outlet area is represented by "A_o" (m^2), and the inlet area is represented by "A_i" (m^2). "W" is equal to the wind velocity (m/s). There are two correction factors represented by letter "f", f_3 and f_4. On the one hand, "f_3" introduces the effect of the nearby constructions and is equal to "g/h", being "h" the height of the nearby obstacles and "g" the distance between them. On the other hand, "f_4" is determined by the floor number where the case study in question is located. "PD" is the subtraction between the windward pressure coefficient (WPC) and the leeward pressure coefficient (LPC) [14] (p. 102).

The dwelling has been modeled in Autodesk Revit 2020®. This model has been exported as a SAT file in order to carry out the corresponding calculations in Autodesk CFD 2021® [15]. These calculations have been developed taking into account the average conditions of the zone where these dwellings were built. This way, in order to obtain representative results, ten locations have been taken, and their wind velocity average has been determined (Figure 5). Just summer wind velocity has been taken into account, since that is the season when ventilation is more important.

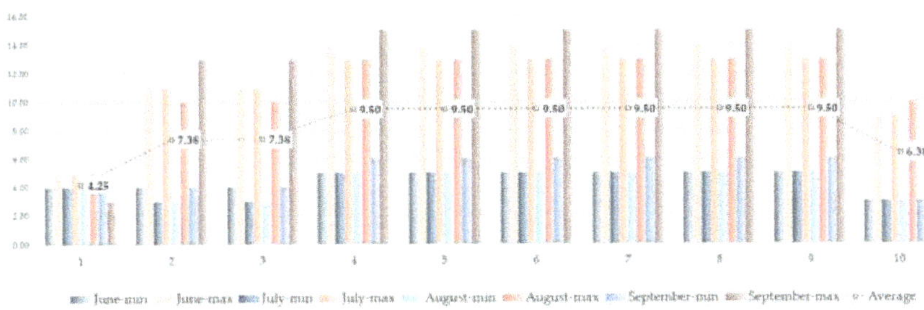

1 Old Kasaan (55.431426°, -132.384658°) (Ketchikan station)

2 Howkan (54.872807°, -132.801788°) (Hydaburg station)

3 Klinkwan (54.890278°, -132.328955°) (Hydaburg station)

4 Kaisun (53.032681°, -132.449980°) (Sandspit station)

5 Kiusta (54.177953°, -133.021781°) (Sandspit station)

6 Kung (54.048427°, -132.572717°) (Sandspit station)

7 Ninstints (52.099875°, -131.216794°) (Sandspit station)

8 Skidegate (53.266872°, -131.990552°) (Sandspit station)

9 Tanu (52.766543°, -131.616467°) (Sandspit station)

10 Hiellan (54.072605°, -131.790643°) (Prince Rupert station)

Figure 5. Average summer wind velocity values (m/s). Ten locations where the Haida built plank houses have been taken. The value that has been used (8.24 m/s) corresponds to the average of all of them based on [16].

As can be seen in Figure 6, the model has been placed in a box. Air material has been assigned to this element, as it has to the gaps between the planks, the smoke hole, and the air mass located inside the house. Wood material has been assigned to the planks, floor, and roof.

As boundary conditions, wind force (8.24 m/s) (Figure 5) has been assigned to one of the sides of the box, normal to the longitudinal axe of the house. Zero pressure has been assigned to the opposite face. This way, the latter absorbed the windflow emitted by the former. The top and longitudinal sides of the box have been characterized as "slim/slippery". The free sides of the gaps have also been assigned zero pressure and the smoke hole has as well (Figure 6).

The dimensions of the air box were established by the software Autodesk CFD®. Its width is five times the width of the dwelling, its length is five times the length of the dwelling, and its height is five times the height of the dwelling.

The solution mode was established as "steady state". The resolution of the mesh was defined as 1, the edge growth rate was defined as 1.1, two points were the minimum on the edges, and 10 was the number of points for the longest edge. These meshes were designed automatically by the CFD software. Calculations have been carried out accepting the Autodesk CFD® criteria, since it limits automatically the optimum amount of iterations to achieve convergence. This way, "Intelligent solution control" and "Automatic convergence assessment" have been selected and set as "tight". The calculation system has been developed from the algorithm SIMPLER, which belongs to one of the best-known CFD tools family. The turbulence was simulated by means of the model k-epsilon, which has been widely developed and studied. The validity of the results that have been obtained are backed by the use of Autodesk CFD®, which is a software that has been tested since 2013. Its calculations have been compared in numerous research studies, such as the one developed in the Victoria University of Wellington by Dr. Jing Li [17]. Autodesk® provides an ample compilation of verification studies; some of them compare test models against experimental results, and others compare them against empirical hand calculations [18].

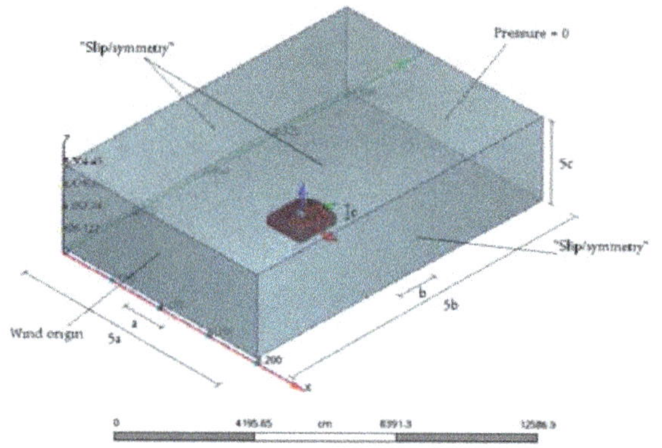

Figure 6. Screenshot in which boundary conditions are indicated (Autodesk CFD®).

Four alternatives have been calculated according to the gaps dimensions: 1, 2, 3, and 4 cm wide. Each case is calculated twice, since the smoke hole can be either open or closed. Five locations have been chosen inside each model, as shown in Figure 7. Each location has been divided into several levels, from level 0 to level 450, in such a way that one point was placed each 50 cm (Figure 7).

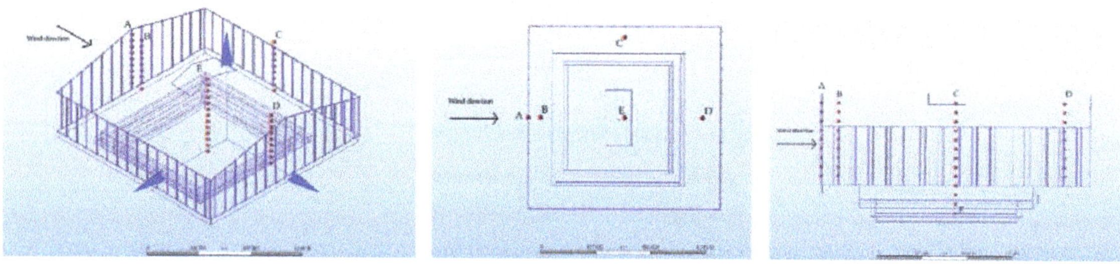

Figure 7. (**a**) Axonometry, screenshot showing the location of the five points analyzed (Autodesk CFD®); (**b**) Floorplan of the same model (Autodesk CFD®); (**c**) Section of the same model (Autodesk CFD®).

3. Results

3.1. Ventilation Rate

As explained before, the ventilation rate of this dwelling has been calculated according to the Second Florida Solar Energy Center method in order to determine if the openings were wide enough to accomplish this minimum requirement. The Spanish Building Technical Code establishes that this value has to be higher or equal to 0.63 ach in abodes.

Calculations have been carried out and, as can be seen, the plank house whose gaps are 1 cm wide accomplished this requirement (Table 1).

Table 1. Ventilation rate for 1 cm wide gaps.

Outlet Area A_0	Inlet Area A_i	Total Area A	WPC	LCP	PD	Volume V	Wind Velocity W	f_3	f_4	Ach
10.49 m^2	0.62 m^2	0.62 m^2	0.7	−0.1	0.8	813.81 m^3	8.24 m/s	0.41	1	8.3

Once it was determined that the ventilation rate was high enough if the gaps between wall planks were equal to or higher than 1 cm wide, the research could be focused on analyzing the way the airflow moved inside the dwelling, varying the dimension of these gaps.

3.2. Airflow

3.2.1. Airflow above the Central Pit

The velocity magnitude has been calculated in five columns, each one composed by nine points from 50 to 450 cm high. Each column has been calculated for two cases, since the smoke hole could be either opened or closed.

As can be seen in Figures 8 and 9, the velocity magnitude increases as the gaps' width does. The thinner the gaps are, the more regular the velocity is along the corresponding column.

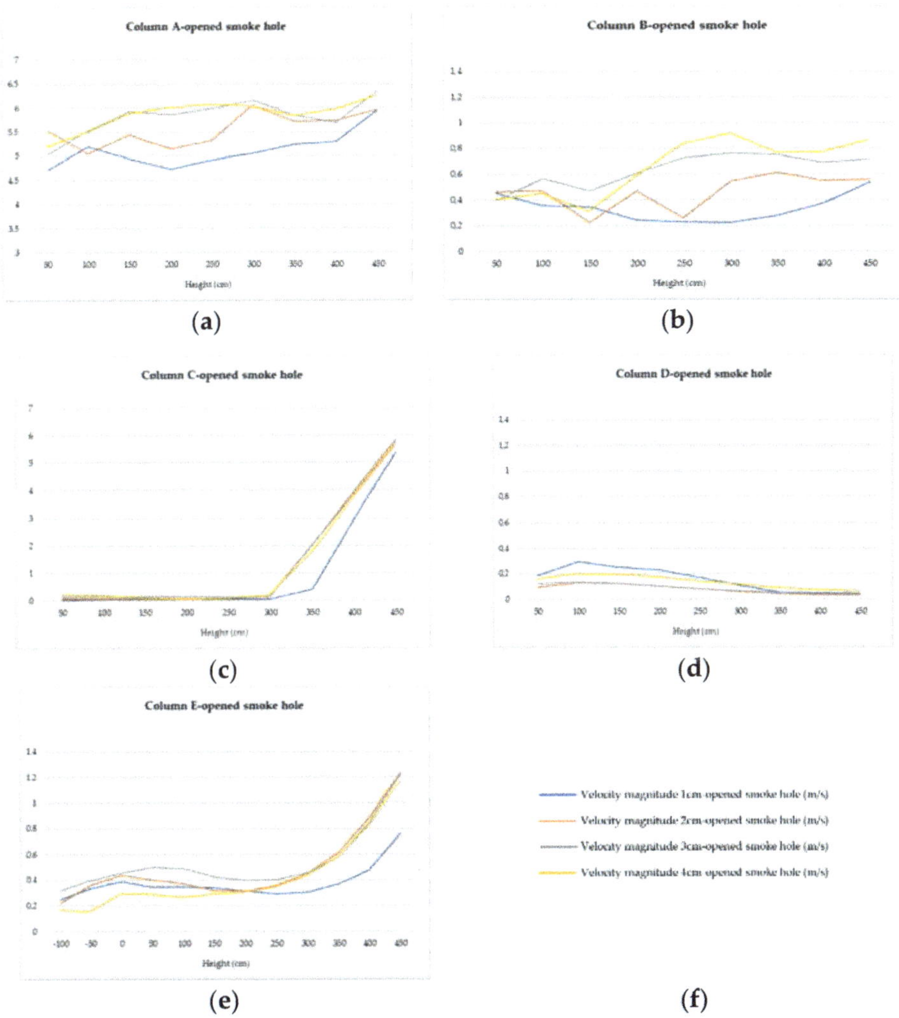

Figure 8. Values of velocity for the five columns of points, which is analyzed when the smoke hole of the plank house is opened (m/s) (Microsoft Excel®; Autodesk CFD®).

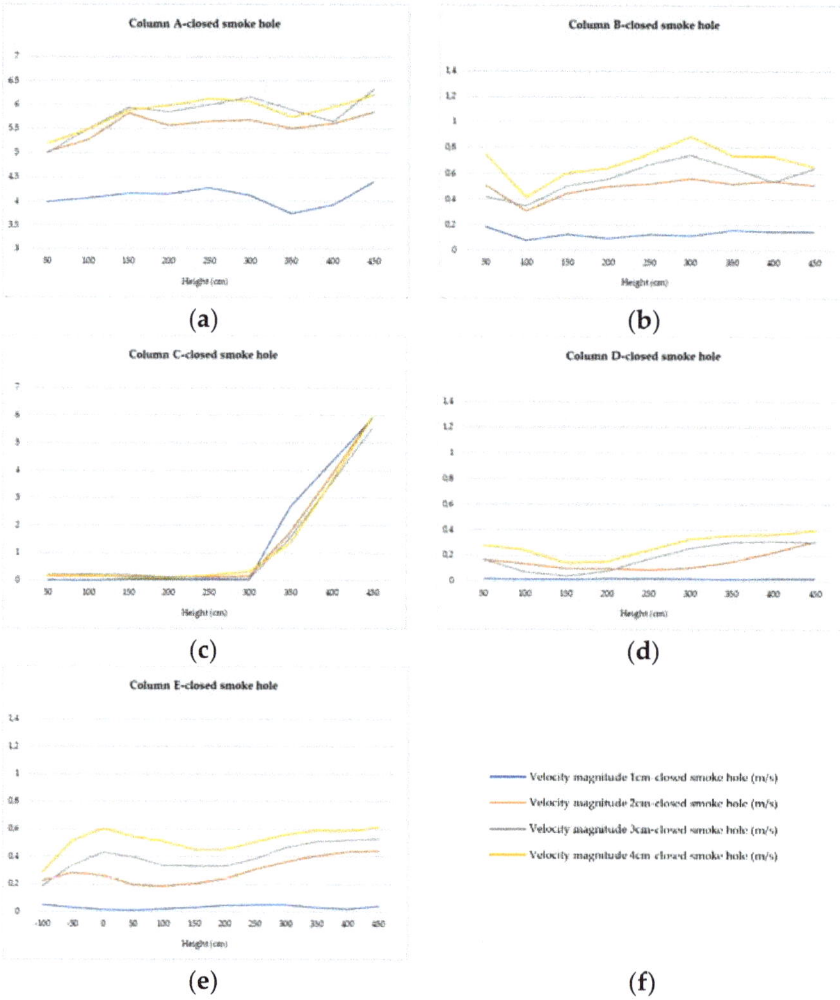

Figure 9. Values of velocity for the five columns of points, which is analyzed when the smoke hole of the plank house is closed (m/s) (Microsoft Excel®; Autodesk CFD®).

"A" is the column that gathers the higher velocity values (Figure 7). The highest values are not obtained by the thinnest gaps, in such a way that wind passes through the gaps that are 1 cm wide more slowly than through the 4 cm gaps. This difference is higher if compared with the closed-smoke hole model. Even though the velocity through the 1 cm gap is lower when the smoke hole is closed, it can be seen that it is almost invariable for the rest of the widths.

The greatest oscillations occur in columns "B" and "D", which are both located near the envelope. The velocity varies along them in an irregular way from the floor to the top of the house. However, their lowest values are reached close to the work plane (about 70 cm high), and the highest ones are found near the ceiling.

The most irregular case is the one of column "C". Its levels are similar to "B" at 50 cm below the floor, and they reach velocities as high as the "A" column when approaching the ceiling. So, the airflow behavior at this point is a combination of both "A" in the highest part of the dwelling and "B" in the lowest one. Whether the smoke hole is opened or closed, the result is the same.

Column "D" is near the outlet gaps. Column "C" is close as well, but the velocity is higher inside the gaps located besides "D". The difference between the results obtained for "C" and "D" is based on the proximity of the former to the smoke hole: that is, the element that triggers the velocity increase in "C" at 300 cm high. Actually, the velocity magnitude is almost null near the ceiling in column "D", which means that the smoke of the hearths could be trapped in this zone. The same velocity level takes place in column "C" below 300 cm high.

The column that is clearly influenced by the state of smoke hole is "E". As can be seen, the airflow is much more regular if this element is closed rather than if it is opened. The most remarkable fact is that the pit stays almost the same. The airflow velocity does not vary in this zone whatever the state of the smoke hole. These graphs also reflect that as expected, the velocity near the smoke hole is higher when it is opened.

In the following diagrams, the way the wind flows along the longitudinal axis of the house, the same as the wind vector, can be observed (Figures 10 and 11). They show how the opened smoke hole creates a quiet zone in the leeward elevation. This zone becomes smaller when the smoke hole is closed, and it turns softer and bigger as long as the gap between the planks becomes bigger. This fact reflects the way Haida transformed these dwellings along the year by moving these boards in order to let the air in. This way, there was no limit between indoors and outdoors, as happened in tipis.

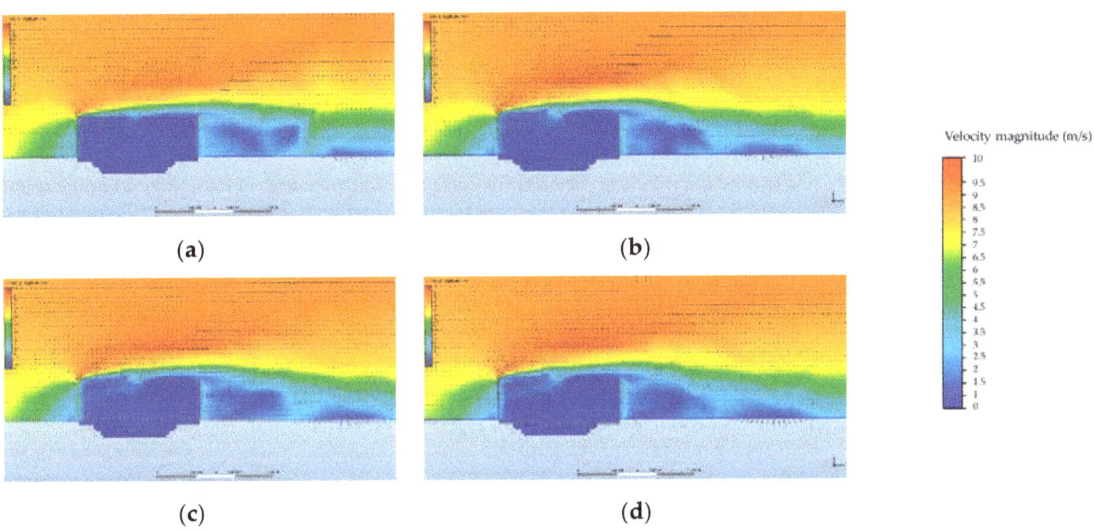

Figure 10. Airflow velocity and direction inside and nearby the plank house, whose smoke hole is opened (**a**) 1 cm gap; (**b**) 2 cm gap; (**c**) 3 cm gap; (**d**) 4 cm gap (Autodesk CFD®).

The airflow velocity inside the dwelling is lower when the smoke hole is closed, but it is higher on the part of the roof surface touched by the airflow from indoors. This way, the roof erosion was more intense when the smoke hole was closed.

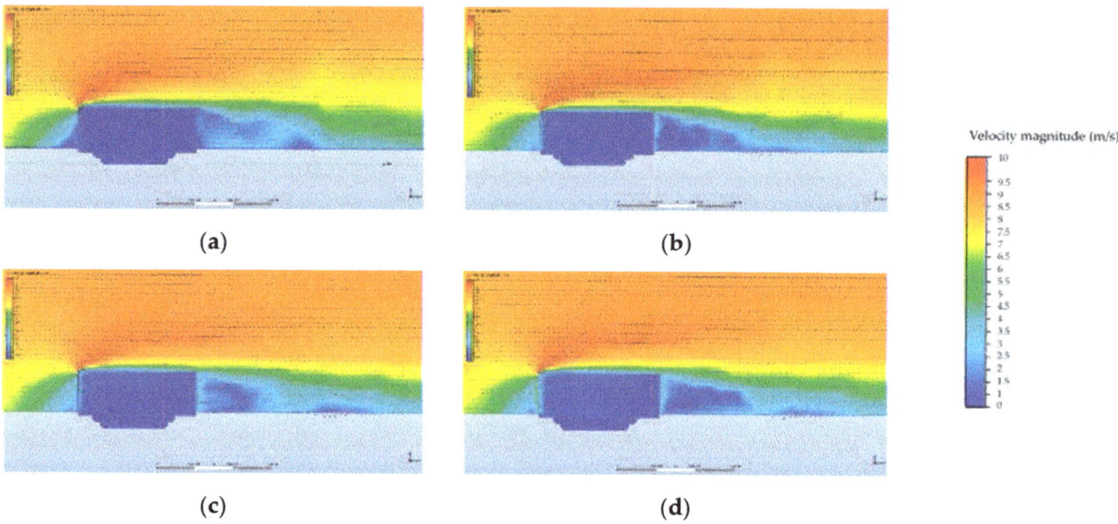

Figure 11. Airflow velocity and direction inside and nearby the plank house, whose smoke hole is closed (**a**) 1 cm gap; (**b**) 2 cm gap; (**c**) 3 cm gap; (**d**) 4 cm gap (Autodesk CFD®).

3.2.2. Airflow in the Central Pit

By means of the following diagrams (Figures 12 and 13), it has been possible to observe how the airflow behaves inside the central pit, where the hearths were lit. As they show, the wider the gaps are between the planks, the bigger the turbulence is inside the pit. They are even more complex if the smoke hole is closed.

As explained before, the airflow velocity was very low in this zone. This turbulence helped to keep the its air clean, but at the same time, the smoke would spread over the place more easily.

When the smoke hole is closed, the airflow in the pit stays in it and cannot exit the house directly. The only case where it becomes possible is the one of the 4 cm gaps (Figure 12f). If the gaps are thinner, the air in the pit has to be dragged by other flows that are able to exit.

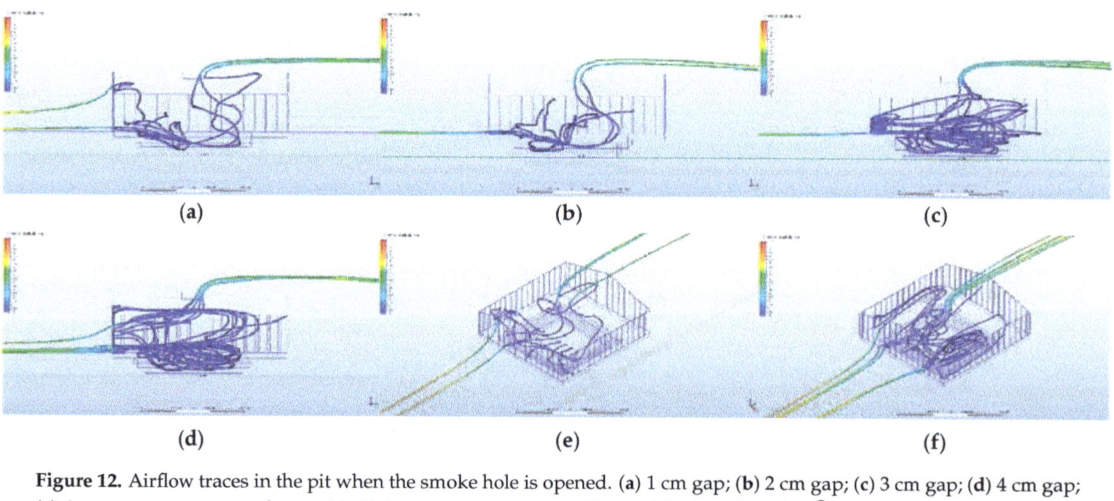

Figure 12. Airflow traces in the pit when the smoke hole is opened. (**a**) 1 cm gap; (**b**) 2 cm gap; (**c**) 3 cm gap; (**d**) 4 cm gap; (**e**) Axonometry corresponding to (**a**); (**f**) Axonometry corresponding to (**d**) (Autodesk CFD®).

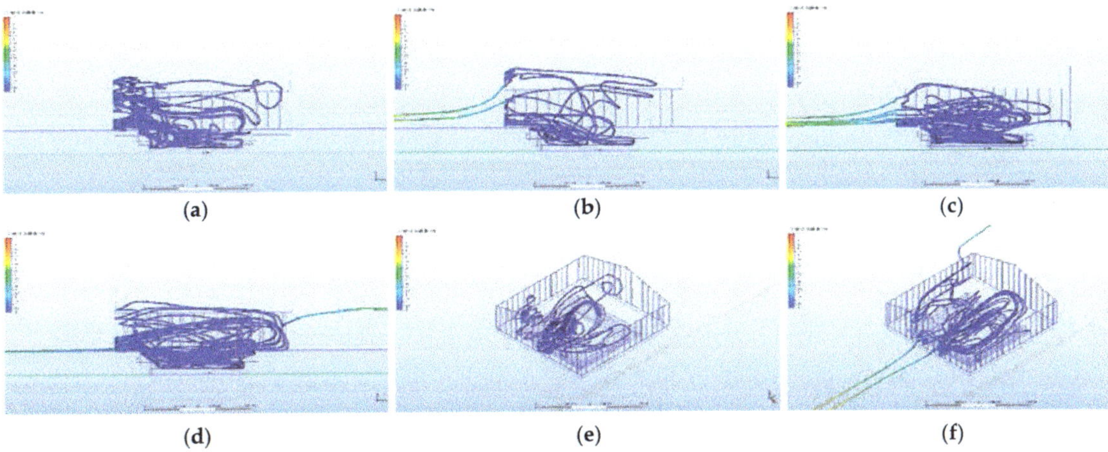

Figure 13. Airflow traces in the pit when the smoke hole is closed. (**a**) 1 cm gap; (**b**) 2 cm gap; (**c**) 3 cm gap; (**d**) 4 cm gap; (**e**) Axonometry corresponding to (**a**); (**f**) Axonometry corresponding to (**d**) (Autodesk CFD®).

Instead, when the smoke hole is opened, the air in the pit usually exits by itself.

In addition to the turbulence, it is important to take into account the velocity of the air particles. Figures 8 and 9 show that the airflow is almost static in the pit if the gaps are 1 cm wide.

4. Discussion

According to the results, it can be asserted that the plank house was a pit covered by a breathable envelope. The pit was the main area of the house, where the hearths were lit and where daily life took place. The origin of this space is lost in time, but it can be found in several Indian dwellings. In different ways, it is the main piece of the Navajo hogans [1] (pp. 325–336), of the pit-houses built by the Thompsons groups [1] (pp. 176–179) and, even, of the sod-houses built by the Yup'ik communities [19] (pp. 119–156). The thermal stability was the target to accomplish in most of these cases, but as shown in previous studies [20], there are some examples in which this function is not so clear. The plank houses are one of them. The depth of the pit is not enough to influence the indoor temperature of the house, so it has to have another aim. As can be seen in the previous graphs, it is a peaceful zone inside the house, where the hearths could be lit up and where the cold air from outdoors decelerated.

The idea of versatility is as important in the plank house as it was in the tipi, which is the most important nomad dwelling. The planks of Haida houses could be removed or displaced as the hide envelope of the tipis could be opened or closed. This way, both models search for the continuity between indoors and outdoors, as nomad tribes do in their daily life [21]. However, the most important point is that these communities analyze conscientiously their environment and take advantage of the opportunities it offers to them [22]. As Atta [23] asserts, an affordable and sustainable housing design requires integrating economic and environmental solutions from a social point of view. Traditional architecture offers a unique opportunity to investigate how these three factors can coexist successfully.

As the research developed by Knowles and Easton established [1,24], Native American architecture is much more complex than expected at first sight. Their research was focused on Acoma dwellings, which are adobe pueblos from New Mexico. Knowles determined that the orientation of Acoma dwellings let the houses take advantage of the solar radiation as much as possible; meanwhile, Easton analyzed the ascending currents of air that occur along the main elevation of the Pueblo houses.

As another example, the tipis could be taken. Their conical design was based on the shape of a Cottonwood tree leaf [25] (p. 7). This specie of tree is well known for being able to resist strong winds, as the tipis do. Its leaves move, adapting to the wind direction in order to survive. This is the way the ventilation system works in the tipis, taking advantage of the main wind in such a way that it takes out the hearth smoke from indoors.

The present research is rooted in the field of virtual archaeology, which is a discipline that offers an enriching amount of opportunities for other areas. The virtual reconstruction of disappeared buildings can show forgotten solutions that are particularly useful nowadays. In this case, the reconstruction of the Haida plank house has been possible thanks to the archaeological records about Ninstints as well as due to the journals written by the first European and Russian explorers. Research studies such as those developed by Trebeleva and Glazov [26] are undoubtedly essential to understand how the ancient buildings were designed and the technical solutions they contained.

The plank house that has been analyzed is the one built by the Haida community in the Haida Gwaii Islands. However, the "plank house" model can be found all along the Pacific coast of North America in different ways. From California, where the Hoopa built their own models, to the coastal Salish in the Vancouver Island and Puget Sound, who dwelled their shed houses, there are enough models to carry out a deeper comparison study.

5. Conclusions

By analyzing the results shown previously, it can be concluded that if the gaps between the planks of the Haida dwelling are wider than 4 cm, the summer airflow comfort limits (1.1 m/s) would be exceeded inside the building [27] (p. 305). However, it should be taken into account that the present research has assigned the same width to every gap in each case, which is a variable that could be studied in further research.

This way, it can be seen that the envelope of the Haida plank house let the indoor area breath. This dwelling is composed of two elements: this envelope and a central pit. The Native American dwellings were usually designed around a central hearth that is protected by an envelope [1]. This envelope can establish a solid limit between indoor and outdoor, as, for instance, happens in the New Mexican adobe pueblos, but there is another group of dwellings that uses this envelope to let the indoor area breath. Among them, the Wichita grass houses can be found. As occurs in the Haida plank houses, their grass envelope is designed to let the smoke out; that is to say, the whole envelope breathes. This type of solution can also be found in the thatched British cottages, confirming the idea that this type of solution is common in high humidity locations. What turns the Haida plank house into a special case is the combination of this breathing envelope and a central pit. As can be seen in Figures 10 and 11, if the smoke hole is closed, this zone is protected from the air currents that take the smoke out and keep the cedar planks dry. Thereby, the smoke hole works as a ventilation trigger and can be used to renovate the air of the dwelling at will.

By means of modern technology, this type of envelope is being developed at the moment by many architecture research centers and offices. For instance, the design developed by the engineer Tobias Becker from Stuttgart can be cited [28]. His envelope is composed of multiple cells controlled by changes in air pressure. This way, they regulate the incident light and the airflow as skin pores and pigments regulate a living being. In addition, the idea behind the Al Bahar towers should be highlighted [29]. Built in Abu Dhabi by Abdulmajid Karanouh and Aedas, their envelopes are integrated by multiple cells inspired in the Arabic lattice. These cells are closed or opened according to the meteorology, since they can let the air and the solar radiation in or they can shield the buildings from them. The prototype developed in Los Angeles by professor Doris Kim Sung also offers an interesting option [30]. Called Bloom, this experiment proposes the use of metal sheets that curl when heated in such a way that the whole structure casts shadows and lets ventilation under specific areas when temperature rises. Obviously, all these projects are technically much more complex than the plank house built by the Haida several centuries ago, but the concept hidden behind all of them was the same. The point was not just to design a breathing envelope but to get the best of every element that composes a building by means of a deep knowledge of the possibilities offered by the environment in order to save the available resources.

Author Contributions: Conceptualization, R.A.G.L. and M.J.M.B.; methodology, R.A.G.L. and M.J.M.B.; software, R.A.G.L. and M.J.M.B.; investigation, M.J.M.B.; resources, R.A.G.L.; writing—original draft preparation, review and editing, R.A.G.L. and M.J.M.B.; supervision, R.A.G.L.; project administration, R.A.G.L.; funding acquisition, R.A.G.L. All authors have read and agreed to the published version of the manuscript.

Funding: The authors wish to thank CEU San Pablo University Foundation for the funds dedicated to the Project Ref. USP CEU-CP20V12 provided by CEU San Pablo University.

Institutional Review Board Statement: Not applicable.

Informed Consent Statement: Not applicable.

Conflicts of Interest: The authors declare no conflict of interest.

References

1. Nabokov, P.; Easton, R. *Native American Architecture*; Oxford University Press: New York, NY, USA, 1989.
2. Spencer, R.; Jennings, J. *The Native Americans: Ethnology and Backgrounds of the North American Indians*; Harper & Row: New York, NY, USA, 1977.
3. Gunther, E. *Indian Life on the Northwest Coast of North America (as Seen by the Early Explorers and Fur Traders during the Last Decades of the Eighteenth Century)*; The University of Chicago Press: Chicago, IL, USA, 1972.
4. Pritzker, B. *A Native American Encyclopedia: History, Culture and Peoples*; Oxford University Press: New York, NY, USA, 2000.
5. Bakeless, J. *America as Seen by Its First Explorers: The Eyes of the Discovery*; Dover Publications: New York, NY, USA, 1989.
6. MacDonald, G. *Ninstints, Haida World Heirtage Site*; University of British Columbia Press in Association with the University of British Columbia Museum of Anthropology: Vancouver, BC, Canada, 1987.
7. Sobel, E.; Gahr, D.; Ames, K. *Household Archaeology on the Northwest Coast*; Berghahn Books: Ann Abour, MI, USA, 2006.
8. Underhill, R. *Indians of the Pacific Northwest*; United States Department of the Interior, Bureau of Indian Affairs, Branch of Education: Washington, DC, USA, 1944.
9. Jarzombek, M. *Architecture of First Societies*; John Wiley & Sons: Hoboken, NJ, USA, 2013.
10. MacDonald, G. *Haida Monumental Art: Villages of the Queen Charlotte Islands*; University of British Columbia Press: Vancouver, BC, Canada, 1995.
11. Stewart, H. *Cedar: Tree of Life to the Northwest Coast Indians*; Douglas & McIntyre: Madeira Park, BC, Canada, 1995.
12. Dawson, G.; Geological Survey of Canada. *Report of Progress for 1878-79. Report on the Queen Charlotte Islands-1878*; Dawson Bros.: Montreal, QC, Canada, 1880.
13. Niblack, A. The coast Indians of southern Alaska and northern British Columbia; based on the collections in the U.S. National Museum, and on the personal observation of the writer in connection with the survey of Alaska in the seasons of 1885, 1886, and 1887. In *U.S. National Museum Annual Report 1888*; Johnson Reprint Corporation: New York, NY, USA, 1890; pp. 225–386. Available online: https://archive.org/details/coastindiansofso00nibl/page/n5/mode/2up (accessed on 5 August 2021).
14. Yarke, E. *Ventilación Natural de Edificios*; Nobuko: Buenos Aires, Argentina, 2005.
15. Autodesk Help-Natural Ventilation. Available online: https://knowledge.autodesk.com/support/cfd/learn-explore/caas/CloudHelp/cloudhelp/2014/ENU/SimCFD/files/GUID-F6011744-B69B-47D0-BB2B-50E1BA8DC32D-htm.html (accessed on 24 May 2021).
16. Windfinder. Available online: https://es.windfinder.com/#3/49.5042/9.5421 (accessed on 24 May 2021).
17. Li, J.; Delmas, A.; Donn, M.; Willis, R. Validation and Comparison of Different CFD Simulation Software Predictions of Urban Wind Environment Based on AIJ Wind Tunnel Benchmarks. In Proceedings of the Symposium on Simulation for Architecture and Urban Design, Delft, The Netherlands, 4–7 June 2018; Available online: https://dl.acm.org/citation.cfm?id=3289777 (accessed on 5 August 2021). [CrossRef]
18. Autodesk Simulation CFD 2021-Verification. Available online: https://help.autodesk.com/view/SCDSE/2021/ENU/?guid=GUID-3DD39B64-5638-4D0D-9175-1FBC95E629FC (accessed on 23 July 2021).
19. Lee, M.; Reinhardt, G. *Eskimo Architecture: Dwelling and Structure in the Early Historic Period*; University of Alaska Press: Fairbanks, AL, USA, 2003.
20. Montero Burgos, M.J.; Sanchiz Álvarez de Toledo, H.; González Lezcano, R.A.; Galán de Mera, A. The Sedentary Process and the Evolution of Energy Consumption in Eight Native American Dwellings: Analyzing Sustainability in Traditional Architecture. *Sustainability* **2020**, *12*, 1810. [CrossRef]
21. Couchaux, D. *Habitats Nomades*; Alternatives: Paris, France, 2011.
22. Rapoport, A. *House form and Culture*; Prentice-Hall: Englewood Cliffs, NJ, USA, 1969.

23. Atta, N.; Dalla Valle, A.; Campioli, A.; Chiaroni, D.; Talamo, C. Construction Technologies for Sustainable Affordable Housing within Fragile Contexts: Proposal of a Decision Support Tool. *Sustainability* **2021**, *13*, 5928. [CrossRef]
24. Knowles, R. *Energy and Form*; The MIT Press: Cambridge, MA, USA, 1980.
25. Goble, P. *Tipi: Home of the Nomadic Buffalo Hunters*; World Wisdom: Bloomington, IL, USA, 2007.
26. Trebeleva, G.; Glazov, K.; Kizilov, A.; Sakania, S.; Yurkov, V.; Yurkov, G. Roman Fortress Pitiunt: 3D-Reconstruction of the Monument Based on the Materials of Archaeological Research and Geological Paleoreconstructions. *Appl. Sci.* **2021**, *11*, 4814. [CrossRef]
27. Neila, F. *Arquitectura Bioclimática en un Entorno Sostenible*; Munilla-Lería: Madrid, España, 2004.
28. Tébe. Available online: https://www.tebe.berlin/innovation/ (accessed on 25 July 2021).
29. Karanouh, A.; Kerber, E. Innovations in dynamic architecture. *J. Façade Des. Eng.* **2015**, *3*, 185–221. [CrossRef]
30. DO | SU Studio Architecture. Available online: https://www.dosu-arch.com/bloom (accessed on 25 July 2021).

energies

Article

Implementation of a Ventilation Protocol for SARS-CoV-2 in a Higher Educational Centre

Alberto Meiss *, Irene Poza-Casado *, Alfredo Llorente-Álvarez, Héctor Jimeno-Merino and Miguel Ángel Padilla-Marcos

RG Architecture & Energy, Universidad de Valladolid, Av. Salamanca, 18-47014 Valladolid, Spain; llorente@arq.uva.es (A.L.-Á.); hector.jimeno@uva.es (H.J.-M.); miguelangel.padilla@uva.es (M.Á.P.-M.)
* Correspondence: alberto.meiss@uva.es (A.M.); irene.poza@uva.es (I.P.-C.)

Abstract: The most recent research confirms that airborne transmission may be the dominant mode of SARS-CoV-2 virus spread in the interior spaces of buildings. Consequently, based on some prescriptions that implemented natural ventilation during face-to-face lessons in a university centre, an experimental characterization of several complementary options aimed at reinforcing the prevention and safety of the occupants was carried out. The action protocol adopted was based on the combination of mandatory natural ventilation, a maximum contribution of outdoor air supply in the air conditioning system, and the use of filtering devices located inside the classroom. All the strategies were incorporated concomitantly with necessary compliance with the basic conditions of social distance, occupation, use of masks and guidelines for use and cleaning within educational buildings. The suitability of this protocol was further evaluated throughout the teaching day with students and teachers by measuring the CO_2 concentration. The results showed that the measures implemented successfully removed the possible pollutants generated inside.

Keywords: indoor air quality; COVID-19; educational buildings; air purifier; airborne transmission

Citation: Meiss, A.; Poza-Casado, I.; Llorente-Álvarez, A.; Jimeno-Merino, H.; Padilla-Marcos, M.Á. Implementation of a Ventilation Protocol for SARS-CoV-2 in a Higher Educational Centre. *Energies* **2021**, *14*, 6172. https://doi.org/10.3390/en14196172

Academic Editor: Christopher Micallef

Received: 1 September 2021
Accepted: 23 September 2021
Published: 27 September 2021

1. Introduction

Throughout the current global pandemic caused by the SARS-CoV-2 virus, there has been a significant evolution of the knowledge of contagion. In the early stages of preventive action, the recognized contagion mechanisms were [1]:

- Direct contact between a healthy person and an infected person;
- Indirect contact or transmission through fomites, in which a healthy individual gets infected through his/her mucosae or conjunctiva after touching a contaminated object or surface;
- Close contact through respiratory droplet transmission and contact routes. "Large droplets" or "ballistic droplets" (greater than approximately 100 μm) are expelled by an infected person when coughing, sneezing and, to a lesser extent, when speaking.

At the beginning of February 2020, it was still not considered necessary for the population to wear face masks. In this way, mandating the population to assume protection mechanisms that might not make sense was avoided [2]. It was in May 2020 when the mandatory use of masks was decreed in Spain [3], following the pattern of the rest of the developed countries, with the predominant aim of blocking the emission of "large droplets" infected with the virus.

Only recently, in April 2021, has the World Health Organization (WHO) begun to as-sess airborne transmission, although emphasis continues to be placed on the fomite path and large droplets [4]. However, there is a current endorsement by a growing number of scientists who do not agree with this limited view [5].

In the field of interior spaces in buildings, the term bioaerosol includes the droplets exhaled by people which can be between 0.4 and 100 μm and remain suspended in the air for a long time [6]. In other contexts, the emission sources can be any living beings in any terrestrial or marine ecosystem.

At an early date (July 2020), the Directorate of the Architecture Technical School in Valladolid developed a supplementary strategy to face a possible airborne transmission method in addition to the strict measures adopted regarding social distance, occupancy, the use of masks and guidelines for the use and cleaning of the building.

It is important to inform society that bioaerosols may be the dominant mode of transmission of the SARS-CoV-2 virus, with some contribution from the fomite pathway and an even smaller contribution of large droplets.

2. State of the Art

2.1. Bioaerosols and SARS-CoV-2

Bioaerosols can contain in their structure:

- Pathogenic bacteria and fungi;
- Virus;
- Allergens and endotoxins;
- Pollen and plant fibres.

The size of an individual SARS-CoV-2 virus is very small (around 0.10 μm), which enables its lodgement in the bioaerosols generated by an infected person (Figure 1).

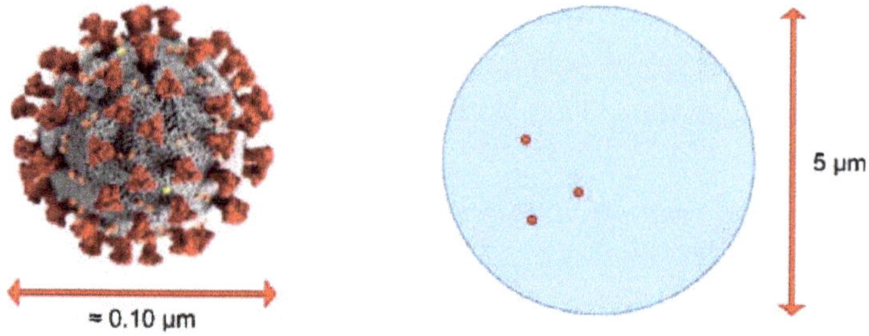

Figure 1. Dimensions of SARS-CoV-2 [7] and its possible distribution within a bioaerosol.

The virus uses the bioaerosol as a means of transport and, depending on the bioaerosol's size, can affect different organs of the recipient body, which in turn influences the viral load that will determine the contagion.

All bioaerosols smaller than 100 μm can be inhaled. There are three main mechanisms for the deposition of bioaerosols in the respiratory tract (Figure 2):

- The larger ones, once aspirated, do not transcend beyond the vicinity of the nostrils and mouth, since they move with difficulty and impact with the walls of those cavities;
- Bioaerosols smaller than 10 μm can penetrate deeper into the respiratory tract and the viruses contained in them can reach its receptors;
- For bioaerosols in the range around 0.3 μm, a singular fact occurs: a significant percentage of bioaerosols are deposited in the respiratory tract when inhaling and are released again when exhaling. This range matches the aerosols of cigarette smoke, and that is the reason why smoke is exhaled when smoking;
- Finally, the smallest bioaerosols are very diffusive, having a high Brownian movement, so that many of them do not transcend beyond the nostrils and mouth.

2.2. Classification of the Droplets Emitted by People

In the case of sedentary or light activity (1.0 to 1.2 met), human emissions through the respiratory tract are limited to events such as breathing, speaking, coughing or sneezing. In each of these events, a mixture of air and droplets of different sizes and at a variable speed is exhaled (Table 1).

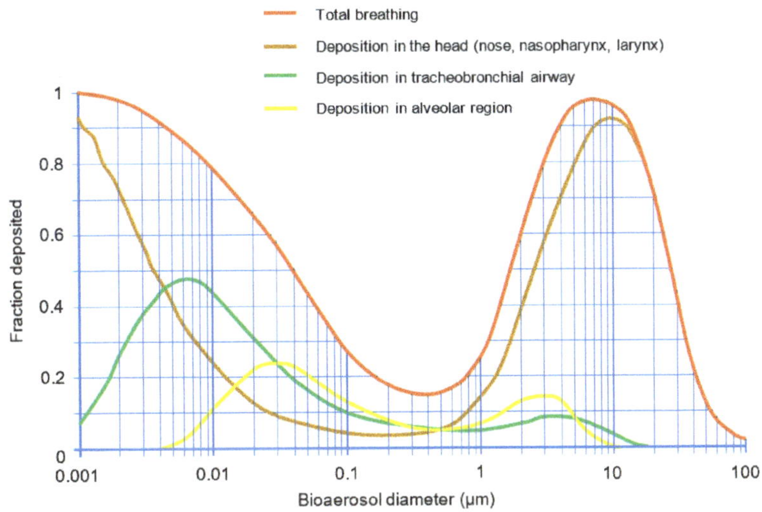

Figure 2. Mathematical model of particle deposition of nasal breathing [7].

Table 1. Average emission per person and event [8].

Event	Droplets Generated (Per Event)	Velocity (m/s)	Exhaled Air (L)
Sneezing	100,000–1,000,000	20–50	1.2
Coughing	3000–10,000	10	1.2
Talking	3900/min	5	700 (1 h)
Breathing	1200/min	1	800 (1 h)

Their distribution based on sizes also varies according to the type of event, which influences the way in which contagion from the infected person could occur (Figure 3, [9–12]). A usual classification establishes three ranges of size:

- Larger droplets (greater than 300 μm);
- Transition droplets (size between 100 and 300 μm);
- Bioaerosols (smaller than 100 μm).

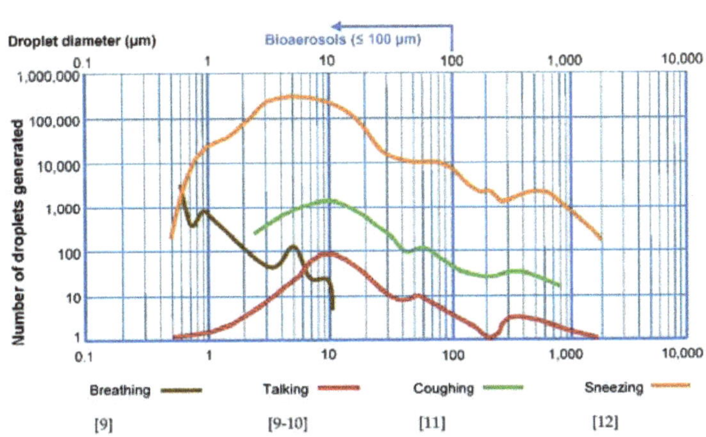

Figure 3. Droplets emitted according to the type of event.

2.3. Evaporation

Evaporation processes affect bioaerosols emitted by people. Thus, the distribution according to droplet size is not only dependent on the event but also a transformation through evaporation (Figure 4). This significantly affects the transition droplets (between 100 and 300 μm).

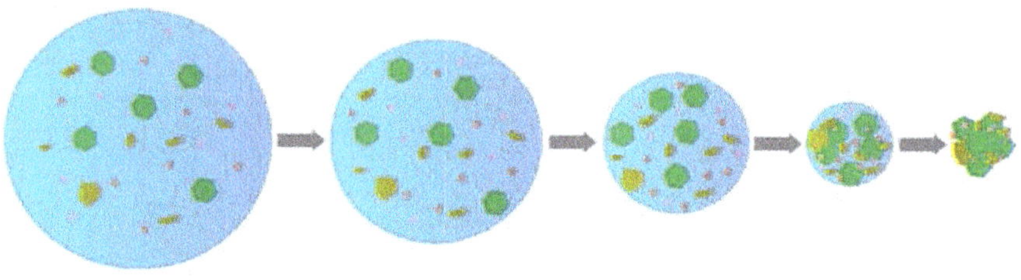

Figure 4. Droplet evaporation process (taken from Verreault, 2008 [13]).

Thus, the rate of evaporation of the droplets depends on the humidity conditions of the environment (the lower the humidity, the higher the evaporation) and the emission speed (depending on the type of event). Based on these parameters, part of the droplets precipitate and others undergo an evaporation process, which allows them to diffuse as bioaerosols (Figure 5). Large droplets are those that could cause infection if they hit the eyes, nose, or mouth of a healthy person. In these cases, social distance greater than 1 m serves as effective protection, as well as the mandatory use of masks.

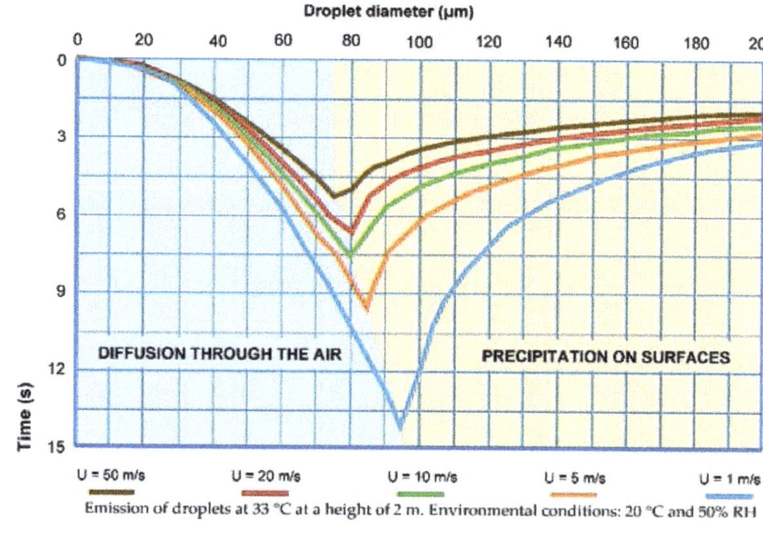

Figure 5. Behaviour of emissions according to their diameter [14].

It is the particles smaller than 100 μm that normally acquire the properties of an air-diffused bioaerosol, which, when inhaled, transmit the disease through the respiratory tract.

A fundamental question is to know how long the SARS-CoV-2 virus remains infectious in bioaerosols emitted by a patient. The life expectancy of the virus is conditioned by the loss of viral infectivity.

However, there is still no consensus on this matter since the time varies between 1 and 3 h under typical temperature and humidity conditions (20 °C and 50% RH) [15–17]. This range of time is important in order to confirm that any virus present in the air of a room has lost its contagiousness after a few hours. On the other hand, it has been demonstrated that the virus remains infectious over much longer periods of time on surfaces (Table 2).

Table 2. SARS-CoV-2 viability in different materials [15].

Material	Time that SARS-CoV-2 Remains Infectious on Surfaces
Plastic	72 h–3 days
Stainless steel	48 h–2 days
Paperboard	4 h
Copper	1–3 h

2.4. Dynamic Behaviour of Bioaerosols

Once the droplet is emitted by the infected person, its dynamic behaviour is determined by the simultaneous and combined action of a system of forces. The acting forces are the gravity force, the buoyancy force and the drag force (Figure 6). The preponderance of one over the others basically depends on the size of the droplet, which determines its diffusion and permanence in the air of the room.

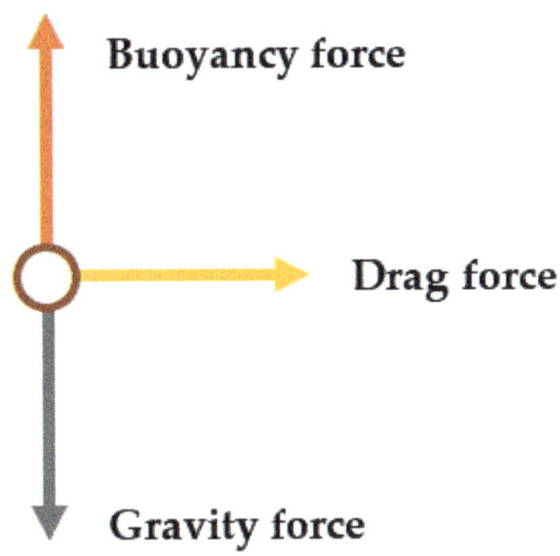

Figure 6. System of forces affecting the bioaerosol.

In the case of the larger droplets, the drag force (from the impulse speed) and the gravity force are dominant, so their impact or their precipitation is favoured after a few seconds. In the case of bioaerosols, buoyancy is the main force, favouring their diffusion and increasing their permanence in the air.

2.5. Standard Protocol

Therefore, it is verified that bioaerosols serve as a vehicle for transporting the SARS-CoV-2 virus when they are emitted from an infected person. Its diffusion capacity is very wide, being able to follow a highly variable pathway from its emission until the bioaerosol leaves the room or the virus loses its infectivity. For this reason, it is necessary to adequately ventilate interior spaces.

The instructions received from the government authorities [18] motivated the development of a first strategy in the classrooms throughout the 2020/2021 school year:

- The frequent opening of the windows was established (without specifying time);
- In air conditioning systems, the supply of outside air was maximized (up to 50% of the circulating flow);

- Low occupancy of the classrooms was set;
- A safe distance between people was determined (1.5 m).

Although the measures were correct as a basic prevention mechanism, a series of drawbacks that could affect the desired level of prevention were identified:

- Natural ventilation was relied upon as the primary mechanism for removing indoor polluted air. Problems derived from its random and chaotic nature were not considered. This caused long periods in which a stable ventilation airflow was not achieved (due to the lack of wind pressure, limited indoor/outdoor thermal gradient, absence of cross flow in the room, little stack effect, etc.);
- Thermal discomfort in the classrooms during the cold season was identified, which entailed the reduction of the cognitive abilities of the students and risk to their health;
- In the case of natural ventilation flows, the appearance of annoying drafts was verified. Thus, the indoor air mixing patterns were altered, and, as a consequence, there were large unventilated short-circuit areas.

Accordingly, in order to supplement the above measures and increase the prevention of possible contagion, an experimental study was carried out [19]. The indoor air quality (IAQ) was assessed assuming strict compliance with the aforementioned provisions, and alter-native proposals based on the use of air purifiers were evaluated.

3. Methods

3.1. Experimental Study in a Typical Classroom

Successive tests were carried out in a representative classroom of the building of the Architecture Technical School of the University of Valladolid (Figure 7) under different ventilation conditions.

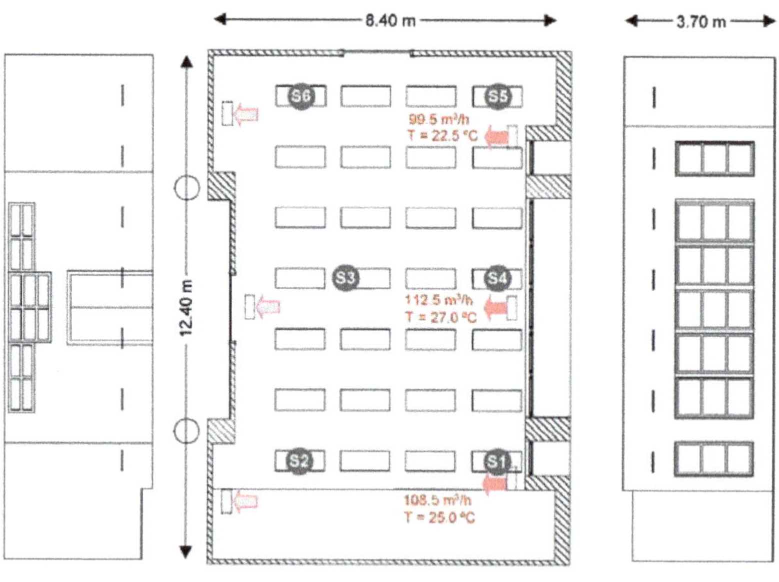

Figure 7. Typical classroom of the Architecture Technical School in Valladolid.

The classrooms have an air conditioner that vertically blows heat-treated air into the room through three inlets in the ceiling and recirculates the indoor air, which is mixed with fresh outside air (50% of the airflow, as a preventive measure against SARS-CoV-2). The windows were modified to allow total opening of their openable surface.

Tests were carried out according to the concentration drop method. In this way, aerosols were released in a controlled way up to a certain concentration, achieving homogeneous distribution by means of fans located inside the classroom. Once the concentration was stabilized, measurements were taken with six sensors (S1 to S6) distributed through-out the room.

To simulate the behaviour of bioaerosols, the emission of smoke was used, since they behave in an analogous way: the smoke contains aerosols with a similar range of sizes (the visible part of the smoke) and gases (recognizable by their smell, which do not constitute a vehicle for virus transport).

The behaviour of smoke emission is analogous to that of bioaerosols released in an environment. When a person smokes, the smoke does not fall to the ground quickly, but rather concentrates in front of the smoker and then mixes with the airflows. If this occurs in a poorly ventilated room, the smoke accumulates. In these situations, only a small fraction, less than 10%, is deposited on the interior surfaces, and the rest remains floating in the environment until the interior air is renewed.

The concept of smoke serves also to illustrate a situation between two people [18] where one person is exhaling smoke and the other wishes to inhale as little of it as possible: this is how the chances of contagion are reduced. This experience carries over to the use of masks: the aerosols are filtered with the masks, but the gases are not. The fact that the odour penetrates the mask does not mean that the mask is not retaining the aerosols.

A smoke machine (1.2 kW model Fz-1200) was used to carry out the tests. Its operation consisted of the combustion of oils that emit fumes with bioaerosols analogous to the composition of the air, allowing its traceability in its airflow path, as well as measuring its concentration by means of multipurpose sensors.

3.2. Filtering Device

As a significant improvement to the initial protocol, it was proposed to incorporate an air filtration device into the classroom. The mechanical capture of particles and aero-sols was done by means of a HEPA H14 filtering surface with an efficiency greater than 99.995%. The efficiency represents the percentage of particles that each type of filter can retain with respect to the most penetrating particle size (MPPS), which is normally be-tween 0.15 and 0.25 μm. Thus, a HEPA H14 filter with an efficiency of 99.995% for a particle size of greater penetration of 0.15 μm implies that for every 100,000 particles of that size, the filter only allows 5 of them to pass through. For particles of other sizes, the filter has an even higher retention capacity.

The flow rates treated depend on the adjustable flow rate (SETPOINT), according to the data supplied by the manufacturer (Figure 8):

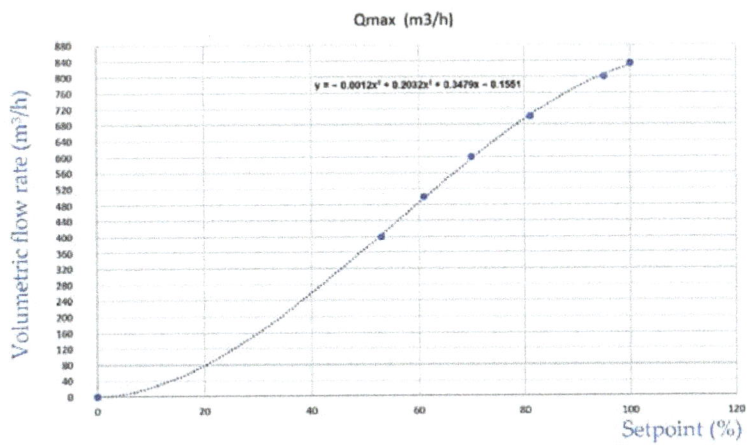

Figure 8. Adjustable flow rate of air filtration device.

A parameter that requires further study is the flow pattern of treated and contaminated air in order to determine the recommended position and height of the air purifiers within the classroom [20]. For the tests conducted for this study, the device was placed in a central position in the space. Since a mixture flow pattern was achieved inside the studied classroom, it was possible to determine that adequate purification of the polluted air was guaranteed [21–25].

4. Results

The experimental study consisted of 12 cases with different configurations regarding the opening of windows and doors, the operation of the air conditioning system with the supply of outdoor air, and the use of a filtration device:

- Case 1: 100% of the windows open and doors closed. Air conditioning system off;
- Case 2: 50% of the windows open and 50% of the doors open. Air conditioning system off;
- Case 3: 100% of the windows open and 100% of the doors open. Air conditioning system off;
- Case 4: 100% of the windows and doors closed. Constant air filtration flow (400 m^3/h) through a HEPA 14 filter. Air conditioning system off;
- Case 5: 100% of the windows and doors closed. Constant air filtration flow (330 m^3/h) through a HEPA 14 filter. Air conditioning system off;
- Case 6: 100% of the windows open and doors closed. Air conditioning system on;
- Case 7: 50% of the windows open and 50% of the doors open. Air conditioning system on;
- Case 8: 100% of the windows open and 100% of the doors open. Air conditioning system on;
- Case 9: 100% of the windows open and doors closed. Constant air filtration flow (550 m^3/h) through a HEPA 14 filter. Air conditioning system on;
- Case 10: 50% of the windows open and 50% of the doors open. Constant air filtration flow (550 m^3/h) through a HEPA 14 filter. Air conditioning system on;
- Case 11: 100% of the windows open and 100% of the doors open. Constant air filtration flow (550 m^3/h) through a HEPA 14 filter. Air conditioning system on;
- Case 12: 100% of the windows open and doors closed. Constant air filtration flow (550 m^3/h) through a HEPA 14 filter. Air conditioning system on.

Throughout all the tests, the outdoor temperature was in the range of 8 to 10 °C, which represents a typical winter situation in Valladolid (Spain).

The results obtained in the tests are summarised in Table 3 and Figure 9, which reflect the average values of the sensors distributed within the room.

Table 3. Aerosol concentration decay time to 10μm/mm^3.

Case	Concentration Decay Time 10 μm Particles Range between 3250 and 200 μg/mm^3
Case 1	>45 min
Case 2	29 min
Case 3	17 min
Case 4	24 min
Case 5	33 min
Case 6	19 min
Case 7	18 min
Case 8	15 min
Case 9	10 min
Case 10	8 min
Case 11	6 min
Case 12	18 min

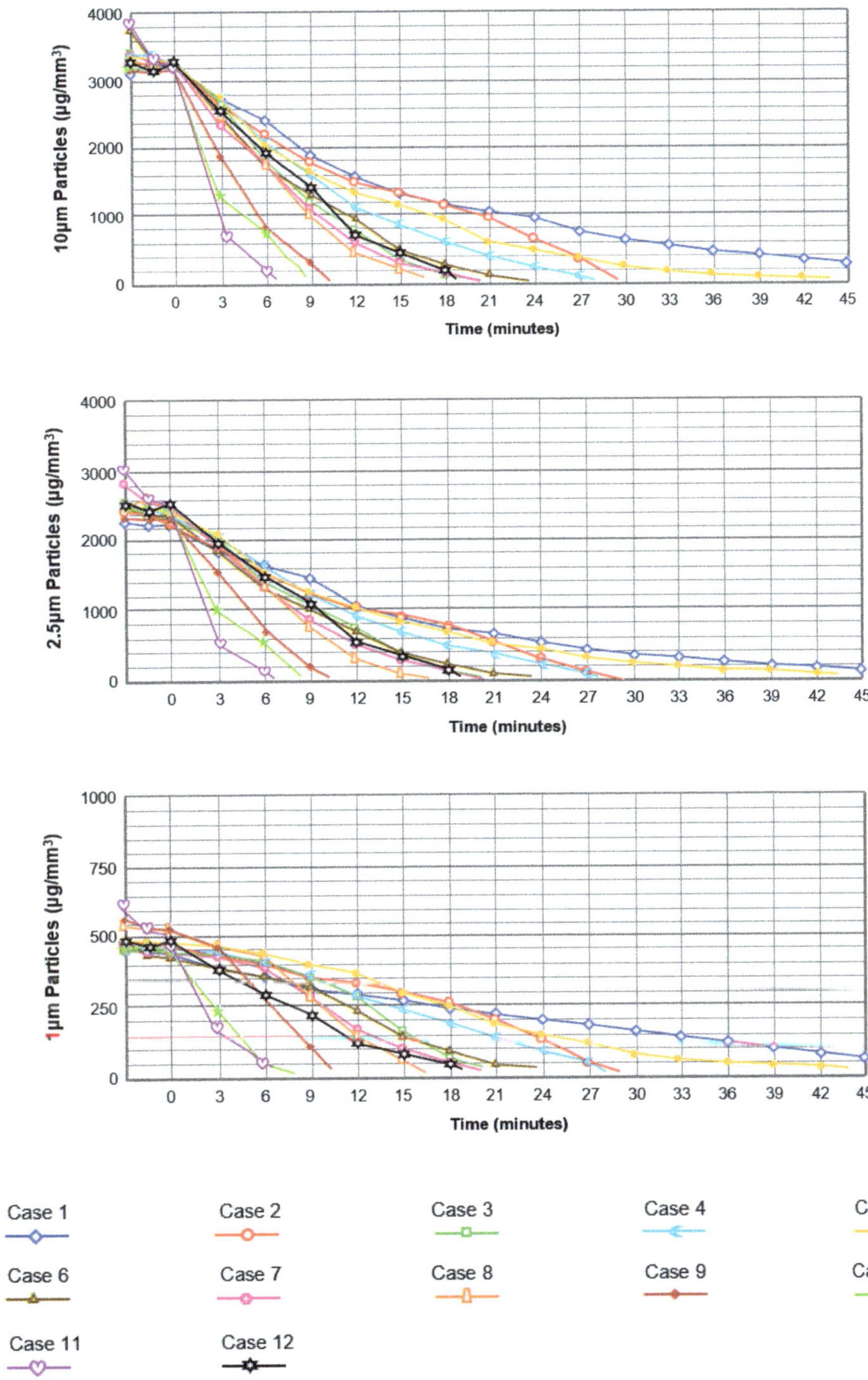

Figure 9. Evolution of the concentration drop of aerosols emitted in the classroom.

5. Discussion

5.1. Conclusions Derived from the Results

From these tests, the following general conclusions can be derived:

- The combined action of natural ventilation, the air conditioning system (with 50% of outdoor air) and the filtering device is the strategy that removes particles and aerosols in the shortest time (Case 11);
- Guaranteeing a mixture flow pattern through the air supply of the air conditioning system enables, in all cases, the best performance (Cases 6 to 11);
- If the exterior and interior conditions allow it, natural ventilation achieves very good results (Case 3). However, natural ventilation does not offer reliable particulate removal conditions due to its chaotic and unpredictable nature. The measurements made were taken at a certain moment and cannot be considered as the average of conditions that can be guaranteed. On the other hand, natural ventilation causes an unacceptable interior temperature drop far from comfort conditions;
- In the event that the windows are closed to prevent the entry of outside air without conditioning in winter (taking into account thermal comfort), the simultaneous use of the filtering device and the air conditioning system with outside air supply (Case 12) reaches average values analogous to a natural ventilation regime, with the added guarantee of controlling the stability of the ventilation process;
- The results obtained when the opening surface of the windows and doors is reduced to 50%, in combination with the operation of the air conditioning system (with 50% outdoor air supply) and the filtering device (Case 10), reach similar quality standards to the best case study. The relative advantage is a better quality of the thermal environment with respect to the complete opening.

5.2. Verification of the Ventilation Conditions

Finally, after the previous comparative analysis, the classroom under study was subjected to a test under typical conditions, that is, with 29 students and a teacher during the usual course of a class. During the test, the outside temperature was 3 °C on average, so the openings were partially closed (50% of their surface) in a similar way to Case 11.

In this case, the tracer gas was the CO_2 concentration. Although CO_2 is not a pollutant since its concentration was far from dangerous levels for health, it is nevertheless a good indicator of the emission of human bioaerosols, because it is strictly related to them.

The classroom was intermittently occupied (from 9:00 to 11:00 on Tuesday and Wednesday, and from 9:00 to 12:00 on Thursday). However, the total CO_2 concentration remained very low (maximum values of 515 ppm with an outdoor CO_2 concentration of 400 ppm), as shown in Figure 10.

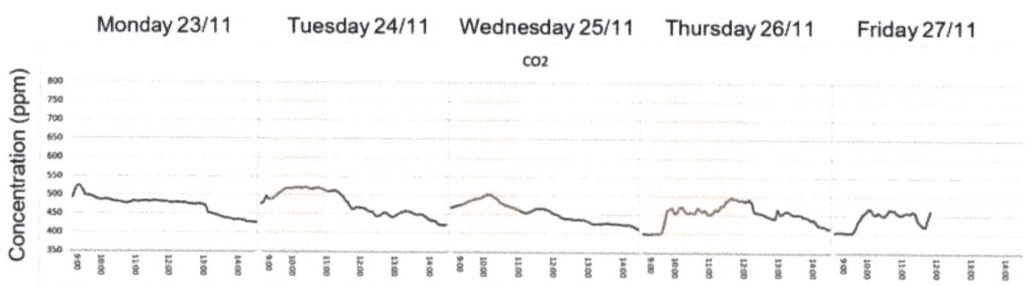

Figure 10. Verification of the ventilation conditions through CO_2 concentration measurement.

Under these circumstances, the combination of these three ventilation mechanisms can be considered as a means to strengthen the prevention measures adopted in the studied classroom.

6. Conclusions

In conclusion, it can be affirmed that it is necessary to strengthen prevention mechanisms against the risk of contagion of SARS-CoV-2 in buildings lacking centralized controlled mechanical ventilation systems, where the main strategy consists of window opening to promote natural ventilation.

The most appropriate methodology needs to address the characterization of each of the possible strategies in the building. In the case studied, the protocol adopted consisted of the combination of natural ventilation, the maximization of the outdoor air supply through the air conditioning system, and the use of filtering devices. Simultaneously, the basic conditions of social distance, occupation, use of masks and guidelines for the use and cleaning of the building must be met.

As a final note, it is necessary to mention that throughout the 2020/2021 school year with 100% face-to-face teaching, there was no case of contagion among students and teachers inside the classrooms of the centre.

Author Contributions: Conceptualization, A.M. and I.P.-C.; methodology, A.M. and I.P.-C.; formal analysis, A.M., A.L.-Á. and M.Á.P.-M.; investigation, A.M.; resources, A.L.-Á.; data curation, A.M. and H.J.-M.; writing—original draft preparation, A.M.; writing—review and editing, A.M. and I.P.-C.; visualization, A.M.; supervision, I.P.-C.; project administration, A.L.-Á.; funding acquisition, M.Á.P.-M. All authors have read and agreed to the published version of the manuscript.

Funding: This research "Metodologías de estudio y estrategias de mejora de eficiencia energética, confort y salubridad de centros educativos en Castilla y León. Investigación básica" was funded by Junta de Castilla y León, Consejería de Educación, grant number VA026G19 (2019–2021).

Institutional Review Board Statement: Not applicable.

Informed Consent Statement: Not applicable.

Conflicts of Interest: The authors declare no conflict of interest.

References

1. World Health Organization (WHO). Modes of Transmission of Virus Causing COVID-19: Implications for IPC Precaution Recommendations. Available online: https://www.who.int/news-room/commentaries/detail/modes-of-transmission-of-virus-causing-covid-19-implications-for-ipc-precaution-recommendations (accessed on 17 August 2021).
2. Europa Press Noticias S.A. Fernando Simón: "No es Necesario Que la Población Use Mascarillas". Available online: https://www.europapress.tv/sociedad/476808/1/fernando-simon-no-necesario-poblacion-use-mascarillas (accessed on 17 July 2021).
3. Boletín Oficial del Estado. *Orden SND/422/2020, de 19 de Mayo, por la Que se Regulan las Condiciones Para el uso Obligatorio de Mascarilla Durante la Situación de Crisis Sanitaria Ocasionada por el COVID-19*; Boletín Oficial del Estado: Madrid, Spain, 2020.
4. World Health Organization (WHO). Coronavirus Disease (COVID-19): How Is It Transmitted? Available online: https://www.who.int/news-room/q-a-detail/coronavirus-disease-covid-19-how-is-it-transmitted (accessed on 17 August 2021).
5. Morawska, L.; Milton, D.K. It Is Time to Address Airborne Transmission of Coronavirus Disease 2019 (COVID-19). *Clin. Infect. Dis.* **2020**, *71*, 2311–2313. [CrossRef] [PubMed]
6. Ministerio de Sanidad. *Gobierno de España Evaluación del Riesgo de la Transmisión de SARS-CoV-2 Mediante Aerosoles. Medidas de Prevención y Recomendaciones*; Ministerio de Sanidad: Madrid, Spain, 2020.
7. Bar-on, Y.M.; Flamholz, A.V.I.; Phillips, R.O.B.; Milo, R.O.N. SARS-CoV-2 (COVID-19) by the numbers. *Epidemiol. Glob. Heal. Infect. Dis.* **2020**, *2*, 1–15. [CrossRef] [PubMed]
8. Ruiz de Adana Santiago, M.; Pastor Pérez, P.; Aledo Vives, S.; Gallardo Salazar, M. *DTIE 2.07. Las Instalaciones de Climatización, SARS CoV 2 y la Calidad del Aire*; Asociación Técnica Española de Climatización y Refrigeración—ATECYR: Madrid, Spain, 2020; ISBN 978-84-95010-71-1.
9. Morawska, L.; Johnson, G.R.; Ristovski, Z.D.; Hargreaves, M.; Mengersen, K.; Corbett, S.; Chao, C.Y.H.; Li, Y.; Katoshevski, D. Size distribution and sites of origin of droplets expelled from the human respiratory tract during expiratory activities. *J. Aerosol Sci.* **2009**, *40*, 256–269. [CrossRef]
10. Prather, K.A.; Marr, L.C.; Schooley, R.T.; McDiarmid, M.A.; Wilson, M.E.; Milton, D.K. Airborne transmission of SARS-CoV-2. *Science* **2020**, *370*, 303–304. [CrossRef] [PubMed]
11. Chen, W.; Zhang, N.; Wei, J.; Yen, H.L.; Li, Y. Short-range airborne route dominates exposure of respiratory infection during close contact. *Build. Environ.* **2020**, *176*, 106859. [CrossRef]
12. Memarzadeh, F. Improved strategy to control aerosol-transmitted infections in a hospital suite. In Proceedings of the IAQ Conference 2012, Houston, TX, USA, 5–10 June 2011.

13. Verreault, D.; Moineau, S.; Duchaine, C. Methods for Sampling of Airborne Viruses. *Microbiol. Mol. Biol. Rev.* **2008**, *72*, 413–444. [CrossRef] [PubMed]
14. Xie, X.; Li, Y.; Chwang, A.T.Y.; Ho, P.L.; Seto, W.H. How far droplets can move in indoor environments—Revisiting the Wells evaporation—Falling curve. *Indoor Air* **2007**, *17*, 211–225. [CrossRef] [PubMed]
15. van Doremalen, N.; Bushmaker, T.; Morris, D.H.; Holbrook, M.G.; Gamble, A.; Williamson, B.N.; Tamin, A.; Harcourt, J.L.; Thornburg, N.J.; Gerber, S.I.; et al. Aerosol and Surface Stability of SARS-CoV-2 as Compared with SARS-CoV-1. *N. Engl. J. Med.* **2020**, *382*, 1564–1567. [CrossRef] [PubMed]
16. Schuit, M.; Ratnesar-Shumate, S.; Yolitz, J.; Williams, G.; Weaver, W.; Green, B.; Miller, D.; Krause, M.; Beck, K.; Wood, S.; et al. Airborne SARS-CoV-2 is rapidly inactivated by simulated sunlight. *J. Infect. Dis.* **2020**, *222*, 564–571. [CrossRef] [PubMed]
17. Smither, S.J.; Eastaugh, L.S.; Findlay, J.S.; Lever, M.S. Experimental aerosol survival of SARS-CoV-2 in artificial saliva and tissue culture media at medium and high humidity. *Emerg. Microbes Infect.* **2020**, *9*, 1415–1417. [CrossRef] [PubMed]
18. Universidad de Valladolid. *Instrucciones de Ventilación y Climatización de Espacios Para Evitar la Propagación del SARS-COV-2*; Universidad de Valladolid: Valladolid, Spain, 2020.
19. Settimo, G.; Indinnimeo, L.; Inglessis, M.; De Felice, M.; Morlino, R.; Di Coste, A.; Fratianni, A.; Avino, P. Indoor air quality levels in schools: Role of student activities and no activities. *Int. J. Environ. Res. Public Health* **2020**, *17*, 6695. [CrossRef] [PubMed]
20. Casals Ventilación Fluid Simulation of REINTAIR 600 Purifier. Available online: https://www.youtube.com/watch?v=5WLPPCNnFzo (accessed on 30 August 2021).
21. Etheridge, D.; Sandberg, M. *Building Ventilation. Theory and Measurements*; John Wiley & Sons Ltd.: Chichester, UK, 1996.
22. Meiss, A.; Feijó-Muñoz, J.; García-Fuentes, M.A. Age-of-the-air in rooms according to the environmental condition of temperature: A case study. *Energy Build.* **2013**, *67*, 88–96. [CrossRef]
23. Meiss, A.; Feijó, J. Influencia de la ubicación de las aberturas en la eficiencia de la ventilación en viviendas. *Inf. Constr.* **2011**, *63*, 53–60. [CrossRef]
24. Mundt, E.; Mathisen, H.M.; Nielsen, P.V.; Moser, A. *Ventilation Effectiveness*; REHVA: Brussel, Belgium, 2004; ISBN 2960046803.
25. Rabanillo-Herrero, M.; Padilla-Marcos, M.; Feijó-Muñoz, J.; Meiss, A. Effects of the radiant heating system location on both the airflow and ventilation efficiency in a room. *Indoor Built Environ.* **2019**, *28*, 372–383. [CrossRef]

energies
MDPI

Review

The Influence of Outdoor Particulate Matter $PM_{2.5}$ on Indoor Air Quality: The Implementation of a New Assessment Method

Dominik Bekierski * and Krystyna Barbara Kostyrko

Thermal Physics, Acoustics and Environment Department, Building Research Institute (ITB), 1 Filtrowa Str., 00-611 Warsaw, Poland; k.kostyrko@itb.pl
* Correspondence: d.bekierski@itb.pl

Abstract: Epidemiological research has shown that there is a positive correlation between the incidence of disease and mortality in humans and the mass concentration of particulate matter. An average 1 g of suspended dust emitted in a room results in the same exposure as 1 kg emitted to the outside air. In this study, the authors described the state of knowledge on dust pollution inside and outside buildings (I/O ratios), and methods of testing the PM infiltration process parameters. According to the law of indoor–outdoor particle mass balance and the physical basis of aerosol penetration theory, a relatively simple but new method for estimating the penetration factor *P* was tested. On the basis of the curve of dynamic changes of internal dust concentration in the process of particle concentration decay and next of the followed curve of dynamic rebound of particle concentration, authors measured penetration factor of ambient $PM_{2.5}$ through building envelope. Authors modification of the method is to be used for determining the value of the particle deposition rate *k* not from the course of the characteristics in the transient state (the so-called particle concentration decay curves) but from the concentration rebound course, stimulated by natural particle infiltration process. Recognition measurements of the mass concentration of suspended $PM_{2.5}$ and PM_{10} particles inside the rooms were carried out. In this study, the choice of the prediction particle penetration factor *P* calculation method was supported by the exemplary calculation of the *p* value for a room polluted by $PM_{2.5}$. The preliminary results of the penetration factors determined by this method $P = 0.61$ are consistent with the *P* factor values from the literature obtained so far for this dimensional group of dusts.

Keywords: particulate matter; dust pollution; IAQ; indoor–outdoor concentration ratio; penetration factor; air quality control

Citation: Bekierski, D.; Kostyrko, K.B. The Influence of Outdoor Particulate Matter $PM_{2.5}$ on Indoor Air Quality: The Implementation of a New Assessment Method. *Energies* **2021**, *14*, 6230. https://doi.org/10.3390/en14196230

Academic Editors: Roberto Alonso González Lezcano and Christian Inard

Received: 5 August 2021
Accepted: 22 September 2021
Published: 30 September 2021

1. Introduction

The aim of the article is to present the authors' proposal for a new method of assessment of the influence of outdoor particulate matter $PM_{2.5}$ on indoor air quality indirectly by determining the value of the parameter that determines the intensity of dust penetration into the interior of the building. This parameter is the penetration factor, i.e., the ratio of the amount of dust entering from the outside, ending the infiltration processes, to the amount of dust retained outside.

In this study, the authors described the state of knowledge concerning dust pollution inside and outside buildings (I/O ratios), and methods of testing the PM infiltration process parameters. Methods known in the literature are based on the observation of the process of dynamic air pumping from a dusty external environment into a room with the use of an exhaust blower or an air cleaner. However, the process is not only dependent on the difference in PM concentration levels but more on the O/I air pressure difference produced. The authors' proposal includes the observation of the PM concentration rebound curve of the dust indoor concentration level increasing caused by the natural process of dust infiltration from the outside, to achieve dust balance in the exterior and interior of the building protected by a dust-permeable envelope.

The authors' field study aimed to estimate indoor air quality, especially particulate matter concentration in relation to the infiltration process through the buildings' envelope. Several analyses based on air quality measurements and calculation procedures were undertaken. On the basis of the curves of dynamic changes of internal dust concentration in the processes of particle concentration decay and next followed curves of dynamic change particle concentration rebound the authors proposed the method of estimating the value of the penetration factor of ambient $PM_{2.5}$ through a building envelope. Information on the value of the P factor can be a reliable parameter that classifies buildings in terms of resistance to dust infiltration into their interior, which partly characterizes the condition of the envelope structure and its resistance to dust penetration into the building. According to research in residential buildings, a large proportion of both groups of particles and their mass concentration are attributed to human activities such as cooking, smoking, vacuuming, gas stoves, the burning of solid fuels and candles, electrical appliances, cleaning, washing and walking. Both sources influence the characteristics of indoor particulate matter, and their concentrations are increased by the resuspension of particles deposited on internal surfaces in the air.

The term PM_{10} is used for particles with an aerodynamic diameter <10 μm. The term $PM_{2.5}$ defines aerodynamic particles with a diameter <2.5 μm. The coarse fraction contains particles with an aerodynamic diameter between 2.5 μm and 10 μm. The "ultrafine particles" fraction has an aerodynamic diameter <0.1 μm. A commonly used indicator describing particulate pollution is the mass concentration of PM_{10}.

Thatcher and Layton's [1] experiments concluded that PM > 5 μm may be resuspended, particles < 5 μm are not easily resuspended and particles < 1 μm show almost no tendency to re-suspend at high human activity.

In the project INDEX PM [2], it was demonstrated that indoor air PM dust pollution mainly consists of salt (ammonium sulphate, ammonium nitrate, sodium and potassium chloride), soot (elemental carbon, EC), minerals (silicon oxides, aluminium, calcium, iron, manganese, titanium, zinc, etc.), organic substances (shredded organic matter, also called organic carbon, OC) and materials of biological origin (bacteria and fungi, dandruff, pollen, fragments of plants and insects). All of these components have different particle size distributions and mechanisms, which are strongly dependent on particle size and are involved in removing particles from the indoor air. Sedimentation is the most important mechanism for the removal of coarse particles (>10 μm), diffusion and the agglomeration/coagulation of ultrafine particles (<0.1 μm).

Particles (0.1–1.0 μm) are most stable in air. A typical input fraction (fraction of the total PM mass emitted from the source that will be inhaled by the whole population) for an indoor PM source is 10^{-2} to 10^{-3}, compared to 10^{-5} to 10^{-6} for outdoor PM sources: An average 1g of suspended dust emitted in a room results in the same exposure as 1kg emitted to the outside air. Consequently, while internal sources are less efficient and their total emission loads are smaller than the PM of external sources, each of these sources, if present, may dominate both the sum of individual exposures and the inhalation risks of PM.

Between 1999 and 2016, the Canadian Census Health and Environment Cohort (CanCHEC) conducted panel studies involving approximately 2.5 million Canadians [3]. The relationship between non-accidental mortality and the concentration of fine particulate matter (≤2.5 μm; $PM_{2.5}$), ozone (O_3) and nitrogen dioxide (NO_2) in the environment was investigated. Models with combinations of these contaminants have also been tested and $PM_{2.5}$ was found to be associated with an increased risk of non-accidental mortality, i.e., lung cancer, diabetes and ischemic heart disease. The results obtained provide evidence that long-term exposure to these three key components of ambient air pollution is associated with an increased risk of non-accidental mortality. Cumulative risk models suggest that exposure to $PM_{2.5}$ alone does not fully characterise the toxicity of a mixed atmosphere and cannot fully explain the mortality risks associated with exposure to environmental contamination. The strongest cumulative risk estimate is for mortality from diabetes (HR = 1.180; 95% CI: 1.125, 1.236). Assuming the additive exposure of individual pollutants,

HR (Health Risk) in 95% CI (Confidence Interval), as estimated using three pollutant models showed that a change in exposure by an average of 5% for all three pollutants together resulted in impact synergies of 1.075. The effect of $PM_{2.5}$ and O_3 was weaker than the effect of NO_2 alone, but HR increased with the combination of all three pollutants, i.e., adding NO_2.

It has also been proved that the strongest correlation between $PM_{2.5}$ concentration and mortality increased by 1.5% when $PM_{2.5}$ concentration increased by 10 $\mu g/m^3$ per day [3], because particulate matter $<PM_{2.5}$ enters into the bloodstream.

The health effects of $PM_{2.5}$ were also studied by Hvelplund [4]. The Brownian motion of particles in the respiratory system is only visible for particles below 0.05 µm. The same tendency was found by Wang et al. [5], who investigated the deposition efficiency of particles with aerodynamic diameters between 0.001 and 300 µm moving through the nasal cavity with an air mass flow of 0.6 m^3/h. Studies in this area have shown that the lowest deposition efficiency is in the range of 0.015 to 10.0 µm due to the low influence of diffusion and the effects of impact and gravity settling.

The number of exhaled particles is debatable, e.g., most particles below 1.0 µm were predicted to escape through the bronchiolar orifices. It is not the case, however, that all particles escape because some of them are absorbed directly into the bloodstream through the bronchioles, which in human airways make up most of the levels of the tracheobronchial tree. Particles larger than 5–10 µm are usually removed from the upper respiratory tract, indicating that the mucosa and nasal hair contribute greatly to the filtration of particles.

According to [4], inhalation exposure to particulate pollutants is one of the main threats to public health. Most existing airway morphometry models are theoretical or semi-empirical in nature; they were designed to predict the deposition fraction for an averaged subgroup of the general population. It is difficult to tailor a quick and accurate prediction to suit individual needs.

Hvelplund et al. [4] aimed to analyse the local particle deposition along an anatomically reconstructed model of the airways, which was developed from computed tomography images of healthy subjects. Computational simulations of airway fluid dynamics show that most particles are deposited in the bronchi. The accumulation of particles (0.1–2.0 µm) is the smallest fraction, about 11%, of sediment in the lower respiratory tract. Increasing the aerodynamic diameter >2.0 µm of the particles increased the fraction of deposited particles. This study determined the size related to the site of particle deposition in the airways. It turned out that most of the particles in the studied bands are deposited in a narrow region of the airway model, such as the BC (bronchi) and larynx.

Already in the 1980s, many epidemiological studies had shown that $PM_{2.5}$ has obvious side effects on human health Lin et al. [6]. Research in China on the relationship between $PM_{2.5}$ and human health has fully demonstrated that $PM_{2.5}$ can increase the incidence of heart and lung disease, respiratory disease, cardiovascular disease, cancer and other diseases, and even the risk of death. Long-term exposure to environmental $PM_{2.5}$ may be an important risk factor for hypertension and is responsible for the significant burden of hypertension in adults in China, as it leads to a reduction in lung function. $PM_{2.5}$ is a risk factor for childhood asthma, attributed to decreased immunoregulation and deterioration in ventilation function. Exposure to $PM_{2.5}$ can also affect reproductive health.

The I/O indicator describing the temporary steady state of dust pollution in a building is related to the process of infiltration of suspended dust from the outside into the building's interior and can be described with the assumption that there are no internal sources generating dust particles:

$$\frac{I}{O}ratio = \frac{C_{in}}{C_{out}} \tag{1}$$

Lee et al. [7] found that room temperature and floor level (expressed by the story where the room is located in the building) are powerful I/O predictors and gave the I/O equation for $PM_{2.5}$, which is:

$$\frac{I}{O} = 0.629 + 0.0102 \cdot T - 0.00654 \cdot FL \quad (2)$$

where T and FL are temperature and floor level, respectively.

The indoor/outdoor (I/O) index is a measure for assessing the difference between the levels of indoor solid concentrations and the current outdoor concentrations, but it may be also recognised as an indicator of the persistence of particulate matter sources inside buildings.PM concentrations are affected by dust infiltration from the outside into buildings and internal sources. I/O ratios can vary considerably due to the building's design, location and the different activities of the occupants. The I/O ratio is also calculated to compare the dynamics of flows between indoor and outdoor PM in different apartments and buildings. As far as health protection is concerned, it is clearly the best when the I/O ratio for each dimensional fraction of dust is less than 1, asthisshows that the structural properties of a residential building reduce the penetration of PM from outside and, accordingly, reduce the residents' exposure to PM.

In epidemiological studies of external contamination, the concentration of external contaminants was previously used as an indicator of the stressor. Thus, the appropriateness of using suspended dust concentrations from the building surroundings as a proxy in personal exposure studies was questioned, and both outdoor and indoor concentrations began to be measured to investigate their correlation with external conditions, as well as the correlations among other pollutants. Positive correlations were found for PM, O_3 and NO_2.

While exposure to indoor air pollution changes under the influence of many factors, such as the type of microenvironment and the source of internal pollution, building characteristics and its location, ventilation parameters and assumed comfort conditions (Branco et al. [8]), as well as individual building occupant activity, studies have shown that only some of these factors are important. Therefore, several parameters of I/O interaction (Kalimeri et al. [9])were selected in order to minimise the error resulting from the use of concentrations in the external environment as a surrogate for estimating the exposure of the PM environment to humans indoors. These parameters are the infiltration coefficient (F_{inf}) (i.e., the equilibrium fraction of particles from the environment that have penetrated into the interior and remain suspended), the P coefficient of penetration efficiency of particles through the leaks of the building envelope (i.e., the fraction of particles that entered the interior from the infiltration path through the external walls of the building, $P < F_{inf}$) and the I/O ratio (the ratio of internal to external concentration of similar PM particles). Lv et al. [10] argued that regression analyses of the ratio of inside and outside particle concentration values confirm that the degree of correlation of changes in these concentrations expresses the infiltration coefficient.

Nadali et al. [11] found that coarse particles, e.g., PM_{10}, have higher falling velocities than fine PM, which leads to lower levels in rooms, because coarse particles (PM) settle under the influence of gravity or settle on doors, window frames and furniture as a result of the effects of electric charges and turbulent diffusion.

In their research, the concentration of solid particles, expressed by the average I/O ratio, also differed significantly in apartments, and did not solely depend on the location of the buildings. This was because, although the air with PM outside can have a significant impact on the concentrations inside, the composition of the air in the room is mostly influenced by the sources (activities and materials) in the rooms, which can be identified as contributing significantly to the internal mass concentration of PM.

Bai et al. [12] addressed the question why the I/O ratio is so important in practice. Everyday health is characterised by the equation of health detriment for each individual, based on an environmental risk assessment from the "receptor" perspective of

$$HHD_{i,j}^{inhalation} = C_{i,j} \cdot \left(\frac{IR \times t_j}{m}\right) \cdot EF \tag{3}$$

where C is the concentration of particulate matter (μg/m^3) with the indices: i = spring or summer or autumn or winter (1:1:1:1 annually), j = inside or outside (ratio 80:20), HHD is damage to human health t is the daily exposure time h, m is an I/O mixing ratio of 1, EF is an influence factor of 78 DALY/kg (or disease cases/kg of chemical intake), IR is the daily inhalation rate of air for each person, which is 13/24 (m^3/h).

In questionnaire studies [12], the daily ratio of the students' exposure time to air polluted with dust inside and outside was calculated, and the mixing coefficient m was determined (the same as the I/O ratio). Due to different daily $PM_{2.5}$ concentrations, health effects at different concentrations were calculated from Equation (3). The daily damage to the health of each person multiplied by the period of a year is the annual damage to the health of one person.

2. The State of Knowledge on Dust Pollution Inside and Outside Buildings

2.1. Characteristics of Actual Air Pollutants in the Internal Environment

According to the European programme INDEX-PM [2], the indoor air sources of PM are classified as:

1. PM from external sources transferred to the premises via ventilated or infiltrated air;
2. PM from external sources, transferred to the premises as settled dust and then re-suspended in the air in the room;
3. suspended dust from indoor combustion sources, such as tobacco smoke, cooking, candles or forest fires;
4. particles of biological origin; and finally;
5. particulate matter produced by indoor air chemistry, e.g., oxidation of cleaning ingredients (e.g., terpenes) by ozone.

Zhang and Duan [13] showed that burning a mosquito on a heating coil can release 626 μg/m^3 $PM_{2.5}$, which is 8.3 times the permissible concentration for a residential environment. Bai et al. [12] examined the concentration of $PM_{2.5}$ in households using coal for cooking and found that it was significantly higher than in households using gas or electricity, and if coal was switched to gas or electricity, the concentration of $PM_{2.5}$ in the kitchen would decrease by 40–70%. Zhang et al. [14] investigated different culinary habits, cooking methods, ingredients and even spices and found that they strongly influenced the composition of particulates. Culinary habits were also the subject of research by another group of researchers, Xue, Zhou et al. [15], who warn against the residential combustion of coal with a high sulphur content, as it can cause the total concentration of air pollutants to contain as much as 11.6% PM_{10} and even 27.5% SO_2 in the winter heating season.Zhou, Liu. [16] indicated that human activities such as walking, dressing and cleaning may increase the concentration of $PM_{2.5}$ indoors by 33%. Chinese researchers conducted experiments on the influence of wet sweeping and dry sweeping in the air in an office. The average levels of $PM_{2.5}$ concentrations in the rooms before cleaning were 47.3 μg/m^3, 40.6 μg/m^3 and 39.4 μg/m^3, respectively. The average levels of $PM_{2.5}$ concentrations inside the premises after cleaning were 109.7 μg/m^3, 97.5 μg/m^3 and 43.3 μg/m^3, respectively. The mean concentrations of $PM_{2.5}$ increased 2.3 times, 2.3 times and 1.1 times, respectively. Therefore, it is recommended to wet sweep as often as possible under ventilation conditions. Printing also plays a role in increasing indoor dust concentration, and the release of $PM_{2.5}$ from printers with different performance varied.

Some systematisation of the main sources of indoor air pollutants in our living environment is a necessary step to facing and reducing the associated health risks. Three different environments are described: Home, school and office, and the pollutants for the home, including particulate matter, are listed in Table 1. These were data published at the Healthy Building conference by Simeone et al. [17] supplemented by data from the literature.

Table 1. Indoor air pollution list based on [16–24].

Sources	Contaminants	
	Bedroom	Office
Furniture	Formaldehyde, VOC, allergens, mites, mould, particulate matter (sink effect)	Formaldehyde, VOC, allergens, mites, mould particulate matter (sink effect)
Walls, floorsand ceilings	Asbestos, formaldehyde, VOC, bacteria, mould, radon, particulate matter (sink effect)	Asbestos, formaldehyde, VOC, bacteria, mould, radon, particulate matter (sink effect)
Environmental tobacco smoke	CO, NO, benzene, formaldehyde, PAH, VOC, particulate matter etc.	CO, NO, benzene, formaldehyde, PAH, VOC, particulate matter etc.
Tapestry	Mites, mould, bacteria, formaldehyde, VOC, particulate matter	Mites, mould, bacteria, formaldehyde, VOC, particulate matter
Clothes (laundry)	Formaldehyde, VOC, PAH, mites, particulate matter	Formaldehyde, VOC, PAH, mites, particulate matter
Air conditioning	Mites, mould, allergens, particulate matter	Mites, mould, allergens, particulate matter
Candles, incenses and deodorants	CO, NO_2, SO_2, particulate matter, benzene, formaldehyde, PAH, VOC	
Personal care products (hairspray)	Aliphatic hydrocarbons, formaldehyde, VOC, acetone, benzene, particulate matter	
Cleaning products	Acetone, terpenes, aldehydes, 1-butanol, hexanal, ultrafine particulate matter	Acetone, terpenes, aldehydes, 1-butanol, hexanal, particulate matter
Cooking, boiling, frying, ovens, toasters *	Aldehydes, terpenes, xylenes, benzene, toluene, nonane, limonene	
Building products	Octanal, formaldehyde, acetaldehyde, benzaldehyde, particulate matter	Octanal, formaldehyde, acetaldehyde, benzaldehyde, particulate matter
Laserprint	O_3, formaldehyde, VOC, breathable particulate matter	O_3, formaldehyde, VOC, breathable particulate matter
Photocopiers		O_3, formaldehyde, VOC, carbon black, benzene, particulate matter
Pets	Allergens, mites, bacteria, fungi	
Outdoor environment	CO, NO_2, SO_2, particulate matter, O_3, pesti-cides, benzene, PAH, pollen, asbestos, noise, electromagnetic fields, radon	CO, NO_2, SO_2, particulate matter, O_3, pesti-cides, benzene, PAH, pollen, asbestos, noise, electromagnetic fields, radon

* Isaxon et al. [19] report the peak number concentration of ultrafine particles for these activities: cooking 1.80×10^5 p cm^{-3}, boiling 5.6×10^4 p cm^{-3}, frying 1.4×10^4 p cm^{-3}, oven 2.3×10^5 p cm^{-3}, toaster 1.6×10^5 p cm^{-3}.

According to Wallace and Ott [20], cooking on gas or electric stoves and electric toaster ovens was a major source of UFP, with peak personal exposures often exceeding 100,000 particles/cm^3 and estimated emission rates in the neighbourhood of 10^{12} particles/min.

He, Morawska and Gilbert [25] investigated submicrometric particle number concentration measured during the cooking test and found that it was in the size range from 0.015 to 0.685 µm.

Jantunen [18], in a study with multiple regression models using data collected from the EXPOLIS from six European cities, determined the internal concentrations of $PM_{2.5}$, starting from the concentration generated by smoking (16%), a gas cooker (1.4%), construction dust and a gas cooker (all < 4%).Smoking and cooking also generated NO_2 emissions. The research focused on the short-term effects of internal sources of PM, which are, e.g., different cooking methods generating concentrations of various solid particles in rooms, with peak concentrations of 30–60 μg/m^3 for $PM_{0.02–0.5}$ and 10–300 μg/m^3 for $PM_{0.7–10}$. Cleaning activities (8 μg/m^3 for $PM_{0.02–0.5}$ and 30 μg/m^3 for $PM_{0.7–10}$) and those related to the mobility of residents (4 μg/m^{3}for $PM_{0.02–0.5}$ and 20 μg/m^3 for $PM_{0.7–10}$) contributed much less. Oven roasting was the most intense source of $PM_{0.5}$ indoors, as well as frying (PM_{10}).

Morawska et al. [26] report that some types of activity of inhabitants result in particularly high concentrations of $PM_{2.5}$ in rooms. These include frying (median peak value: 745 μg/m^3), grilling (718 μg/m^3), candle evaporating eucalyptus oil (132 μg/m^3) and smoking (79 μg/m^3). The high maximum concentrations caused by these activities may result in the US EPA $PM_{2.5}$ standard being exceeded for 24 h at 65 μg/m^3 in homes where such activities are carried out, provided that the activity is carried out for more than 24 h.

The reported emission rate of 0.99 mg/min^{-1} from smoking is comparable with the results reported in the literature. For example, Klepeis et al. [27] measured the emissions factor of the respirable particle ($PM_{3.5}$) emitted in an apartment where smoking took place. It turned out that the average $PM_{3.5}$ emission index ranged from 0.98 mg/min^{-1} (cigar) to1.9 mg/min^{-1} (Marlboro cigarette). Brauer et al. [28] measured cigarette smoking using a nephelometer positioned in an environmental chamber and found that the emission rate of $PM_{2.5}$ particles was 1.67 mg/min^{-1}.

In a US indoor/outdoor air study, it was determined that smoking and cooking were the predominant activities associated with an elevated concentration of "fine particles". Smoking can add 20 μg/m^{-3} (24 h average) of particles per household smoker [29] with short-term peaks of 300 μg/m^{-3} that may persist for up to 30 min after the cigarette is extinguished. Home cooking generated particles (0.1 μm) accounted for 30% of the particle volume [30]. The cooking method, and particularly frying, variably increased particle concentration, and the consensus states that the I/O ratios were somewhat higher in houses with gas cookers than in houses without such cookers and heating sources. Large particles (>2.5 μm in diameter) are generated in homes with activities such as home cleaning (vacuuming and sweeping) [30,31], which can lead to the re-suspension of particles embedded in horizontal surfaces such as floors, carpets and furniture. The ratio between indoor and outdoor particle concentrations gives an indication of whether particles found indoors are the result of indoor generation.

2.2. Limit Values for Indoor PM Concentrations

The data taken from the Health-Based Ventilation Guidelines: Principles and Framework, which were published in [32,33], can be considered as setting the latest limit values for the exposure of humans to indoor pollutants, see Table 2 The guidelines adopted the data from the WHO Air Quality Guidelines, Global Update 2005 published in 2006 by the Regional Office for Europe (Copenhagen) [34] concerning the most life-threatening pollutants NO_2, PM_{10}, $PM_{2.5}$ and O_3, but also SO_2, considered by some researchers to be an equally hazardous pollutant.

Table 2. Current air quality guidelines (figures in brackets indicate the average time for which the guideline values apply) [32,33].

Time of Exposure \ Air Pollution [μg/m^3]	NO_2	SO_2	PM_{10}	$PM_{2.5}$	O_3
Short-term	200 (1 h)	500 (10min)	50 (24 h)	25 (24 h)	100 (8 h)
Long-term	40 (1 year)	20 (24h)	20 (1 year)	10 (1 year)	-

2.3. Characteristics of Actual Air Pollutants in the Outdoor Environment in Statistical Terms

Primary particles from the industry and agriculture are usually larger than 10 μm, and their share in primary PM emissions is usually lower than that of PM_{10} and $PM_{2.5}$. The same applies to non-road exhaust emissions from traffic sources (road, tyre and brake wear), which are a secondary source of particles smaller than 2.5 μm. Soot (BC–black coal) emissions mainly come from combustion processes in the transport sector (diesel vehicles) and from small domestic boilers The projected share in primary PM_{10} and $PM_{2.5}$ emissions in the EU (Poland and 14 other countries) was included in the basic scenario of the European CAFE programme (European Commission's Clean Air for Europe) implemented in the CAFE programme for 2020, as illustrated in Figure 1.

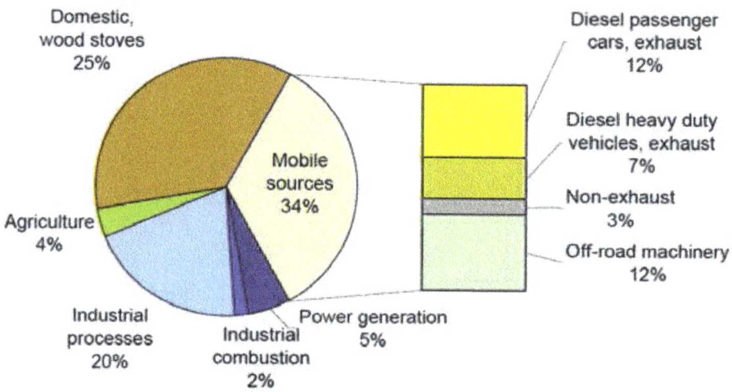

Figure 1. Sectoral emissions of $PM_{2.5}$ projected in the years 2000 to 2020 in the CAFE programme [35].

Overall, the CAFE baseline estimated that PM_{10} emissions were expected to decrease by around 40% in new Member States between 2000 and 2010. For 2020, it is planned that the sources of $PM_{2.5}$ emissions from diesel exhaust gases will drop from 12% to 6% and that the largest source of primary $PM_{2.5}$ emissions will reduce their emissions by 40%. On 21 November 2011, the European Commission took Poland to the European Court of Justice, accusing the country of lack of progress in the implementation of Directive 2008/50/EC of the European Parliament and of the Council of 21 May 2008 on ambient air quality and cleaner air for Europe (CAFE).

According to the WHO recommendation, the standards of the average daily concentration of PM_{10} may not be exceeded more than 35 times a year. Unfortunately, in Poland, it happens that during the winter months, days when the concentration of suspended dust is within the norm can be counted on the fingers of one hand Samek [36]. In China, from the statistics of $PM_{2.5}$ and PM_{10} on the pollution levels of2015–2017, it was observed that the annual average of $PM_{2.5}$ mass concentration was decreased Wang [37]. In Shenyang, $PM_{2.5}$ decreased from 72 to 51 $\mu g/m^3$ and PM_{10} mass concentration decreased from 115 to 88 $\mu g/m^3$ Yu [38].

This has disastrous consequences for the health of the inhabitants of large cities (Figures 2 and 3).

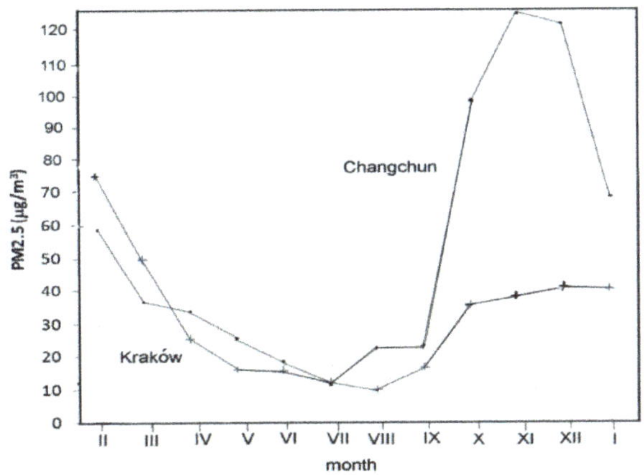

Figure 2. Measured by Samek et al. [36] average concentrations of $PM_{2.5}$ in µg/m^3 in Krakow (2014) and average concentrations of $PM_{2.5}$ in µg/m^3 measured in Changchun [12].

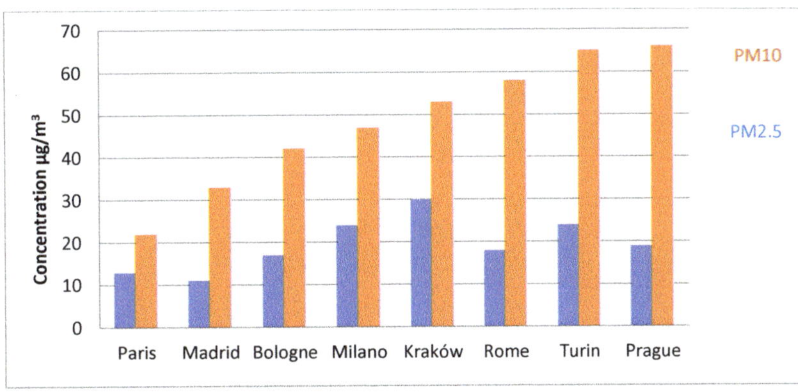

Figure 3. Average concentrations of suspended dust PM_{10} (red) in µg/m^3 measured in European cities in 2004 [39] and $PM_{2.5}$ (blue) in µg/m^3 observed in the same cities in 2016 [40].

The World Health Organization (WHO) has established specific standards for particulate matter concentrations in the outside air. Currently these are:

- the standard of the average daily concentration of PM_{10} dust: 50 µg/m^3;
- the standard of the average annual concentration of PM_{10} dust: 20 µg/m^3;
- the standard of the average daily concentration of $PM_{2.5}$ dust: 25 µg/m^3;
- the standard of the average annual concentration of $PM_{2.5}$ dust: 10 µg/m^3.

2.4. *Research on Actual Dust Pollution of the External Environment in Polish Cities*

The study was conducted in Warsaw as part of a project conducted by Juda-Rezler et al. [40] from the Warsaw University of Technology. The average concentration of $PM_{2.5}$ in Warsaw in 2016 was 18.8 µg/m^3 (measured with standard deviation ±11.9 µg/m^3), when the EU annual limit value for $PM_{2.5}$ was 25 µg/m^3; WHO's annual air quality (10 µg/m^3) was not met. PM pollutants in the atmospheric air differ in size and composition and according to an author [40], it is a mixture of primary sources (emitted from anthropogenic and natural sources) and secondary sources (formed in the atmosphere compounds as a result of the reaction of primary pollutants). The combustion of fossil fuels in the energy and smelting industry and the housing and road transport sector produces the most significant anthropogenic source of both solid and gaseous precursors of secondary particles, including sulphur dioxide (SO_2), nitrogen oxides (NOx) and volatile organic compounds (VOCs). Most of the particulate mixture in the atmospheric air are mineral dust organic matter, secondary inorganic aerosols (including nitrates, sulphates, and ammonia) and water. Many other components, however, are associated with PM, including micronutrients such as, for example, silicon (Si), magnesium (Mg), aluminium (Al), calcium (Ca), potassium (K), titanium (Ti) and trace metals including detrimental heavy metals, such as copper (Cu), arsenic (As), cadmium (Cd), chromium (Cr), lead (Pb) and zinc (Zn). The chemical composition of particulate matter, as well as other PM characteristics, may vary widely in different areas.

According to researchers Samek et al. [36] from the AGH University of Science and Technology in Kraków, in industrial areas of Poland such as Silesia or urban areas such as Kraków and some other cities, the levels of pollution cleanings often exceed air quality standards. The dust masses (PM) are the most important component of atmosphere pollution. The research was carried out in 2014 with an extension to the 2015 heating season. During this period, approximately 200 samples were collected, showing a daily variability of $PM_{2.5}$ concentration in the air. The AGH University of Science and Technology team determined the lowest monthly concentration value for August 2014, amounting to about 10 $\mu g/m^3$, the highest for February 2014 (70 $\mu g/m^3$), while the annual mean value was approximately 31 $\mu g/m^3$. Additionally, the samples were analysed for content and particle size distribution. Research by the AGH team also included the determination of the black carbon content (BCN).

The team of Ścibor et al. [41] from the Jagiellonian University in Kraków conducted a study in 2014–2015 of the percentage of $PM_{2.5}$ dust in relation to the mass of PM_{10} dust indoors, with open and closed windows, and found that this share reached about 70% for both types of weather conditions: Good—at high wind speed; and bad—at low wind speed. The total share of PM_{10} and $PM_{2.5}$ dust masses penetrating into the interior of the rooms from the outside was higher by about 10% in good weather (high winds) than in bad weather. Opening the window had no significant effect here. Due to the lower degree of dilution and chemical transformations, and also with a larger number of occupants indoors, the impact of exposure per unit mass of $PM_{2.5}$ emitted in a room is two to three orders of magnitude greater than that of exposure from the same emissions in the outdoor environment.

The results obtained in Kraków by the Samek and Ścibor teams can be compared, for example, with the work of Morawska et al. [26] who studied the relationship between indoor/outdoor airborne particles in 16 localised residential homes in a suburb of Brisbane, Australia. By measuring the mass concentration of particles smaller than 2.5 μm, the Australian team showed that, while the periodic values of the I/O index varied widely from $PM_{0.2}$ to $PM_{2.5}$ with both less and more effective ventilation, their average I/O ranged from 1.01 to 1.08, so that the I/O ratios were much higher than in Kraków. Ścibor et al. [41] showed that the mean I/O values for PM_{10} and $PM_{2.5}$ were higher in the rooms when the windows were closed under bad weather conditions. In good weather conditions, the relationship was the same for PM_{10}, while for $PM_{2.5}$, opening the window increased the value of the I/O ratio.

The results for Kraków show that I/O values were significantly below 1, but the fact that indoor concentrations are much lower than outdoors should be juxtaposed with the concentration level situation in this city where there are much higher outdoor and indoor dust concentrations than in many other cities. Therefore, it can be assumed that in Kraków there are much larger amounts of PM dust migrating inside buildings.

In the case of Kraków, I/O values in good weather conditions were much higher (0.92 for cases with open windows and 0.79 for closed windows) than values in bad weather conditions (0.46 and 0.47, respectively), which was consistent with the results of the regression analysis. Indoor PM concentrations were then significantly lower than outdoors. High I/O values for $PM_{2.5}$ occurred in good weather, when the concentrations of

$PM_{2.5}$ indoors were two to four times higher than outdoors. The t-test showed that weather conditions and window opening are factors with a statistically significant impact on I/O.

2.5. Influence of Meteorological and Building Parameters on the Level of Indoor Dust Pollution

Weather conditions are considered to be the main factor influencing the quality of atmospheric air because of their role in dispersing, transforming and removing pollutants from the atmosphere. In urban environments, episodes of severe pollution are mainly caused by unfavourable weather conditions. Precipitation has a large impact on the concentration of particulate matter and washes away mainly coarse particles, but has little effect on changes in the concentration of fine particles. Therefore, atmospheric conditions are important for the level of indoor air pollution.

In a study on Hong Kong, Chan [42] has researched how changes in temperature, humidity, pressure, atmospheric stability, building insulation, wind, etc., influence constantly changing outdoor conditions and can modify the infiltration of outdoor air into a building in a built environment, regardless of which airflow systems one refers.

Chan [42] investigated the I/O ratios for respirable suspended particles (PM) and nitrogen oxides (NO_x) in various meteorological conditions. At higher outside temperatures, pollutants in the air are pushed inside through the doors and gaps in the windows, while the reverse is true for lower outside temperatures. This also explains the small slope of the relationship with the windows closed. Conversely, it has also been found that this effect is slightly more significant for the gaseous NO_x pollutant compared to the effect size for the particulate matter. This is because when the room is relatively well sealed, most of the air comes from the air conditioning vent, which effectively filters a large proportion of particulate matter but likely does not filter NO_x. Chan also noticed that in most cases the I/O for particulate matter was slightly lower than for NO_x, especially at higher temperatures. Overall, it can be seen that the I/O ratio for PM increases with increasing temperature (although at a slightly slower rate than for NO_2).

Chan [42] also investigated the dependence of the I/O index on relative air humidity. In general, the I/O ratio increased with increasing outdoor humidity, which explains the fact that both the PM and NO_x pollutants are readily absorbed or washed away by the water vapour in the atmosphere. Presumably, for the particulate form, this was of greater importance. The I/O ratios for PM increased slightly more for PM than for NO_x, which may confirm that the leaching effect to the outside of the building was more significant for PM than for NO_x.

The results of Chan's research were partially confirmed by Klaic et al. [43],who studied winter correlations between the spread of $PM_{1.0}$ dust indoors and outside weather conditions in Zagreb. Xu et al. [44] showed a decrease in PM_1 with a simultaneous increase in outdoor temperature, rainfall and horizontal wind speed, as well as an increase in indoor $PM_{1.0}$ concentration with increasing relative humidity outside the building.

The question of the correlation of the PM concentration level in interiors with humidity and rainfall outside of buildings posed the most doubts, as shown by further research. Zheng et al. [45], in Beijing, addressed this concern by demonstrating the obvious impact of precipitation on the removal of solid particles from atmospheric air. Researchers observed that the mean concentration of $PM_{2.5}$ decreased by 56.3% as a result of precipitation, and the mass concentration of $PM_{2.5}$ fell below 60 g/m^3 in 72 h after precipitation. Within one hour following rain, the $PM_{2.5}$ concentration level remained almost unchanged, but during the next 12 h it decreased. The large thickness of the mixed layer and the unstable structure of the atmospheric layers help to reduce the mass concentration of $PM_{2.5}$.

A different look at the existing correlations between the PM concentration in interiors and meteorological parameters is presented by a team of Iranian researchers in the article by Nadali et al. [11]. They built, like Klaic et al. [43], a symmetric linear correlation matrix in which they analysed the possibilities of correlation of C_{in} concentration with air temperature and relative humidity, as well as wind speed. Since these newer results were based on a large number of buildings and because they partially contradict earlier results, it is worth mentioning that the correlation coefficients (r) between the PM concentration in the interior and the parameters were as follows: With temperature the correlation is negative; with relative air humidity, however, the correlation is $r = 0.07$, which means that no significant relationship was observed between the concentrations of solid particles and air humidity. This problem merits further investigation.

The positive correlation between the concentration of solids and temperature may be caused by the impulse to the phenomenon of thermal diffusion. At higher temperatures, external PMs are forced into buildings through windows and gaps in doors, while the opposite is true at lower ambient temperatures. The results of Nadali et al. [11] confirmed the positive correlations between I/O ratios and wind speed.

The existing correlations between the indoor concentrations of PM_{10}, $PM_{2.5}$ and $PM_{1.0}$ in the parameters of a building were presented by Nadali et al. [11]. A symmetrical linear correlation matrix was built, which examined the possibilities of correlating the concentration of particles with a given particle diameter size (PM_{10}, $PM_{2.5}$ and $PM_{1.0}$), i.e., indoor concentration C_{in} with building parameters: Age of building, building type, number of windows, ventilation, indoor smoking and particles of other dimensions. A less significant correlation was found between the age of construction and the number of windows, ventilation and cigarette smoke in rooms, and more significant correlations ($p > 0.05$) between the concentrations of PM particles of different sizes.

2.6. Sources of Air Pollution by Particulate Matter (PMF Numerical Methods of Outdoor Sources Apportionment)

The first databases on separate sources of air pollution (outside) of buildings with PM_1 and $PM_{2.5}$ dusts were created in2014 by the World Health Organization. These data were organised and collected by Karagulian et al. [46].

In order to reduce the health impacts of air pollution, it is important to understand the sources of pollutants contributing to environmental exposure. This study systematically reviewed and analysed the distribution of available sources for particulate matter surveys (10 and 2.5 microns in diameter, PM_{10} and $PM_{2.5}$) performed in cities in order to estimate typical source contributions by country and region.

The percentage of separate sources of $PM_{2.5}$ and PM_{10} dust in the urban environment for the Central and Eastern Europe region per source category according to the World Bank's list of economies (2012) for Central and Eastern Europe, provided by [46], is shown in Figure 4.

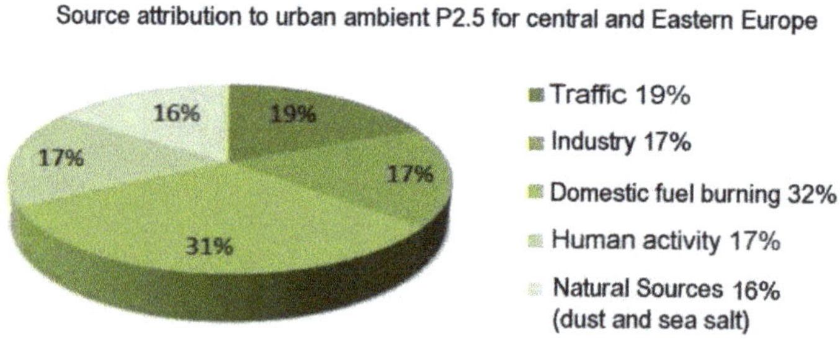

Figure 4. First source attribution to urban ambient particulate matter pollution by source category for Central and Eastern Europe according to [42].

The available records regarding the distribution of sources, however, show significant heterogeneity in the assessed source categories and incompleteness in some countries/regions.

In the years 2014–2017, receptor models tested using the PMF (Positive Factorised Matrices) method were increasingly used to isolate the sources of pollution in the outside air. The basis for the application of such a model by Samek et al. [36] in urban areas such as Kraków, where the levels of pollution often violate air quality standards, was the collection of measurement data on air pollution, especially $PM_{2.5}$, considered to be the most hazardous to health. In 2014, approximately 200 samples were collected.

The lowest monthly concentration values were found for August 2014—around 10 $\mu g/m^{-3}$.The highest was for February 2014—70 $\mu g/m^{-3}$—while the annual average value was around 31 $\mu g/m^{-3}$. Using the X-ray fluorescence method to test PM concentration, 15 elements were determined for each sample and eight inorganic ions were analysed by ion chromatography. Additionally, the samples were analysed for soot (PNE). The average concentration of $PM_{2.5}$ in Kraków (31 $\mu g/m^{-3}$) was twice as high as in 2009–2013 in Genoa, Barcelona and Florence, and similar to Milan.

A receptor model adapted to the research with the EPA PMF 5.0 program was developed, which was used to identify sources of air pollution with suspended dusts and chemical components (elements and ions) generated by separate sources. As a result of the modelling, six sources were identified. The sources of dust and their quantitative share in the total mass of $PM_{2.5}$ was determined. The following sources have been identified: (1) Combustion, (2) secondary nitrates and sulphates, (3) biomass burning, (4) industry, (5) the soil and (6) road traffic. Monthly deviations of the efficiency of the source of pollution with $PM_{2.5}$ dust are presented.

In the Warsaw area, the Juda-Rezler et al. [40] team also applied the positive matrix factorisation (PMF) method, supported by the analysis of enrichment factors, in order to identify the six main sources of $PM_{2.5}$ dust: Residential combustion (fresh and old aerosols) (46% by weight of $PM_{2.5}$), road fumes (21%) and unused exhausts (10%), mineral dust/construction works (12%), high temperature processes (8%) and steel processing (3%). In this analysis, primary organic carbon (POC) and secondary organic carbon (SOC) were classified as two separate components that helped distinguish between primary and secondary aerosol sources. Identification of the sources was also supported by the study of their annual and weekly profiles, as well as the analysis of the correlation of particulate components with meteorological conditions. Comparison of the attributed sources of air pollution with dust in Warsaw and Kraków is shown in Table 3.

The most expressive elements of PM sources in Warsaw are SOC, Cl^- and As for combustion in apartments, NH_4^+, Sb and POC for road transport, Ca and Mg for construction works, and SO_4^{2-} for PM transport over long distances.

The analysis of the collected data, including carbon species, eight main water-soluble ions, 21 small trace elements and local meteorological conditions were used to assess the nature and seasonal variability of PM aerosols in Warsaw, as well as to identify the contribution of natural and anthropomorphic sources to the aerosol levels.

The method of studying the active sources of dust pollution from which the pollutants originate in school buildings located near Athens, Greece, was investigated by [17]. The tests using the PMF method were carried out in 2017 and they referred to the mass concentrations (the highest was 72.02 $\mu g/m^{-3}$) and the chemical analysis of PM_{10}. Seasonal fluctuations were also examined.

Lv et al. [10] studied the concentration of solid particles in Daqing China, from the outside C_{out}, and their percentages expressed in mass concentrations in buildings of various applications, as well as the correlations of internal pollution with the locations of buildings exposed to external sources of dust of various origins (office: C_{out} = 22 $\mu g/m^3$ and ρ = 79.4% (ρ–contribution rate I/O of indoor and outdoor particle sources to the concentration of indoor particulate matter); classroom: 20 $\mu g/m^3$ and 87.6%; city residence: 30 $\mu g/m^3$ and 75.0%; rural house: 34 $\mu g/m^3$ and 90.2%). The researchers found that the share of external sources in the concentration of PM indoors exceeds 70%, and in classrooms and in the countryside it exceeds 90%.

Table 3. Comparison of the attributed sources of air pollution with dust in Warsaw and Kraków.

Warsaw (Juda–Rezler et al. [40])			Kraków (Samek et al. [36])		
Sources	Percentage %	Concentration μg/m^3	Sources	Percentage%	Concentration μg/m^3
Residential combustion (fresh andaged aerosols)	45.6	6.7	Combustion	22.9	27.3
			Secondary nitrate	17.1	11.4
			Secondary sulphate	19.3	10.5
Exhaust traffic emissions from gasoline and diesel engines	21.1	3.1	Traffic	8.3	1.6–4.0
Non-exhaust traffic emissions from abrasion of road, brake pads and tyres	10.2	1.5			
Mineral dust/construction works	12.2	1.8	Biomass burning	15.6	10.0
High temperature industrial processes–ferrous and non-ferrous metal processing	8.2	1.2	Industry and/or soil	2.5	0.5–1.2
Steel processing	2.7	0.4			
			Unidentified	14.3	

3. Materials and Methods

3.1. Preliminary Measurements of Mass Concentrations of Dust Pollutants Inside Rooms

There are two types of PM measurement instruments: Those that provide average concentrations over the sampling period and those that provide real-time instantaneous concentration monitoring. Instruments based on the gravimetric method are considered as reference methods and must meet the requirements of PN-EN 12341:2014 [48]. The principle of the reference method is to collect the particulates on a circular filter. This must be a glass fibre filter (GFF), a quartz fibre filter (QFF), polytetrafluoroethylene (PTFE) or PTFE-coated glass fibre, and must be thoroughly conditioned before and after collection. The filter itself should be weighed before and after collection.

Optical instruments based on light scattering, absorption or occultation of light by particles are used to measure dust concentration in real time. They must have certificates confirming their equivalence with the reference method.

A detailed description of the measurement strategy for airborne particles and legal measuring equipment can be found in ISO 16000-34:2018 [49] and ISO 16000-37:2019 [50]. Measurement of $PM_{2.5}$ mass concentrations describes the strategies and procedures for measuring the mass concentration of $PM_{2.5}$ indoors.

Tests of dust air pollutants in selected rooms were carried out from 27 January 2020 to 10 March 2020 in a low office building (two-story high), located in the centre of Warsaw in an urban development. Three office rooms with a different area and a different proportion of external and internal walls were selected for the study of air pollution. The rooms were empty, the leaks had been identified and taped over, leaving only airflow through the window vents and the undercutting of the front door. Localisation of the measurement devices in rooms is shown in Figure 5. The cubic volumes and number of windows of the analysed rooms were as follows:

Room 1–23.48 m^3, 1 window

Room 2–48.75 m^3, 2 windows

Room 3–93.14 m^3, 3 windows

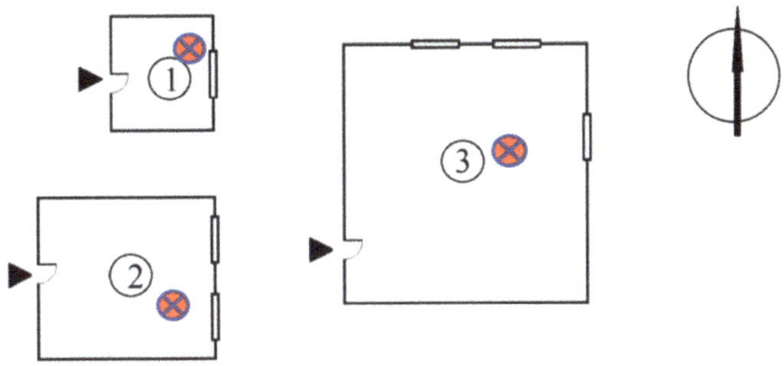

Figure 5. Rooms 1,2,3 where measurements were performed with indicated point of measurement devices.

The rooms were situated on the ground floor of the building (approximately 2 m above ground level and 104 m above sea level). At the time of the test, the exhaust ventilation was turned off, and the ventilation was turned off one hour before the start of the measurements. In the immediate vicinity of the building in question, there are other office buildings and multi-occupancy residential buildings connected to the municipal central heating. In terms of height, the adjacent buildings can be characterised as low (height above ground level up to 12 m), medium (height above ground level up to 25 m) and tall (height above ground level up to 55 m). There are local access roads with low traffic around the office building in question. At a distance of approximately 120 m from the building in question, there is a main road with heavy traffic.

3.2. Measurement Methods (Measuring Instruments and Procedures of Particle Concentration Measurements Inside and Outside the Building)

Measurements of the mass concentration of dust inside were carried out by the AEROCET 831 Handheld Particle Counter (Met One Instruments Inc, Grants Pass, OR, USA) operating on the optical principle with a laser diode, and by the TSI QUEST EVM-7 optical-gravimetric environment monitor, indicating the mass concentration of particles, but also their size fractions $PM_{2.5}$, PM_4, PM_{10} (in terms of measuring the particle size distribution, the optical device is classified as a laser light scattering aerosol spectrometer—LSAS—for measuring particle sizing). The TSI QUEST EVM-7 environment monitor (TSI INCORPORATED, Shoreview, MN, USA) also measures the CO_2 concentration inside the rooms. Used equipment is shown in Figure 6. Used devices are under the supervision of the environmental laboratory and are regularly checked and periodically calibrated. Each device was checked and calibrated before the experiments. Meters from Table 4, the dust and environmental condition measuring devices AEROCET and the TSI were set to zero prior to the measurement using the "filter 0" in order to "reset" the measurement path.

Figure 6. Set of measuring devices for indoor environment testing.

Table 4. Equipment for measuring the thermal parameters of the internal environment and the mass concentration of PM_{10} and $PM_{2.5}$ particles in selected rooms and outside the building.

<table>
<tr><th colspan="2">Type of Measuring Device</th><th>Measured Quantities</th><th>Units</th><th>Resolution</th><th>Measurement Range</th><th>Accuracy</th></tr>
<tr><td colspan="2" rowspan="3">TSI QUEST EVM-7</td><td>Mass concentration of particles</td><td>µg/m³</td><td>1</td><td>0–20,000</td><td>±15%</td></tr>
<tr><td>Particle sizing</td><td>µm</td><td>N/A</td><td>0.1–10</td><td>±2.5%</td></tr>
<tr><td>CO_2 concentration</td><td>ppm</td><td>1</td><td>0–5000</td><td>±100 ppm</td></tr>
<tr><td colspan="2">AEROCET 831</td><td>Selective mass concentration of particles</td><td>µg/m³</td><td>0.1</td><td>0–1000</td><td>±10%</td></tr>
<tr><td rowspan="4">ALMEMO 2690-8AKSU (for microclimate parameters)</td><td rowspan="3">FHAD46C sensor</td><td>Temperature</td><td>°C</td><td>0.1 K</td><td>−20–+80 °C</td><td>±2K</td></tr>
<tr><td>Relative Humidity (RH)</td><td>% RH</td><td>1</td><td>5–98</td><td>±2% RH at 23 °C</td></tr>
<tr><td>Atm. Pressure</td><td>hPa</td><td>0.1</td><td>300–1100</td><td>±2.5 hPa at 23 °C</td></tr>
<tr><td>FV A605-TA1 Thermo-anemometer</td><td>Air flow</td><td>m/s</td><td>0.001</td><td>0.001–1</td><td>± 3 %</td></tr>
</table>

The operating principle of a LSAS relies on particles being guided individually through an intensely illuminated volume. Commercially available equipment enables the particle mass concentrations to be estimated and displayed by the monitor, using evaluation software. Normally, these programs assume the ideal spherical form of particles and convert them to mass concentration. Calibration of the TSI QUEST EVM-7m is required and usually performed with PM latex particles of a defined diameter as the test aerosol. A comparison with other gauges should, however, always be made carefully because ultimately what is determined by LSAS is the particle equivalent optical diameter determined by calibration with monodisperse spherical latex particles.

The lower detection limit depends very much on particle size: The larger the particles, the lower the limit of quantification. The characteristics of the devices used for measuring the internal environment, including PM_{10} and $PM_{2.5}$ dust, are shown in Table 4 below.

Outdoor concentrations of PM_{10} and $PM_{2.5}$ on both measurement days will be based on CIEP-Chief Inspectorate of Environmental Protection, pol. GIOŚ archival data and the results of measurements made with the AEROCET 831 m.

At CIEP, dust concentrations were measured using an automatic method, by a CIEP environmental monitoring station located closest to the building selected for the study. CIEP provides access to hourly monitoring data on air quality in Poland, produced as part of the State Environmental Monitoring and collected in the JPOAT2.0 database.

The CIEP pollution measurement stations taken into account during the tests differ from each other in terms of their location, as well as their location in relation to communication routes affecting the measured pollution levels.

At the same time, measurements of the mass concentration of PM_{10} and $PM_{2.5}$ were performed using the AEROCET 831 m described in Table 4, placed outside, right next to the window opening of the monitored room, i.e., outside the building. Due to the use of different devices at the Building Research Institute and CIEP, the not fully established measurement procedure of "outside the window" (a critical increase in the AEROCET 831m indications was noticed when the cleaner was turned on), different distances and positioning of the monitored rooms in relation to the sides of the world from the official air quality control points, the results of the present study's measurements are not fully comparable with the results presented by CIEP. Self-readings of outdoor concentrations should be more reliable for comparison because they were carried out at the same time points as the indoor pollutant concentration. The outdoor environmental parameters were observed and recorded during experimental campaign. In these preliminary tests, the air velocity in the immediate external environment was not measured.

During the measurements of the mass concentration of dusts, the concentration of CO_2 was also measured with the TSI QUEST EVM-7 m in each test room.

The measurements of the time curves of decay (measuring particle decay curve) and the particle concentration rebound curve in the interiors were carried out in time cycles of 150 min. The concentration of suspended dusts was rebuilt as a result of natural dust penetration through the external wall of the building and the gaps near windows and doors in the time period of from −50 to 0 min, as the cleaner was turned on in the time interval from 0 to 150. At point 0 (on the x-axis), when the cleaner was switched off, the indoor dust concentration was measured and had a minimum value. The second part of the transient curve is the concentration rebound curve.

3.3. *Procedures for Measurements of $PM_{2.5}$ Mass Concentration in the Indoor Mode Performed at ITB*

During the tests, which were carried inside each room on three days, when the air pollution by $PM_{2.5}$ outside was different, the following weather conditions were recorded, as shown in Table 5.

Table 5. Weather conditions during air quality and dust concentration measurements (source: www.weather.com and www.meteomodel.pl, accessed on 27 January 2020, 28 January 2020, 10 March 2020).

	Temp (°C)	Atm (hPa)	Wind (km/h)	RH (%)	Precipitation (mm)
Day 1 (27 January 2020)	3.0	1013.75	12.5	87.1	0.0
Day 2 (28 January 2020)	1.0	1002.50	15.0	93.1	0.0
Day 3 (10 March 2020)	9.0	1008.75	16.0	75.0	0.0

Indoor air pollution tests in three selected rooms were carried out according to the following scheme:

1. An initial measurement of the concentration of PM_{10}, $PM_{2.5}$ and CO_2 was performed.
2. The Samsung AX90R7080WD/EU air cleaner was turned on and PM mass concentration measurements were performed until the minimum dust concentration was achieved.
3. After turning off the air cleaner, measurements of the concentration of PM_{10} and $PM_{2.5}$ dust were taken until reaching "saturation"—the state where indoor PM concentration no longer decreases.
4. After achieving the point of a steady state of PM concentration the measurements of the concentration of PM_{10} and $PM_{2.5}$ dust were going on until reaching equilibrium of I/O concentration.

Thus, this study measured the transient concentration decay curves of the decrease in dust concentration in the room after the air cleaner was turned on until it was turned off at point $C_{in,0}$, and the transient PM concentration rebound curves of the recovery of dust concentration in specific time intervals determined on a real-time basis. The particle concentration rebound curves not the particle concentration decay curves were used in our research to measure the penetration factor. This modification of the method is to be used for determining the value of the particle deposition rate k not from the course of the characteristics in the transient state, the so-called particle decay curve, but from the course of the further time stimulated by infiltration, the so-called particle concentration rebound curve. This method was used to determine the value of k and then P was recognized as correct by He [25] andDiapuoli [51] but it has not been used so far.

One person stayed in the room during the measurements, hence the constant increase in the concentration of carbon dioxide, which was also measured with the TSI QUEST EVM-7 m.

The CO_2 concentrations marked as c_0 and c_t were measured on each measurement day at the beginning and at the end of the measurements of $PM_{2.5}$ mass concentrations.

The results of CO_2 concentration measurements in room 1 (with one window) and in room 3 (with three windows) in ppm are given in Table 6 (assumed CO_2 concentration outside equal to 500 ppm); according to the statement of the Minnesota Department of Health: 'The outdoor concentration of carbon dioxide is about 400 parts per million (ppm) or higher in areas with high traffic or industrial activity. The present authors' assumption is determined by localisation of the measurements points and their own experience in this area.

Table 6. The results of indoor CO_2 concentration measurements (ppm).

Room 1; One Window		**Room 3; Three Windows**	
Initial measurement	Final measurement	Initial measurement	Final measurement
1600 ppm	3400 ppm	1000 ppm	1400 ppm

Dust concentration tests in individual rooms were carried out sequentially (not simultaneously); therefore, the external environment for rooms 1–3 may differ slightly.

4. Results and Discussion

4.1. Results and Discussion of Preliminary Measurements of Indoor $PM_{2.5}$ and PM_{10} Concentrations

The concentrations of PM_{10} and $PM_{2.5}$ outside were taken from published data on the basis of measurements made by the CIEP Environmental Protection Station closest to the studied rooms (Warsaw, Wokalna 7and Al. Niepodległości 227/233). Measurements of air quality stations are reported with an hourly interval and shown in Table 7.

Table 7. Outdoor air quality conditions during air quality and dust concentration measurements (source: GIOŚ).

	Station	**Hour Interval**	**PM_{10} (µg/m^3)**	**$PM_{2.5}$ (µg/m^3)**	**NO_2 (µg/m^3)**	**CO (µg/m^3)**	**Benzene (µg/m^3)**
Day 1 (27 January 2020)	Wokalna	9:00–10:00	60.8	46.2	33.0	-	-
		10:00–11:00	61.0	45.9	-	-	-
		11:00–12:00	58.8	44.0	-	-	-
		12:00–13:00	55.1	40.4	30.7	-	-
		13:00–14:00	48.6	33.8	34.8	-	-
	Niepodległości	9:00–10:00	65.6	62.5	66.9	1.5	3.0
		10:00–11:00	65.1	60.9	-	-	-
		11:00–12:00	68.8	61.2	-	-	-
		12:00–13:00	82.0	62.6	47.0	1.1	2.4
		13:00–14:00	87.6	54.8	58.2	1.1	2.1
Day 2 (28 January 2020)	Wokalna	8:00–9:00	58.9	43.5	35.6	-	-
		9:00–10:00	58.7	42.8	37.9	-	-
		10:00–11:00	51.3	36.6	34.3	-	-
		11:00–12:00	40.5	29.2	28.3	-	-
	Niepodległości	8:00–9:00	64.3	58.1	55.7	1.3	2.9
		9:00–10:00	64.2	59.6	51.7	1.1	2.7
		10:00–11:00	60.6	54.5	59.6	1.2	2.7
		11:00–12:00	52.3	44.0	60.5	1.1	2.5
Day 3 (10 March 2020)	Wokalna	6:00–7:00	41.5	37.1	36.7	-	-
		7:00–8:00	51,3	36.6	34.3	-	-
		8:00–9:00	31.7	28.9	39.1	-	-
		9:00–10:00	30.4	28.1	38.7	-	-
	Niepodległości	6:00–7:00	51.8	44.0	47.2	0.3	2.0
		7:00–8:00	60.6	54.5	59.7	1.2	2.7
		8:00–9:00	43.6	41.0	67.4	0.6	1.6
		9:00–10:00	40.7	39.9	84.5	0.8	1.9

Based on the data on dust concentrations obtained from the hourly monitoring of the Chief Inspectorate of Environmental Protection for PM_{10} and $PM_{2.5}$ dusts on Day 1 (27/01/2020), Day 2 (28/01/2020) and Day 3 (10/03/2020), the curves of dust concentration changes outside the building were determined. The curves of dust concentration inside the building in the indicated rooms were drawn on the basis of measurement data made with the AEROCET 831 m. Figure 7a–c show changes in the mass concentration of PM_{10} and $PM_{2.5}$ on the first, second and third days of measurements in rooms 1 and 3.

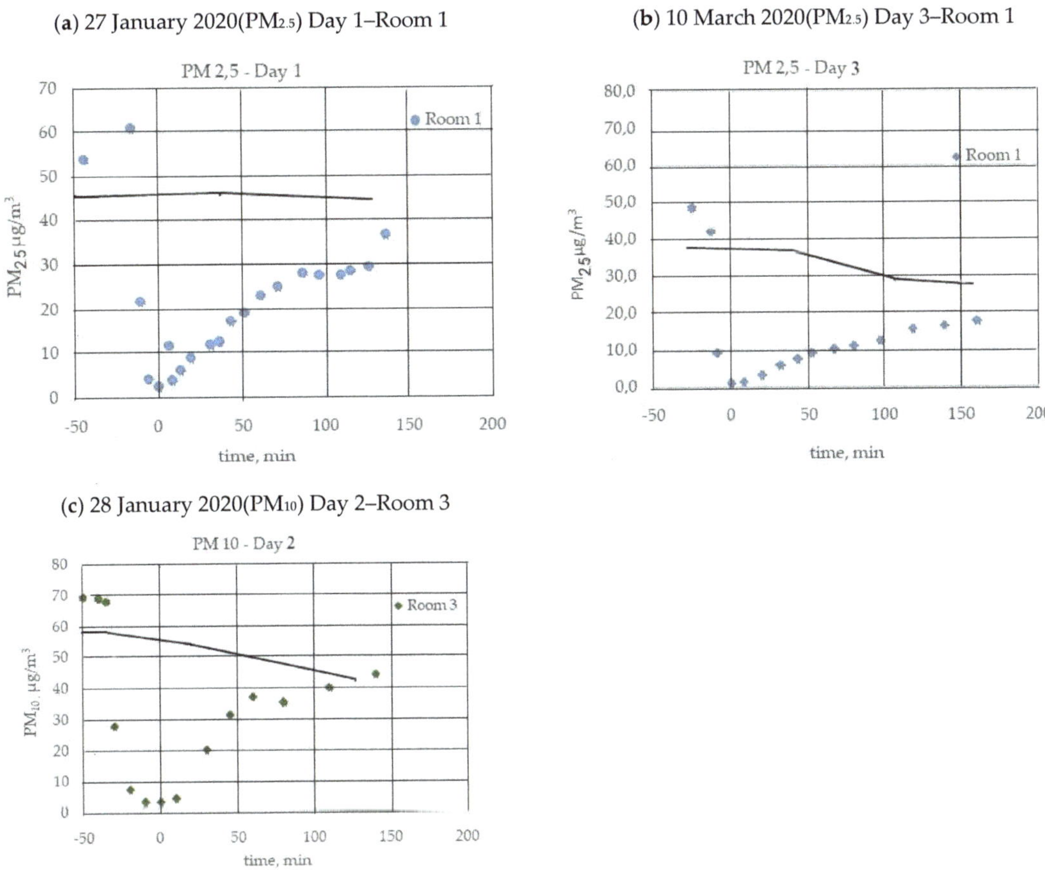

Figure 7. Forced changes in the concentration of PM_{10} and $PM_{2.5}$ dust inside the tested rooms (points) and outside the building (solid line) on three working days of dust concentration rebound curve measurements.

Comparing the measurements of $PM_{2.5}$ dust on day 1 (Figure 7a) and day 3 (Figure 7b) in room 1, it can be concluded that the rate of dust concentration level increase was higher on day 1, and the indoor pollution equilibrium state was reached by approximately100 min. The rate of dust concentration level was much lower on day 3, and the indoor pollution equilibrium state was reached by approximately150 min. This may be due to the smaller and continually decreased PM concentration gradient (C_{out}–C_{in}) (although at the same time during the measuring cycle, the temperature and wind speed were slightly higher). On day 3, the ambient concentration of dusts C_{out} measured by CIEP decreased in measuring time from 38 to 28 µg/m^3.

By analysing the course of the measurement of PM_{10} on day 2 (Figure 7c) in room 3, it is possible to notice also a uniform decrease in pollution outdoors caused by high relative air humidity; therefore, the concentration gradient (C_{out}–C_{in}) also decreased. According to theoretical assumptions, however, the equilibrium level of PM_{10} concentration inside the room is set at a level about 40% higher than the level of $PM_{2.5}$ concentration (Figure 7a). The measurement of the PM_{10} concentration rebound curve (Figure 7c) was completed when rainfall occurred. Authors found that the compliance between the CIEP (Chief Inspectorate of Environmental Protection) results with our $PM_{2.5}$ measurements outside the building is of the order of uncertainty in the measurement of PM concentrations inside regarding the used field instruments in accordance with Table 4.

Obtained results of preliminary experiments have shown that a system for measuring the concentration of $PM_{2.5}$ directly outside the window should be designed, which would be protected against disturbances caused by the operation of the air cleaner in the interior and changes by the meteorological parameters.

4.2. Theoretical Basis and Possibilities of Predicting Dust Penetration through the Building Envelope. A Dynamic Model of the Mass Balance Equation of Indoor Particle Concentration Levels

The process of penetration of dust from the outside into the interior of the building and the generation of PM by internal sources is a dynamic process [46]; therefore, all findings emphasise the importance but also the difficulty of determining separately the time courses of changes in the concentration of suspended dust in the interior, and the share in their mass of suspended particles from outdoors and the proportion of particles generated indoors. Such data are necessary for the interpretation of the results of epidemiological studies. In this framework, recent scientific research has focused on the study of factors influencing the penetration of particles through the building envelope and the quantification of the relative indoor proportion of particles coming from outside and remaining as dust suspended inside the room. The basic form of the dynamic mass balance equation is given in [25,51–56]; it describes the profiles of changes in the temporal concentration of PM during the particle decay curve or during the reconstruction of this concentration (particle rebound curve) in the room with Equation (4)

$$\frac{dC_{in(t)}}{dt} = a \cdot P \cdot C_{out}(t) - (a + k) \cdot C_{in}(t) + \frac{Q_{is}}{V} \tag{4}$$

where $C_{in}(t)$ and $C_{out}(t)$ are the concentrations of particles inside and outside at time t, respectively (mg/m^3); a is the multiplicity of air changes (h^{-1}); P is the dimensionless coefficient of particle penetration efficiency; k is the particle deposition rate (h^{-1}); V is the volume of the interior (m^3); and Q_{is} is the rate of generation of particles by the indoor sources(mg/h). Equation (4) assumes perfect mixing of indoor air. It also ignores particle mass losses or gains due to differences in the gas-phase concentrations of condensable substances and changes in temperature/relative humidity conditions between indoor and outdoor spaces. The penetration efficiency factor (P) and particle deposition loss rate (k) are also related to building characteristics and indoor/outdoor conditions, but these parameters also depend on particle size, composition and their electric charge.

The particle penetration factor (P) is defined as the mass fraction of the particles in the infiltrated air passing through the building envelope, which depends on particle diameter (d_p). This simple definition is

$$P = N_{escape} / N_{total} \tag{5}$$

where N_{escape} are the particles escaping through the leak outlet and N_{total} are the particles collected at the entrance of the leak in the building. The penetration factor is the most appropriate parameter for describing the mechanism of particle penetration through cracks and leaks in the building envelope, and at the same time the parameter describing the functionality of the building and, thus, its balance.

The main difficulty in the solution to this equation lies in the separate calculation of the penetration efficiency factor (P) and deposition rate (k). The values for these two parameters that are reported in the literature vary significantly. Deposition rate k presents the widest range of values in terms of size fractions in the literature. The penetration efficiency factor P seems to be more accurately calculated through the application of dynamic models.

Bennett and Koutrakis [57] developed a method for calculating the unknowns (P) and (k) using time-dependent indoor and outdoor particle concentrations and air exchange rate (a). Assuming there are no indoor particle sources, when using discreet time steps Δt, expressed as Equation (4), Equation (6) can be rewritten as follows

$$C_{in,t} = \frac{a_i P_i C_{out,t-\Delta t}}{(k+a_i)} \times \left(1 - e^{-(k_i+a_i)\Delta t}\right) + C_{in,t-\Delta t} \times e^{-(k_i+a_i)\Delta t} \tag{6}$$

Assuming steady-state conditions inside the building, no internal sources of suspended dust generation and no mechanical ventilation with a filter, Equation (6) can be transformed into a Equation defining the dust infiltration coefficient F_{inf}. The infiltration of air with particles suspended indoors is also described by the infiltration coefficient (F_{inf}), which determines the fraction of external particles that enters the internal microenvironment and remains suspended. The particulate matter infiltration process is defined by the following equation

$$F_{inf} = \frac{C_{in}}{C_{out}} = \frac{P \cdot a}{a+k} \tag{7}$$

where F_{inf} is the PM infiltration factor (-).

"When indoor particle emissions cannot be avoided, the method developed by Long et al. [58] which is based on the linear regression approach can be used for F_{inf} determination".

The analytical solution of Equation (6), based on field measurements of mass concentrations of $PM_{2.5}$ inside and outside the building was used by Chao Chen et al. [59] in their study of particle penetration through window gaps. They used Equation (6), where the Δt is time step Δt = 1 h; and a_i, P_i, k_i are hourly air exchange rate, penetration factor and deposition rate, respectively. Therefore, Equation (6) can be resolved when $C_{in,t}$ and $C_{out,}$ (i = 1, 2 … n) are known for a period Δt of time, where hourly indoor and outdoor $PM_{2.5}$ mass concentrations could be determined by field measured data, so the unknowns are air exchange rate (a_i), penetration factor (P_i) and deposition rate (k_i). As such, the number of equations is ($n-1$) and the unknowns are 3(n−1) in the new Equation (4) ($n-1$, a_i, P_i and k_i).

The experimental solution of the dynamic mass balance equation of indoor PM concentration is recommended by [25,52,53]. They argued that a dynamic variation of the PM concentration level in the sample room should be performed while simultaneously measuring the PM particle decay curve in real buildings.

When the test time is longer than the time constant, it is implied that the indoor $PM_{2.5}$ concentration has reached a steady-state condition, then

$$\frac{dC_i(t)}{dt} = 0 \tag{8}$$

Therefore Equation (4) can be summarized by Equation (9)

$$C_{in,t} = C_{in(0)} = \frac{P \cdot a \cdot C_{out}}{a+k} \tag{9}$$

where $C_{in,t}$ is the steady-state indoor $PM_{2.5}$ concentration.

Upon determining the "decay term" (a + k) and the air exchange rate a, the penetration efficiency factor P is described by the following equation, which assumes the absence of internal sources PM.

$$P = \frac{(a+k)}{a} \cdot \frac{C_{in,0}}{C_{out}} \tag{10}$$

where $C_{in,0}$ is the concentration C_{in} of the PM particles already at a steady state with time.

By estimating the coefficient of particle infiltration from outside into the room, F_{inf}, from the dynamic mass balance equation, it is difficult to determine independently the values of P and k. Over the last two decades, many methods have been used to estimate P

and k and calculate the relative contributions of particles from indoor and outdoor sources in the measured mass concentrations of given indoor particles. Diapouoli et al. [51] describe a broad overview of the methods for determining the values of these parameters, presenting different approaches, which they grouped into four categories according to the principles of their determination: (1) Steady-state assumption using the steady state of the mass balance equation; (2) dynamic solution of the mass balance equation using complex statistical techniques; (3) experimental studies using conditions that simplify model calculations (e.g., decreasing the number of unknowns); and (4) infiltration surrogates using a particulate matter (PM) constituent with no indoor sources to act as surrogate of indoor PM of outdoor origin. As they claim, however, the analysis of various methodologies and results shows that the estimation of particle infiltration parameters is still difficult.

The penetration factor P measurement principle, according to the authors' proposal, covers the realisation of the dynamic mass balance equation of indoor PM concentration, which is presented in Figure 8. Air exchange rate a is measured at first using the properly documented principle and therefore it is not described in Figure 8.

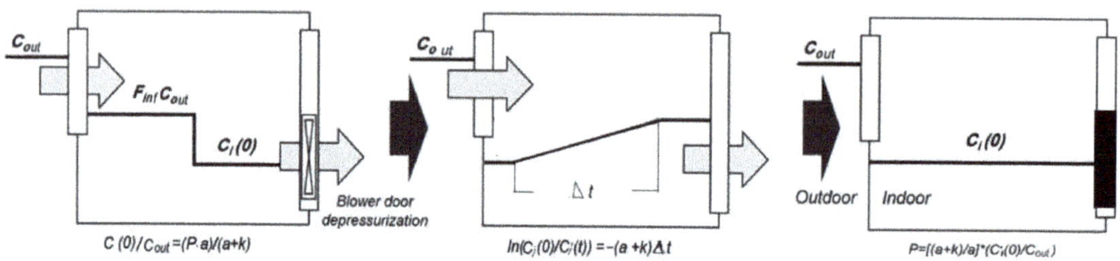

Figure 8. Schematic diagram of the P factor test procedure (with reference to Dong et al. [24]).

4.3. Method of Determination of the Penetration Efficiency Factor P

Air exchange rate in the internal microenvironment (a)

The rate of air exchange in the room (a) due to infiltration of air from the outside is a measurable parameter, and is mainly influenced by the construction and technical condition of the external wall of the building and the ventilation system, the activity of residents and meteorological conditions.

The size of (a) in the room where people are present can be estimated on the basis of the time curve of the decay of CO_2 concentration in the room at night. The air exchange rate can be estimated from the exponential curve of decay of CO_2 concentration over time, using Equation

$$a = \frac{1}{t - t_0} \ln\left(\frac{c - c_{out}}{c_0 - c_{out}}\right) \quad (11)$$

where a is the air change rate (h^{-1}), and the values of time t and t_0 are read at the end and the beginning of the concentration decay curve (h^{-1}), respectively. The values of c and c_0 are CO_2 concentrations (ppm) measured at times t and t_0, respectively, and c_{out} (ppm) is the concentration of PM outside at time t. The air change rate estimated for different measurement cycles shows seasonal fluctuations in the multiplicity of air changes. In naturally ventilated buildings, air flow is caused by differences in temperature or pressure inside and outside the building.

Equation (11) was used by the Czech–Norwegian team of Chatoutsidou et al. [54] in a study of the infiltration process in The Baroque Library Hall building in Prague with a natural ventilation system by infiltration in order to calculate the number of air changes. In this study, a dynamic mass balance model was used, taking into account the penetration of particles from external and internal losses (deposition, ventilation). The model was used to determine the particle deposition rate k and the penetration efficiency P. As a result of the three-year research, the values of a for spring were $a = 0.13$, for summer $a = 0.11$ and for winter $a = 0.15$. Equation (11) was also used in a simplified form by [25,60].

$$a = \frac{1}{t} \ln \frac{c_t}{c_0} \quad (12)$$

Deposition rate of airborne particles (k).

One of the simpler methods of determining the particle deposition coefficient is recommended by [25,51], who specified the factor of Equation (6) as the "decay term"—($a + k$) in the time interval Δt, using the regression analysis of the experimentally determined particle concentration decay curve. The temporary disappearance of the dust concentration in the interior, i.e., the C_{in0} concentration, to the C_{int} highest concentration (e.g., C_{in} level caused, for example, by the operation of the air cleaner) in accordance with equation

$$\ln\left(\frac{C_{in,0}}{C_{int}}\right) = -(a + k)\Delta t \quad (13)$$

is used to then derive the value of k from the value ($a + k$). An increasing trend was found for increasing particle sizes. At the maximum particle size, particle deposition rates are of the same order or even higher than their removal rates due to blown air exchange. According to the results of [1], higher deposition rates were found for larger particles, so the deposition coefficient k of PM_{10} particles is greater than k for $PM_{2.5}$ particles. They also found large differences in deposition rates among the six residences studied as a result of a discrepancy in surface material texture and roughness as confirmed by Abadie et al. [61]. They also demonstrated that the deposition effect is temperature dependent.

Particle penetration efficiency through the external walls of the building (P).

There are several different models in the literature for predicting the penetration efficiency P of particles through the gaps in the building envelope. The equation resulting from the Lagrangian model (Chen, Zhao [52]) is particularly interesting. It calculates the trajectory of each particle by integrating the equilibrium forces acting on individual particles.

The practical significance of the research conducted by Chao et al. [62] from the Hong Kong Technical University concerned the determination of the P and k coefficients in residences with natural ventilation. Both parameters were dependent on particle size but showed a different upward and downward inversion profile with respect to particle size. The main causes of loss by deposition and penetration effects are diffusion, inertial impact/interception and deposition (showing a different up-and-down inversion profile against particle size).

The efficiency of P is related to particle size, as demonstrated by many researchers who compared the penetration of particles of different sizes [1,25].

Chao et al. [62] studied six buildings and argued that 'the penetration coefficient showed a hill-shape with respect to particle sizes and there was a peak (0.79) at the size range of 0.853–1.382 µm'.

It follows that the deposition coefficient k of PM_{10} particles is greater than k of $PM_{2.5}$ particles while the penetration coefficient P of PM_{10} particles is less than P of $PM_{2.5}$ particles. The P particle penetration factor is, therefore, another important factor determining the value of the I/O ratio. If the building is ventilated by infiltration, however, the expansion of the particle penetration coefficient is a strong function of the air exchange coefficient a, particle size and fracture geometry in the building envelope.

Chao et al. [62] also described a simplified way to analyse the model of indoor particle behaviour.

This simplified method has been developed for users of particle counters, who need to express concentrations in the form of a particle number concentration, i.e., ($\times 10^3$) p cm^{-3}, so that there is a need to convert the units in which the particle concentrations are recorded, i.e., convert the particle mass concentration units µg/m^{-3} per unit of particle number concentration ($\times 10^3$) p cm^{-3}, assuming a spherical shape and uniform particle density and packing. Such conversions are performed for each particle size range.

Returning to the dynamic mass balance Equation (6), it can be said that all parameters in this equation are known values, except for the penetration efficiency coefficient P and the deposition rate k. Chao et al. [62] transformed Equation (6) into a transient form. For each particle size range, the concentration of particles in the interior during the forced change in dust indoor concentration is equal to Equation (14)

$$C_{in} = \frac{PaC_{out}}{(a+k)} + \left(C_{int} - \frac{PaC_{out}}{a+k}\right)e^{-(a+k)t} \tag{14}$$

where C_{int} the is the initial dust pollution on the transient concentration decay curve.

Equation (14) consists of two parts: $P{\cdot}a{\cdot}C_{out}/(a + k)$ and $(a + k)$, which are the steady-state particle concentration and the time constant of the transient term, respectively.

Chao et al. [62] performed an experiment of the process of particle concentration decay of C_{in} at time $(t\text{–}t_o)$ caused by intensive cleaning of the room from particles using an air cleaner. A typical profile curve of such a process is given by [62] and is also cited by [63]. The experiment consisted in artificially increasing the PM concentration to the C_{int} highest concentration value in the room in order to obtain a "drive" for the increase in the deposition rate, and then quickly reducing the concentration of solid particles to the steady state C_{in0} by inducing ventilation present in the building. Substituting all measured particle concentrations and the air exchange rate a obtained in the tracer gas CO_2 decay test, the deposition rate from the particle decay curve can be expressed as follows:

$$k = \left(\frac{1}{t}\right)\ln\left(\frac{C_{in} - C_{in0}}{C_{int} - C_{in0}}\right) - a \tag{15}$$

The difficulty in applying this equation, however, lies in the correct choice of the time point in particle concentration decay curve at which the k value is to be calculated, since the chosen value of t will determine the value of $C_{in(t)}$ from the decay curve or from the particle rebound curve. The penetration rate P can be found from the steady-state particle concentration upon determining the deposition rate k, which is shown as follows:

$$P = \left(1 + \frac{k}{a}\right)\frac{C_{in0}}{C_{out}} \tag{16}$$

The penetration factor consists of the ratio of the concentration found inside to the level of particle concentration outside (I/O ratio) and from the expression to the ratio of the particle deposition rate to the air exchange rate. In the case where the deposition rate k is much slower than the air exchange rate a, the penetration coefficient is equivalent to the I/O ratio.

The particle decay method can be replaced by the particle concentration rebound method as mentioned by [25,51,52]. Then, it will be a reversed process, the transition curve of which will reflect more closely the ease of particle penetration. This is because the experiment takes place under conditions of natural pressure difference I/O, and reflects better the efficiency of penetration of particles through the building envelope.

4.4. Exemplary Determination of the Penetration Efficiency Factor Based on the Results of Tests of the Profile of Forced Changes in the Concentration of PM Particles in a Selected Room

As an example of a measurement cycle, the results of which were used to calculate the value of the penetration efficiency factor P of $PM_{2.5}$ particles through the external walls of the building, we chose the mass concentration C_{in} and C_{out} measurement cycle of 27 January 2020, made in room 1 (with one window), and presented in Figure 7a.

Assuming that for further calculations of the penetration coefficient, an experiment must be conducted that will force the concentration decay process, and then the process of restoring the concentration level of $PM_{2.5}$ in the selected room (as in experiments in [26,62]), the authors present the transient curve of this dynamic process with the penetration of $PM_{2.5}$ particles through the building envelope in Figure 7.

The first step was to determine the value of the air changes rate a in the selected room and under the thermal conditions of the measurement cycle. The CO_2 concentrations marked as c_0 and c_t were measured on each measurement day at the beginning and at the end of the measurements of $PM_{2.5}$ mass concentrations. The values of CO_2 concentration outside the building, c_{out} = 500 ppm, were assumed in accordance with ISO 16000-26:2012 [64]. The a value calculated from Equation (9) was for the selected room and measurement date a = 0.28.

The value a (ACH air changing rate h^{-1}) is calculated from equation

$$a = \frac{1}{(t - t_0)} \cdot \ln\left(\frac{c_t - c_{out}}{(c_0 - c_{out})}\right)$$

In order to use Equations (14)–(16), the authors had to convert the results of particle concentration measurements in the selected measurement cycle (Figure 9) from the mass concentration units μg/m^{-3} to the numerical concentration units ($\times 10^3$) p/cm^{-3} (number of particles to cubic cm). The method of conversion was taken from the article by Morawska et al. [26], assuming (for the size range of $PM_{2.5}$ particles) the conversion factor of 21.3/18 of the values of mass concentration units to units of numerical concentration.

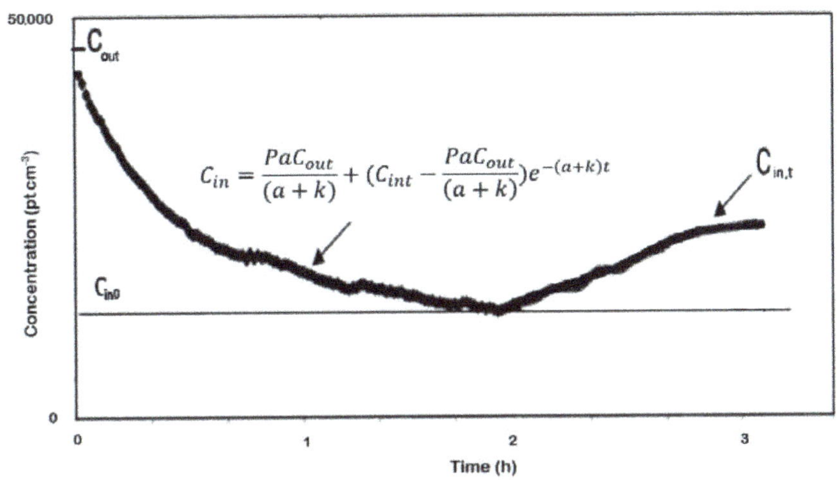

Figure 9. Coupled dynamic processes: Particle concentration decay curve combined with particle concentration rebound curve. Decay curve starts at the concentration point C_{int}.

The second step is to determine the value of particle deposition rate k (from the concentration decay curve C_{in} of $PM_{2.5}$ particles or from the particle concentration rebound curve (see Figure 7) in the selected time interval Δt (then, it is a reverse cycle) in the room under the conditions of the determined air change rate a (see Table 8). The k factor will be calculated for the measurement cycle shown in Figure 7, using Equation (7) simplified to the form (13) with the data from Table 9.

Table 8. Data for calculating the a value (air changing rate h^{-1}).

c_{out} = 500 ppm	Room 1 (one window)
c_0 = 1600 ppm	c_t = 3400 ppm
c_0–c_{out} = 1100 ppm	c_t–c_{out} = 2900 ppm
ln 2.64 = 0.97	a = 0.28 h^{-1}

Table 9. Data for calculating the k value (deposition rate h^{-1}).

C_{in0} = 3.8 μg/m^{-3} = 4.5 × 10^3 p/cm^{-3} in time t = 0 h	$C_{in,t}$ = 26 μg/m^{-3} = 30.7 × 10^3 p cm^{-3} in time t = 1 h
ln(4.5/30.7) = ln(0.15) = (−1.90)	(−1.90) = −(a + k) × 1
1.90 = (a + k)	
a = 0.28	k = 1.62 h^{-1}

From the $PM_{2.5}$ particle concentration decay curve measured in Room 1 on 27 January 2020 (Figure 8) curve segment $\Delta t = t\text{-}t_0$ of the particle concentration rebound curve was selected. $C_{in0} = 4.5 \times 10^3$ p cm^{-3} was calculated for time t_0 and $C_{in} = 30.7 \times 10^3$ p cm^{-3} was calculated after 60 min for break point t of the particle level restoration curve. Based on Equation (13), the value $(a + k) = 1.90$ was determined, which gives, after subtraction, $a = 0.28$ and $k = 1.62$

Calculation of the value of k (deposition rate h^{-1}) from Equation (13)

$$\ln\left(\frac{C_{in0}}{C_{in,t}}\right) = -(a+k)\Delta t$$

In the literature, however, there is a large discrepancy in the values of the deposition rate factors for the size range of $PM_{2.5}$ particles, which are reported with the values of the air exchange rate.

In the range of a from 0.45 to 0.61 h^{-1}, i.e., in conditions similar to our air changing rate $a = 0.28$ h^{-1}, the value of $k = 2.01 \pm 1.11$ $h^{-1}$was given in He, Morawska and Gilbert [25].The authors decided to compare the obtained value with others, e.g., Chao et al. [62] ($k = 0.53$ h^{-1} to $k = 1$ h^{-1}). We chose our value of $k = 1.62$ h^{-1} for further calculations because it is close to that given by [25] and because previously a similar value of $k = 1.3$ h^{-1} for $PM_{2.5}$ was obtained by the team Thatcher et al. [65] from the Lawrence Berkeley National Laboratory.

The third step is to calculate penetration P from Equation (16) with the data from Table 10.

$$P = \left(1 + \frac{k}{a}\right)\frac{C_{in0}}{C_{out}}$$

Table 10. Data for calculating penetration factor P.

C_{in0} = 3.8 μg/m^{-3} = 4.5 × 10^3 p/cm^{-3}	C_{out} = 44 μg/m^{-3} = 52 × 10^3 p/cm^{-3}
[(a + k)/a] [4.5/52] = (1.90/0.28) (4.5/52) = 0.61	
P = 0.61	

Thus, after converting the values of $C_{in0} = 4.5 \times 10^3$ p/cm^{-3} and $C_{out} = 52.0 \times 10^3$ p/cm^{-3}, and substituting all parameter values into Equation (16), the value $P = 0.61$ is obtained. From the course of the measurement cycle (Figure 7a), it can be seen that the value of P will be higher because the concentration level of C_{out} during the particle concentration rebound curve measurement was constantly decreasing.

The value is verified by the data collected by Chao et al. [62] who, based on their research, compiled the values of P factors for particles in wide particle size ranges.

For $PM_{2.5}$, it is possible to determine the P value of $PM_{2.5}$ particles by interpolating the P factor value from the chart prepared by Chao et al. [62]. The calculated value $P = (2.458 + 2.642)/2 = 0.63$ is roughly consistent with our value $P = 0.61$.

According to a recent study by Yu et al. [38], in diverse pressure differences, the 0.25–2.5 μm particles within the scope of penetration factor P is close to 1, which increased along with the pressure difference between the two sides of the gap. Four penetration factor models compared Chen and Zhao [52] and present much lower P values for $PM_{2.5}$.

5. Conclusions

Predicted infiltration factors F_{inf} in urban residences were collected, tested and classified by Baxter et al. [66] from the Harvard School of Public Health in Boston. They studied various factors related to occupant behaviours and home characteristics, which might influence ventilation patterns or infiltration factors. These include age of construction, housing type (multi- vs. single-occupancy homes), floor levels, opening windows, seasonal differences in F_{inf} and air conditioning or HVAC system use.

The transport of outdoor particles across the building envelope (i.e., penetration) is an important physical factor that contributes to particle concentration and size distribution inside buildings. The penetration factor can be equalised when doors and windows are open, but under different conditions the p values ought to be measured.

Outdoor pollution sources, rather than indoor pollution sources, are the principal contributors causing indoor particulate pollution in modern office buildings. When there are no substantial indoor particle pollution sources, 60–70% of the indoor $PM_{2.5}$ mass density comes from outdoor particulate pollution, according to the study.

This paper presents a modified method of determining the parameters of the process of penetration of $PM_{2.5}$ particles through leaks in the external walls of the building with the use of the processes of natural recovery of the levels of particle concentration. So far, dynamic tests of PM infiltration have been performed using the blower-door depressurization procedure by introducing the n-th power disturbance in the air flow caused by increasing the pressure gradient in the I/O system. The modified method promotes the measurement of C_{in} concentration during the process of equalizing the level of $PM_{2.5}$ concentrations inside and outside the building.

The preliminary results of the penetration factors determined by this method are consistent with the P factor values from the literature obtained so far for this dimensional group of dusts. Information on the P factor value can be a reliable parameter that classifies buildings in terms of resistance to dust infiltration into their interior, which may have a significant impact on indoor air quality. Obtained results of preliminary experiments have led to the conclusion that a system for measuring the concentration of $PM_{2.5}$ directly outside the window should be designed, which would be protected against disturbances caused by the operation of the air cleaner in the interior and changes by the meteorological parameters.

The authors have tried to describe the current state of knowledge in the field of research on interior pollution by dust and establish a roadmap for launching and improving measurement and test methods so that in the future a dynamic indoor particle model could describe indoor air quality, when exposure to $PM_{2.5}$ takes place. This model will be further tested using the Monte Carlo method before being included in the Combined IAQ model (Piasecki, Kostyrko [67,68]).

Author Contributions: Conceptualization, K.B.K. and D.B.; methodology, K.B.K. and D.B.; validation, K.B.K. and D.B.; formal analysis, K.B.K. and D.B.; investigation, K.B.K. and D.B.; resources, K.B.K. and D.B.; data curation, K.B.K. and D.B.; writing—original draft preparation, K.B.K. and D.B.; writing—review and editing, K.B.K. and D.B.; visualization, K.B.K. and D.B.; supervision, K.B.K. and D.B.; project administration, K.B.K. and D.B.; funding acquisition, K.B.K. and D.B. Both authors have read and agreed to the published version of the manuscript.

Funding: This research received no external funding.

Institutional Review Board Statement: Not applicable.

Informed Consent Statement: Not applicable.

Data Availability Statement: Data supporting reported results can be found in the Thermal Physics, Acoustics and Environmental Protection Department, Instytut Techniki Budowlanej, Warsaw, Poland.

Conflicts of Interest: The authors declare no conflict of interest.

References

1. Thatcher, T.L.; Layton, D.W. Deposition, resuspension, and penetration of particles within a residence. *Atmos. Environ.* **1995**, *29*, 1487–1497. [CrossRef]
2. Arvanitis, A.; Kotzias, D.; Kephalopoulos, S.; Carrer, P.; Cavallo, D.; Cesaroni, G.; Brouwere, K.D.; de Oliveira-Fernandes, E.; Forastiere, F.; Fossati, S. The index-pm project: Health risks from exposure to indoor particulate matter. *Fresenius Environ. Bull.* **2010**, *19*, 2458–2471.
3. Crouse, D.L.; Peters, P.A.; Hystad, P.; Brook, J.R.; van Donkelaar, A.; Martin, R.V.; Villeneuve, P.J.; Jerrett, M.; Goldberg, M.S.; Pope, C.A., III; et al. Ambient PM2.5, O3, and NO2 exposures and associations with mortality over 16 years of follow-up in the canadian census health and environment cohort (CanCHEC). *Environ. Health Perspect.* **2015**, *123*, 1180–1186. [CrossRef]
4. Hvelplund, M.H.; Liu, L.; Frandsen, K.M.; Qian, H.; Nielsen, P.V.; Dai, Y.; Wen, L.; Zhang, Y. Numerical investigation of the lower airway exposure to indoor particulate contaminants. *Indoor Built Environ.* **2019**, *29*, 575–586. [CrossRef]
5. Wang, S.; Inthavong, K.; Wen, J.; Tu, J.; Xue, C. Comparison of micron- and nanoparticle deposition patterns in a realistic human nasal cavity. *Respir. Physiol. Neurobiol.* **2009**, *166*, 142–151. [CrossRef] [PubMed]
6. Lin, Y.; Zou, J.; Yang, W.; Li, C.-Q. A Review of Recent Advances in Research on PM2.5 in China. *Int. J. Environ. Res. Public Health* **2018**, *15*, 438. [CrossRef] [PubMed]
7. Lee, J.Y.; Ryu, S.H.; Lee, G.; Bae, G.-N. Indoor-to-outdoor particle concentration ratio model for human exposure analysis. *Atmos. Environ.* **2016**, *127*, 100–106. [CrossRef]
8. Branco, P.; Alvim-Ferraz, M.D.C.; Martins, F.; Sousa, S. Children's exposure to indoor air in urban nurseries-part I: CO2 and comfort assessment. *Environ. Res.* **2015**, *140*, 1–9. [CrossRef] [PubMed]
9. Kalimeri, K.K.; Bartzis, J.G.; Sakellaris, I.A.; Fernandes, E.D.O. Investigation of the PM2.5, NO2 and O3 I/O ratios for office and school microenvironments. *Environ. Res.* **2019**, *179*, 108791. [CrossRef]
10. Lv, Y.; Wang, H.; Wei, S.; Zhang, L.; Zhao, Q. The Correlation between Indoor and Outdoor Particulate Matter of Different Building Types in Daqing, China. *Proc. Eng.* **2017**, *205*, 360–367. [CrossRef]
11. Nadali, A.; Arfaeinia, H.; Asadgol, Z.; Fahiminia, M. Indoor and outdoor concentration of PM10, PM2.5 and PM1 in residential building and evaluation of negative air ions (NAIs) in indoor PM removal. *Environ. Pollut. Bioavailab.* **2020**, *32*, 47–55. [CrossRef]
12. Bai, L.; He, Z.; Li, C.; Chen, Z. Investigation of yearly indoor/outdoor PM2.5 levels in the perspectives of health impacts and air pollution control: Case study in Changchun, in the northeast of China. *Sustain. Cities Soc.* **2019**, *53*, 101871. [CrossRef]
13. Zhang, S.; Duan, Y. Determination the PM2.5 concentration in the room of Liting mosquito-repellent incense and cigarette. *Inn. Mong. Environ. Sci.* **2013**, *25*, 184–185.
14. Zhang, N.; Han, B.; He, F.; Xu, J.; Zhao, R.; Zhang, Y.; Bai, Z. Chemical characteristic of PM2.5 emission and inhalational carcinogenic risk of domestic Chinese cooking. *Environ. Pollut.* **2017**, *227*, 24–30. [CrossRef] [PubMed]
15. Xue, Y.; Zhou, Z.; Nie, T.; Wang, K.; Nie, L.; Pan, T.; Wu, X.; Tian, H.; Zhong, L.; Li, J.; et al. Trends of multiple air pollutants emissions from residential coal combustion in Beijing and its implication on improving air quality for control measures. *Atmos. Environ.* **2016**, *142*, 303–312. [CrossRef]
16. Zhou, Z.; Liu, Y.; Yuan, J.; Zuo, J.; Chen, G.; Xu, L.; Rameezdeen, R. Indoor PM2.5 concentrations in residential buildings during a severely polluted winter: A case study in Tian, 2016. *China. Renew. Sustain. Energy Rev.* **2016**, *64*, 372–381. [CrossRef]
17. Simeone, M.G.; Ubaldi, V.; Lepore, A.; Cirillo, M.C. A Web graphig tool for travelling through a virtual home, school or office to improve our awareness on Indoor air risk. In Proceedings of the 8th International Conference and Exhibition on Healthy Buildings, Lisboa, Portugal, 4–8 June 2006; p. 315.
18. Jantunen, M.J. Indoor Air Exposure. In Proceedings of the 8th International Conference and Exhibition on Healthy Buildings, Lisboa, Portugal, 4–8 June 2006; p. 23.
19. Isaxon, C.; Gudmundsson, A.; Nordin, E.; Lönnblad, L.; Dahl, A.; Wieslander, G.; Bohgard, M.; Wierzbicka, A. Contribution of indoor-generated particles to residential exposure. *Atmos. Environ.* **2015**, *106*, 458–466. [CrossRef]
20. Wallace, L.; Ott, W. Personal exposure to ultrafine particles. *J. Expo. Sci. Environ. Epidemiol.* **2011**, *21*, 20–30. [CrossRef] [PubMed]
21. Wallis, S.L.; Hernandez, G.; Poyner, D.; Birchmore, R.; Berry, T.-A. Particulate matter in residential buildings in New Zealand: Part I. Variability of particle transport into unoccupied spaces with mechanical ventilation. *Atmos. Environ.* **2019**, *2*, 100024. [CrossRef]
22. Hänninen, O.; Lebret, E.; Ilacqua, V.; Katsouyanni, K.; Künzli, N.; Srám, R.; Jantunen, M. Infiltration of ambient PM2.5 and levels of indoor generated non-ETS PM2.5 in residences of four European cities. *Atmos. Environ.* **2004**, *38*, 6411–6423. [CrossRef]
23. Lai, H.; Bayer-Oglesby, L.; Colvile, R.; Götschi, T.; Jantunen, M.; Künzli, N.; Kulinskaya, E.; Schweizer, C.; Nieuwenhuijsen, M. Determinants of indoor air concentrations of PM2.5, black smoke and NO2 in six European cities (EXPOLIS study). *Atmos. Environ.* **2006**, *40*, 1299–1313. [CrossRef]
24. Choi, D.H.; Kang, D.H. Infiltration of Ambient PM2.5 through Building Envelope in Apartment Housing Units in Korea. *Aerosol Air Qual. Res.* **2017**, *17*, 598–607. [CrossRef]
25. He, C.; Morawska, L.; Gilbert, D. Particle deposition rates in residential houses. *Atmos. Environ.* **2005**, *39*, 3891–3899. [CrossRef]
26. Morawska, L.; He, C.; Hitchins, J.; Gilbert, D.; Parappukkaran, S. The relationship between indoor and outdoor airborne particles in the residential environment. *Atmos. Environ.* **2001**, *35*, 3463–3473. [CrossRef]
27. Klepeis, N.; Apte, M.G.; Gundel, L.; Sextro, R.G.; Nazaroff, W. Determining Size-Specific Emission Factors for Environmental Tobacco Smoke Particles. *Aer. Sci. Technol.* **2003**, *37*, 780–790. [CrossRef]

28. Brauer, F.M.B.L.H.; Brauer, L.H.; Behm, F.M.; Lane, J.D.; Westman, E.C.; Perkins, C.; Rose, J.E. Individual differences in smoking reward from de-nicotinized cigarettes. *Nicotine Tob. Res.* **2001**, *3*, 101–109. [CrossRef]
29. Dockery, D.W.; Spengler, J.D. Indoor-outdoor relationships of respirable sulphates and particles. *Atmos. Environ.* **1981**, *15*, 335–343. [CrossRef]
30. Kamens, R.; Lee, C.-T.; Wiener, R.; Leith, D. A study of characterize indoor particles in three non-smoking homes. *Atmos. Environ. Part A. Gen. Top.* **1991**, *25*, 939–948. [CrossRef]
31. Chao, C.Y.; Tung, T.C.; Burnett, J. Influence of Different Indoor Activities on the Indoor Particulate Levels in Residential Buildings. *Indoor Built Environ.* **1998**, *7*, 110–121. [CrossRef]
32. Carrer, P.; Fernandes, E.D.O.; Santos, H.; Hänninen, O.; Kephalopoulos, S.; Wargocki, P. On the Development of Health-Based Ventilation Guidelines: Principles and Framework. *Int. J. Environ. Res. Public Health* **2018**, *15*, 1360. [CrossRef]
33. World Health Organization. *Air Quality Guidelines for Particulate Matter, Ozone, Nitrogen Dioxide and Sulfur Dioxide. Global Update 2005: Summary of Risk Assessment*; WHO: Geneve, Switzerland, 2006.
34. World Health Organization. *Air Quality Guidelines for Europe. Copenhagen*; WHO Regional Office for Europe: Copenhagen, Denmark, 1987; p. 426, (WHO regional publications. European series no.23).
35. World Health Organization. *Health Risks of Particulate Matter from Long-Range Transboundary Air Pollution. 2006 Joint WHO/Convention Task Force on the Health Aspects of Air Pollution*; European Centre for Environment and Health: Bonn, Germany, 2006.
36. Samek, L.; Stegowski, Z.; Furman, L.; Styszko, K.; Szramowiat, K.; Fiedor, J. Quantitative Assessment of PM2.5 Sources and Their Seasonal Variation in Krakow. *Water Air Soil Pollut.* **2017**, *228*, 1–11. [CrossRef] [PubMed]
37. Wang, R.; Wang, Y. Study on the Influence of Thermal Properties of Building Envelope on Indoor Pollutant Diffusion. *Environ. Sci. Eng.* **2020**, 1113–1122. [CrossRef]
38. Yu, L.; Kang, N.; Wang, W.; Guo, H.; Ji, J. Study on the Influence of Air Tightness of the Building Envelope on Indoor Particle Concentration. *Sustainability* **2020**, *12*, 1708. [CrossRef]
39. Vestreng, V.; Goodwin, J.; Adams, M. *Inventory Review 2004. Emission Data Reported to CLRTAP and Under the NEC Directive*; Norwegian Meteorological Institute: Oslo, Norway, 2004.
40. Juda-Rezler, K.; Reizer, M.; Maciejewska, K.; Błaszczak, B.; Klejnowski, K. Characterization of atmospheric PM2.5 sources at a Central European urban background site. *Sci. Total Environ.* **2020**, *713*, 136729. [CrossRef] [PubMed]
41. Ścibor, M.; Bokwa, A.; Balcerzak, B. Impact of wind speed and apartment ventilation on indoor concentrations of PM10 and PM2.5 in Kraków, Poland. *Air Qual. Atmos. Health* **2020**, *13*, 553–562. [CrossRef]
42. Chan, A.T. Indoor–outdoor relationships of particulate matter and nitrogen oxides under different outdoor meteorological conditions. *Atmos. Environ.* **2002**, *36*, 1543–1551. [CrossRef]
43. Klaić, Z.B.; Ollier, S.J.; Babić, K.; Beslic, I. Influences of outdoor meteorological conditions on indoor wintertime short-term PM1 levels. *Geofizika* **2015**, 237–264. [CrossRef]
44. Xu, X.; Zhang, T. Spatial-temporal variability of PM2.5 air quality in Beijing, China during 2013–2018. *J. Environ. Manag.* **2020**, *262*, 110263. [CrossRef] [PubMed]
45. Zheng, X.; Zhao, W.; Yan, X.; Zhao, W.; Xiong, Q. Spatial and temporal variation of PM2.5 in Beijing city after rain. *Ecol. Environ. Sci.* **2014**, *23*, 797–805.
46. Karagulian, F.; Belis, C.A.; Dora, C.F.C.; Prüss-Ustün, A.M.; Bonjour, S.; Adair-Rohani, H.; Amann, M. Contributions to cities' ambient particulate matter (PM): A systematic review of local source contributions at global level. *Atmos. Environ.* **2015**, *120*, 475–483. [CrossRef]
47. Barmparesos, N.; Saraga, D.; Karavoltsos, S.; Maggos, T.; Assimakopoulos, V.D.; Sakellari, A.; Bairachtari, K.; Assimakopoulos, M.N. Chemical Composition and Source Apportionment of PM_{10} in a Green-Roof Primary School Building. *Appl. Sci.* **2020**, *10*, 8464. [CrossRef]
48. *PN-EN 12341: 2014-07 Atmospheric Air—Standard Gravimetric Measurement Method for Determining the Mass Concentrations of PM10 or PM2.5 Suspended Dust*; Polski Komitet Normalizacyjny: Warszawa, Poland, 2014.
49. ISO 16000-34:2018. *Indoor Air—Part 34: Strategies for the Measurement of Airborne Particles*; International Organization for Standardization: Geneva, Switzerland, 2018.
50. ISO 16000-37:2019. *Indoor Air—Part 37: Measurement of PM2.5 Mass Concentrations Describes the Strategies and Procedures for Measuring the Mass Concentration of PM2.5 Indoors*; International Organization for Standardization: Geneva, Switzerland, 2019.
51. Diapouli, E.; Chaloulakou, A.; Koutrakis, P. Estimating the concentration of indoor particles of outdoor origin: A review. *J. Air Waste Manag. Assoc.* **2013**, *63*, 1113–1129. [CrossRef] [PubMed]
52. Chen, C.; Zhao, B. Review of relationship between indoor and outdoor particles: I/O ratio, infiltration factor and penetration factor. *Atmos. Environ.* **2011**, *45*, 275–288. [CrossRef]
53. Chen, C.; Zhao, B.; Weschler, C.J. Assessing the Influence of Indoor Exposure to "Outdoor Ozone" on the Relationship between Ozone and Short-term Mortality in U.S. Communities. *Environ. Health Perspect.* **2012**, *120*, 235–240. [CrossRef]
54. Chatoutsidou, S.E.; Mašková, L.; Ondráčková, L.; Ondráček, J.; Lazaridis, M.; Smolík, J. Modeling of the aerosol infiltration characteristics in a cultural heritage building: The Baroque Library Hall in Prague. *Build Environ.* **2015**, *89*, 253–263. [CrossRef]
55. Chatoutsidou, S.E.; Ondráček, J.; Tesar, O.; Tørseth, K.; Ždímal, V.; Lazaridis, M. Indoor/outdoor particulate matter number and mass concentration in modern offices. *Build. Environ.* **2015**, *92*, 462–474. [CrossRef]

56. Kopanakis, I.; Chatoutsidou, S.; Torseth, K.; Glytsos, T.; Lazaridis, M. Particle number size distribution in the eastern Mediterranean: Formation and growth rates of ultrafine airborne atmospheric particles. *Atmos. Environ.* **2013**, *77*, 790–802. [CrossRef]
57. Bennett, D.; Koutrakis, P. Determining the infiltration of outdoor particles in the indoor environment using a dynamic model. *J. Aer. Sci.* **2006**, *37*, 766–785. [CrossRef]
58. Long, C.M.; Sarnat, J.A. Indoor outdoor relationships and infiltrationbehavior of elemental components of outdoor PM2.5 for Boston-area homes. *Aer. Sci. Technol.* **2004**, *38*, 91–104. [CrossRef]
59. Chen, C. A methodology for predicting PM2.5 penetration and deposition based on the air infiltration through the window gaps. Healthy Housing 2016. In Proceedings of the 7th International Conference on Energy and Environment of Residential Buildings, Queensland University of Technology, Brisbane, Australia, 20–24 November 2016. [CrossRef]
60. Nowak-Dzieszko, K.; Kisilewicz, T. Internal particulate matter pollution in educational building. In Proceedings of the E3S Web Conference, 12th Nordic Symposium on Building Physics (NSB 2020), Tallinn, Estonia, 7–9 September 2020; Volume 172, p. 06008. [CrossRef]
61. Abadie, M.; Limam, K.; Allard, F. Indoor particle pollution: Effect of wall textures on particle deposition. *Build. Environ.* **2001**, *36*, 821–827. [CrossRef]
62. Chao, Y.H.C.; Wan, M.P.; Cheng, E.C. Penetration coefficient and deposition rate as a function of particle size in non-smoking naturally ventilated residences. *Atmos. Environ.* **2003**, *37*, 4233–4241. [CrossRef]
63. Chen, C.; Zhao, B.; Zhou, W.; Jiang, X.; Tan, Z. A methodology for predicting particle penetration factor through cracks of windows and doors for actual engineering application. *Build. Environ.* **2012**, *47*, 339–348. [CrossRef]
64. ISO 16000-26:2012(E). *Indoor Air—Part 26: Sampling Strategy for Carbon Dioxide (CO_2)*; International Organization for Standardization: Geneva, Switzerland, 2012.
65. Thatcher, T.L.; Lai, C.K.A.; Moreno-Jackson, R.; Sextro, R.G.; Nazaroff, W. Effects of room furnishings and air speed on particle deposition rates indoors. *Atmos. Environ.* **2002**, *36*, 1811–1819. [CrossRef]
66. Baxter, L.K.; Suh, H.H.; Paciorek, C.J.; Clougherty, J.E.; Levy, J.I. Predicting infiltration Factors in Urban Residences for a Cohort Study. In Proceedings of the 8th International Conference and Exhibition on Healthy Buildings, Lisboa, Portugal, 4–8 June 2006; p. 194.
67. Piasecki, M.; Kostyrko, K.B. Combined Model for IAQ Assessment: Part 1—Morphology of the Model and Selection of Substantial Air Quality Impact Sub-Models. *Appl. Sci.* **2019**, *9*, 3918. [CrossRef]
68. Piasecki, M.; Kostyrko, K. Development of Weighting Scheme for Indoor Air Quality Model Using a Multi-Attribute Decision Making Method. *Energies* **2020**, *13*, 3120. [CrossRef]

MDPI
St. Alban-Anlage 66
4052 Basel
Switzerland
Tel. +41 61 683 77 34
Fax +41 61 302 89 18
www.mdpi.com

Energies Editorial Office
E-mail: energies@mdpi.com
www.mdpi.com/journal/energies

www.ingramcontent.com/pod-product-compliance
Lightning Source LLC
Chambersburg PA
CBHW061547210926
51850CB00009B/2028
* 9 7 8 3 0 3 6 5 2 5 0 6 8 *